The Birds
of County
Donegal

The Birds
of County
Donegal

Residents, Regulars and Rarities

Ralph Sheppard

Illustrations by Robert Vaughan
Maps by Cóilín MacLochlainn

EST
1925
CUP
CORK
UNIVERSITY
PRESS
CLÓ OLLSCOILE CHORCAÍ

First published in 2025 by
Cork University Press
Boole Library
University College Cork
Cork T12 ND89
Ireland

Authorised representative: Sinead Neville.
Email: corkuniversitypress@ucc.ie

Library of Congress Control Number: 2025933544

ISBN: 9781782050230

Distribution in the USA: Longleaf Services, Chapel Hill, NC, USA

Book design and typesetting by Studio 10 Design, Cork

Printed by Gutenberg in Malta

www.corkuniversitypress.com

FSC
www.fsc.org

MIX
Supporting
responsible forestry
FSC® C022612

All photos have been taken within County Donegal
Photographers (other than the author) are named in the captions

Comhairle Contae
Dhún na nGall
Donegal County Council

An Chomhairle Oidhreachta
The Heritage Council

NPWS
An tSeirbhís Páirceanna
Náisiúnta agus Fiadhúlra
National Parks and Wildlife
Service

CONTENTS

ACKNOWLEDGEMENTS

This book would not have been possible without the enthusiasm for birds of a great many people. Many observations, perhaps even the majority, are routine and predictable, but have nonetheless been essential to the success and value of all sorts of lists, databases, projects and publications, and are greatly valued. Then there are the observers who have submitted their more notable sightings to the relevant experts and authorities of the day. Their observations make up the bulk of this book and I owe my thanks to all of them.

They are: Joe Adamson, Dave Allen, J. Baird, Chris Batty, Michael Bell, Keith Bennett, W.K. (Bill) Bigger, Jamie Bliss, Tamsin Bliss, Helen Boland, R. Bonser, Michael Boyle, Dermot Breen, Derek Brennan, J.J. Breslin, Chris Brewster, Victor Brook, M. Brush, Boyd Bryce, Martin Burke, D. Burns, K. Burns, Majella Callaghan, Geoff Campbell, Helen Campbell, Oscar Campbell, Theo Campbell, Brian Carruthers, W.J. Carlyle, Vittorio (Victor) Caschera, J. Cassidy, Derek Charles, B. Clarke, Kendrew Colhoun, John Collins, Henrietta Cooke, Anthony Cooper, H. Copeland, John Coveney, Archdeacon Cox, John Cromie, Fionnbarr Cross, Michael Cunningham, Noel Curran, Gordon D'Arcy, Lee Dark, Amanda Doherty, Brendan Doherty, Neil Doherty, Joe Donaldson, James Dowdall, P.J. Doyle, Nick Duff, Dave Duggan, Frank Egginton, Andy Ellard, Kieran Fahey, Padraig Farrelly, Wilton Farrelly, Seamus Feeny, Andrew Ferguson, Desmond Fitzgerald, S. Fleming, Ralph Forbes, Charlie Gallagher, Tony Gallagher, Martin Garner, J.W. and W.M. Gentleman, A. Gibbs, M. Golley, George Gordon, Kieran Grace, Bill Gregg, G. Griffin, John Hamilton, Hazlett Harkness, P. Harrison, Henry Chichester Hart, Dennis Hawke, Dermot Hughes, David Hunter, Chris Ingram, Emmet Johnston, A. Kelly, A.A. Kelly, Paul Kelly, Aengus Kennedy, Kevin Kiely, Bill Laird, A.A.K. (Tony) Lancaster, J. Larkin, J.R. Leebody, Nigel Lloyd, John K. Lovatt, Paddy Mackie, Colman Mac Lochlainn, E. Masterson, Ben McAuley, Pete McBride, Brian McCloskey, F. McDaid, Larry McDaid, Willie McDowall, Anthony McGeehan, Grainne McGettigan, Neal McGregor, Stuart McKee, Liz McKenna, Dermot McLaughlin, Jim McLaughlin, Orla McLaughlin, Rónán McLaughlin, Andrew McMillan, Anton Meenan, Eamon Meenan, Clive Mellon, E. and R. Millar, Roger Millington, Daniel Moloney, C.C.

Moore, Aoife Moroney, J.C. Mortimer, R. Moss, Killian Mullarney, R. Mundy, Chris Murphy, Noel Murphy, Tim Murphy, Gerard Murray, Richard Nairn, B. Naughton, Noel Newell, E. Max Nicolson, I.C.T. (Ian) Nisbet, Hannah Northridge, Robert Northridge, A. Ó Dónaill, Michael O'Cleary, John O'Connell, M. O'Donnell, D. O'Hara, James O'Neill, N. O'Neill, Danny O'Sullivan, Kenneth W. Perry, R.G. Pettitt, Peter Phillips, Simon Prientnall, D.J. Radford, Brad Robson, J. Russell, Robert Salter, Mariona Sarda, P.J. Selby, Liz Sheppard, William Sinclair, Andrew Speer, R.D. Steele, J.V. Stewart, C.V. Stoney, G. Thomas, Matthew Tickner, R.J. Ussher, D.I.M. (Ian) Wallace, D. Watts, Dennis Weir, Robert Wheeldon, W.E. Wheeler, D.R. Wilson and Richard Woods.

The band of seawatchers is small, but without them the true picture of what happens just offshore would never have been revealed. They are: Derek Brennan, Oscar Campbell, Theo Campbell, John Cromie, Brendan Doherty, Joe Donaldson, Stephen Foster, Tony Gallagher, Chris Ingram, Rónán McLaughlin, Andrew McMillan, Oscar Merne, Eric Randall and Philip Redman.

Dedicated counters who have done valuable service monitoring water-fowl populations have been: Tony Berry, Jamie Bliss, Derek Brennan, Martin Burke, David Cabot, Kendrew Colhoun, John Cromie, Fionnbarr Cross, Kieran Cunnane, Ethna Diver, Nick Duff, Dave Duggan, Andy Ellard, Mervyn and Susanne Guthrie, Hazlett Harkness, Brian Hegarty, John Hennigan, Chris Ingram, Aengus Kennedy, Lee McDaid, Graham McElwaine, Dermot McLaughlin, Gearoid Mac Lochlainn, Brian Magee, Peter McCarron, Paddy McCrossan, Gerard McGeehan, Dermot McLaughlin, Clive Mellon, Daniel Moloney, Martin Moloney, David Norriss, John O'Boyle, Brad Robson, Andrew Speer, Noel Teague, Matthew Tickner, Ann and Jim Toland, Robert Vaughan, Alyn Walsh and Gary Wilkinson – with apologies to the many who filled in once or twice, or just tagged along.

The names of many field workers and experts appear only in the pub-lications listed in the reference section. Their work has given us the under-standing of what all the individual observations add up to, and point to the important tasks needed to secure the future of our birds, and their habitats.

My thanks are also due to the photographers who have generously allowed their work to be displayed here. They are: Michael Bell, Molly Bell, Dermot Breen, Derek Brennan, Victor Caschera, Derek Charles, John Cromic, Tony Gallagher, Aidan Kelly, Niall T. Keogh, Rónán McLaughlin,

Anton Meenan, Eamon Meenan, Grace Meenan, Kim Periera, Richard Smith, Andrew Speer, Gerry Studd and Robert Vaughan.

Robert Vaughan has painted the evocative bird-filled Donegal scenes that will bring back special days in special places for so many of us, and for which we all owe him our thanks. And I am very grateful to Cóilín MacLochlainn for so expertly drawing the maps. Micheál Mac Gloinn's help with the Irish names was invaluable.

I'm very grateful to Joseph Gallagher (Heritage Officer) and Julie Corry (Biodiversity Officer) with Donegal County Council for their encouragement, and for supporting this publication as an action of the County Donegal Heritage Plan and the Local Biodiversity Action Fund, with funding from Donegal County Council, The Heritage Council and National Parks & Wildlife Service. Thanks also to Maria O'Donovan and Sinead Neville at Cork University Press for guiding this book through the publication process. The reviewers who read various parts of the book made many helpful suggestions, and spotted many mistakes, and I am greatly indebted to them all. They are: Joe Hobbs, Olivia Crowe, John Cromie, Chris Ingram, Robert Vaughan and Derek McLoughlin. Most of all, I could not have done this book without the constant help and vigilance of Liz. She has been my severest critic and my most enthusiastic supporter throughout.

LIST OF FIGURES, TABLES AND MAPS

Front cover: Stonechat on Greenan Mountain, overlooking Inch and Lough Swilly.

Frontispiece: Buzzard, Tory.

Page vi: top, Red-throated Diver, Malin; bottom, Wren, Tory.

FIGURES

TABLES

MAPS

Chapter 1

Introduction

A S A CHILD I knew about birds from books, but I had never seen one in flight. So at the age of seven, when I got my first pair of spectacles, a flock of Rooks against the sky opened up a whole new world. The bond with nature was forged, and there was no going back.

My long-suffering father ferried me around hidden corners where I could indulge my passion. He took me to places like Blanket Nook – with a name like that, it just had to be explored. In the 1950s, to reach the lagoon, we had to cross marshy fields with as many wintering waterfowl as in the lagoon itself – and Grey Partridges in the weedy margins. The summers took us to the Inishowen coast where the numerous tern colonies were the main attraction.

Featuring strongly in the teenage years that followed were three elderly gentlemen who gave me great encouragement. Two of them published little more than the occasional record in the *Irish Bird Report*, but they were both excellent ornithologists. Michael Franklin was a cathedral organist in Derry. His eyesight wasn't great, so it was his musical ear that guided his initial approach to a bird – I still recall his rendition of the song of the Linnet. He also had a special interest in the then disappearing Tree Sparrow.

Bill Bigger was a surgeon who had stitched up wounded soldiers in Palestine during the First World War, taking his breaks with a pair of binoculars. He remained there for the rest of his career, mastering the birds of the Middle East without the benefit of an illustrated field guide. Franklin couldn't drive, and some would say neither could Bigger. He would keep his foot on the accelerator regardless of the condition of the road – a technique honed on the tracks and trails of the Middle East, and still useful on the Donegal roads of the 1960s. But we survived many memorable visits to great places like Horn Head and Tormore, and what was then the only known, and very secret, nesting haunt of the Red-throated Diver.

My third great mentor was Major R.F. Ruttledge – formally Robert, Robin to his friends, but to every Irish birdwatcher he was simply, and in every sense, 'The Major'. He was an exemplary correspondent, and I still have about 150 of his letters. If replying to one of mine, his comments always ran to several pages, and they were always posted by return. Other letters exhorted me to provide data for whatever bee was in his bonnet at the time. Our correspondence continued until not long before he died in 2002, at the age of 102.

Those early years merged into a short period when I could borrow my father's car before breakfast for a quick check on the migrant waders at Donnybrewer Lake, a small lagoon on the polders at the Northern Ireland side of Lough Foyle. It wasn't long before the lake was drained, and with it went the Spotted Redshanks and Ruff that had been daily fare through each autumn. And the Black-tailed Godwits that mobbed me when I approached their nests – they went too. I never did get better proof of breeding, as the lake was drained after only two seasons of incomplete evidence. Donnybrewer was 'my patch', and its loss was something that went very deep, turning me overnight into a conservationist as well as a birdwatcher.

Growing up in Derry/Londonderry in the 1950s and '60s, I was able to make the most of the wildlife on offer in Donegal, Derry and Tyrone, but there was no question that the size and diversity of Donegal had the greater pull. At that time, its spectacular coastline, mountains and wetlands were relatively undeveloped and offered enormous scope for exploration.

Life moved me on to other places, but my frequent visits home ensured that the links were never broken, and I continued to keep a close eye on the county's birdlife. In 1978, I returned with my wife Liz to take over her family's farm in the heart of Donegal, so I have now spent more than half my life equidistant from all points on the coast – useful for keeping tabs on the whole county.

The account that follows is an attempt to gather together all that is known about the birds that have been found here from the very earliest records up to the present. Nowadays there is a steady flow of information from the growing band of local and visiting birdwatchers/birders/ornithologists/ ecologists.

It is very encouraging now to see awareness growing of the deep connection between humans and nature, and the urgent need to protect wild habitats. The conservation movement touches many lives in different ways, and in my case has included planting trees, mapping wildflowers and trapping moths. But it was discovering the joys of birdwatching that started my lasting relationship with the natural world. So if you are reading, or browsing, through this book, and have not yet encountered the amazing world of birds, this could well be the first crucial step that will lead you to a lifetime of enchantment.

Chapter 2

Early Days

THE EARLIEST EVIDENCE of birds contemporaneous with humans in Donegal is from bone fragments of the now extinct Great Auk. These were identified in the kitchen middens of our Mesolithic (middle Stone Age) ancestors of 6,000 or more years ago. They were found in the sandy machair zones of Donegal and other coastal localities around Ireland. No other archaeological evidence exists to challenge or support what we currently know of our avifauna. Capercaillie bones have only recently been confirmed from elsewhere in Ireland, but not yet from Donegal.[1]

Apart from casual comments in the writings of visitors from the seventeenth to the early nineteenth centuries, what was known about birds wasn't committed to paper in Donegal until landowners first adopted the shooting of game for entertaining their guests, as well as diversifying their diet. Raptors, being the perceived enemies of partridges, grouse, wild ducks, etc., were also fair game, and the better specimens of all these were mounted by taxidermists. It was a short step then to use these weapons to bring other birds close enough to identify. One clue to the extent of this practice is found in a bird list included in the *Ordnance Survey Memoirs* from the 1830s.[2] The known wild species in the county make up the bulk of the list, but there are also the various farmyard fowls, and other birds for which there is no other record of them having occurred naturally (or in captivity) in Donegal – for example, the Eurasian Nuthatch. Along with personal observations and local information, the list would have had to include birds in collections of mounted specimens and skins. The most valuable gem among the 277 species listed is the Capercaillie – the best evidence we have that it was a native species in the county in historical times (see Chapter 18).

The first person to critically assemble the available information was William Thompson, whose three-volume *Natural History of Ireland* (published from 1849 to 1851) was devoted to birds – the incomplete fourth volume, published posthumously, dealt with the remaining vertebrates and the invertebrates. Although he visited Donegal himself in 1832, Thompson's references are mainly supplied by correspondents. One such was John Vandeleur Stewart (1804–72), who lived at Rockhill House, near Letterkenny, and published his notes in the *Magazine of Natural History*. He also owned land at Horn Head, and his observations on Golden and White-tailed Eagles were related in person when Thompson visited the area. The gamekeeper at Horn Head had apparently killed only one Golden Eagle, as against twelve or thirteen White-tailed Eagles, during the four years he had worked there prior to Thompson's visit.

Stewart apparently shot a Bee-eater, in about 1830 or 1831, but the record seems to have been rejected or overlooked since it was first reported by William Sinclair to Ussher and Warren.[3] Sinclair lived in Belfast, but spent the summer months at Drumbeg House, Inver, and supplied other interesting observations, not least that his son, Major Sinclair, shot one Red-necked Phalarope out of a party of three at Inver.

However, the most authoritative source of information from the nineteenth century was Henry Chichester Hart (1847–1908). A Shakespearean scholar by profession, his first degree was in experimental and natural science, and he was a hill-walking botanist by inclination and the author of the definitive flora of the county.[4] His summer home was in Carrablagh House, at Portsalon on the Fanad peninsula, but he also travelled widely on various expeditions among the mountains of Ireland, to Palestine and to the North Polar regions. He published his 'Notes on the Birds of Donegal' in 1891, in the *Zoologist* magazine – the forerunner of *British Birds*.[5] It was clearly his intention to cover all species, as he apologised later for omitting some common species. His own observations are mainly from the summer months. His most intriguing claim is of a Marsh Tit at his well-wooded home in November 1889. He says, 'I observed it carefully: there was no white on the nape.' If that was all he observed, it could explain why the record was not reported by Ussher and Warren or subsequently. His views on Bean Geese are outlined in the species entries. Otherwise his observations are convincing. He reports from J.E. Stewart (probably a misprint for J.V.) in 1830 that Bitterns were rare residents but had been common thirty years previously. Treecreepers were also thought to be very scarce – very likely reflecting the nadir of the county's woodland cover following the Famine of 1845 to 1852. Similarly, he had never himself encountered a Quail in Donegal, their decline probably also preceding him.

J. Steele Elliott was another observer who published notes, almost all from the Dawros area, and including some from the Tormore stack and from Roaninish.[6]

John Robinson Leebody (1840–1927), professor of mathematics and natural philosophy at Magee College in Derry from 1865, made a useful contribution in his 'Notes on the Birds of Lough Swilly', which was published in *The Irish Naturalist* in December 1892. Despite his academic background in natural sciences, Leebody seems to approach this subject in the manner of a typical ornithologist of his time, and his observations are peppered with references to the success or otherwise of shooting his quarry. However, his conclusions ring true, and almost all concur with what we would expect. The main differences between then and the present, as revealed in his notes, include Bewick's Swans being common but Whoopers rare, only thirty to forty White-fronted Geese wintering on Inch Water, but Scaup were the most plentiful duck after Wigeon. He noted that Brent Geese covered several acres of water between Inch Island and the Leannan estuary. This would have been well before the disease that wiped out the great beds of eel grass (the scientific names of species other than birds are given in Appendix 2) on which they traditionally fed.[7]

Most of these later individuals, after Thompson's time, reported their observations for inclusion in the volume that summed up the century, *The Birds of Ireland* by Ussher and Warren, published in 1900.

The early twentieth century was in some ways a lean period in ornithological activity. Shooting was going out of fashion among the ornithological community, but the guns were not replaced by decent optics until mid-century, and the sharing of information was still no quicker than it had been for centuries. An interim list was published by G.R. Humphries in 1937.[8] But it was a full half-century following the Ussher and Warren book before leaders again emerged who could repeat that exercise, with inputs from a network of competent observers. They were Fr P.J. Kennedy, Major R.F. Ruttledge and Lt-Col C.F. Scroope, who published the fruits of the collective labour in 1954, in *Birds of Ireland*.[9] It is tempting to wonder if the people-organising skills implicit in the titles of the three authors were crucial to the success of the project.

In that period the main focus of interest shifted to the offshore islands and their breeding seabirds. Many attempts were made to locate colonies of Storm Petrels and estimate their numbers. Robert Ruttledge was himself the driver in this activity, with contributions from R.S.R. Fitter and C.V. Stoney.[10] But it was S.P. Cummins and I.M. Goodbody in 1948 who finally came up with an estimate of 1,000 pairs on Roaninish.[11] Stoney also found Red-necked Phalarope breeding there, and on the mainland the artist Frank Egginton took that story further.

Chapter 3

Recent Times

JUST BEFORE THE PUBLICATION of *Birds of Ireland* in 1954, Ruttledge started the annual *Irish Bird Report*, in 1953, on behalf of the Irish Ornithologists' Club. This formula has stood the test of time, but has expanded. The report was essentially a list of rare bird sightings, and is now the *Irish Rare Bird Report*, published on behalf of the Irish Rare Birds Committee (IRBC). The extra items that accompanied the annual list grew, in number and rigour, to the point where the journal *Irish Birds* had to be started to accommodate them, and encourage more of the same. The Irish Ornithologists' Club went through some mergers and re-namings to become the Irish Wildfowl Conservancy (IWC), then the Irish Wildbird Conservancy, and finally BirdWatch Ireland (BWI), a professional organisation carrying out research, conservation and advocacy with, and on behalf of, its nationwide membership of amateur birdwatchers.

The *Irish Bird Report* made us all more aware of other birdwatchers – or in the case of Donegal, the lack of them. In those first couple of decades following 1953, there was very little happening in Donegal. The observatories at Tory and Malin Head in the early 1960s were hugely important, but they were mostly manned by people from the ornithologically more advanced counties to the south, and from Northern Ireland and Great Britain.

A particularly useful training ground for local birdwatchers has been the succession of bird atlases – roughly every ten or fifteen years – run jointly by the British Trust for Ornithology (BTO), the Scottish Ornithologists' Club and BWI (including its previous incarnations). Fieldwork for the first of these took place between 1968 and 1972, but when the most recent one was carried out between 2007 and 2011, professionals took most of the pressure off the local birdwatchers.

There have always been a few people based in Derry/Londonderry who have regarded Donegal as their hinterland (I was one of them in my youth), and Ken Perry's *The Birds of the Inishowen Peninsula* in 1975 was a significant contribution from a Derry resident.[1] We might have expected that a fairly large city just a few miles across the county boundary would dominate the local scene in Donegal, but this has not been the case, although it has always provided a trickle of individuals to boost the local effort.

In 1982, the IWC became involved in a campaign to save a threatened goose habitat on the west coast, culminating in the purchase of some land. This started a process of land acquisition by the National Parks and Wildlife Service (NPWS) across what is now the Sheskinmore Nature Reserve. The campaign also prompted the formation of the Donegal branch of IWC in 1983. The branch continues, under the BirdWatch banner, to cater for birdwatchers throughout the county, with regular meetings, usually outdoors, and by making its voice heard when necessary on conservation and planning issues. In such a large county, small independent groups in Inishowen and the south-west have helped ensure that everyone has an active local group to join.

Monitoring the waterfowl in our bays and estuaries has been very effective at challenging birdwatchers into greater activity. These counts started in 1984 with the Winter Wetlands Survey, and continue now with the ongoing Irish Wetland Bird Survey (I-WeBS). Some of our keenest birdwatchers cut their teeth on the monthly Lough Swilly censuses.

Since 1998, monitoring 1 x 1 km squares for the Countryside Bird Survey has involved a small number of dedicated volunteers, and over the years it will increasingly provide us with vital information on the fortunes of the commoner non-colonial breeding bird species – those which account for most of the individual birds in Ireland.

In the 1980s and ’90s, bird reporting advanced from the leisurely annual ritual of meeting the publishing deadline of the *Irish Bird Report*. *Flightline*, manned by Pat and George Gordon in Northern Ireland, provided daily telephone updates, keeping birdwatchers on the move, including in Donegal. The rate at which rare birds were sighted increased accordingly. The process advanced in the early millennial years with *Flightline*'s successor in Northern Ireland, the *nibirds* blog, which still carries occasional reports from Donegal, thanks mainly to Wilton Farrelly. Joe Doolan and Eugene Archer's Irish Birding website (irishbirding.com) now provides national coverage. But the development that has been most useful to the Donegal-based birdwatchers for the last five or six years has been a WhatsApp group run by Chris Ingram. For its forty-seven current members, on-the-spot sharing of observations has kept us all on our toes. Most importantly, it continues to draw in new participants.

Responding to this book's subtitle, we can now say that there are ninety permanent year-round resident taxa (species and subspecies), ninety-two are regular visitors (winter, summer, or as passage migrants), and so far there have been 174 rarities – some appearing most years, but all of them here by accident. The total of 376 includes a further twenty taxa that don't fit into the above groupings. Appendices 5 and 6 identify which is which, and the back-story for every one of them can be found in Chapter 18. The preponderance of rarities shows just what can be found in a coastal county at the corner of the island, but it is the residents and regulars that will give greatest pleasure to the majority of birdwatchers. After all, we have only to step outside our doors or look through our windows to start engaging with them.

Chapter 4

The County

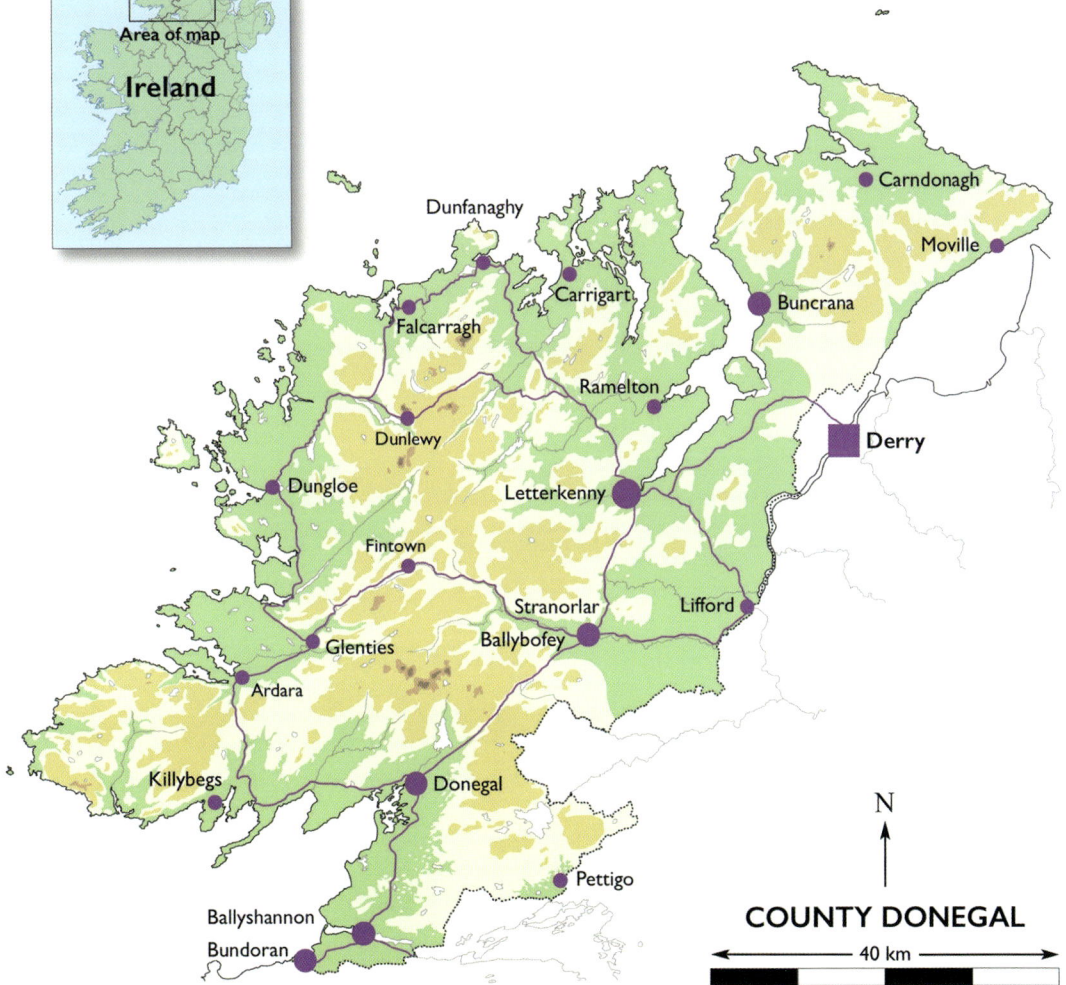

Map 4.1: County Donegal, with the main roads, towns and villages

ONEGAL IS BOTH a county and a region (Map 4.1). Its isolation from the other counties in the Republic gives its people a sense of difference. Birdwatchers are not immune to such influences, and our expeditions south across the narrow county boundary to enjoy the birds of Leitrim or Sligo are less frequent than they might be. To the east, the broad River Foyle with no crossing between Derry/ Londonderry and Strabane-Lifford, and the barren forests and bogland on the Tyrone and Fermanagh borders also serve to limit our vision – even without regard for the different jurisdictions. The landscapes of Donegal and the four adjacent counties are also all markedly different. Although the bird communities, as everywhere in Ireland, are less contrasting, differences do exist and they heighten that sense of distinctiveness.

At 4,858 km², Donegal is one of the largest counties in Ireland. From Malin Head in the north-east to Malin Beg in the south-west – a day-trip that a keen birdwatcher might well want to make – is 168 km. The county is largely upland, with the highest peak being Errigal, at 752 m high. The coastline is arguably the longest in the country. The county council put it at 1,134 km, but of course it all depends on how fine your scale is. Also arguable is whether or not the 601 m sea cliffs of Slieve League are the highest in Europe, or even in Ireland. At any rate, the upper slopes are (or were) good enough to harbour Ring Ouzels, but significant seabird colonies are limited to Cormorants.

At the north-west corner of Ireland, and one of a very select group of areas that could claim to be at the north-west corner of Europe, Donegal has to be of special interest as either the start or the end of a spectrum of bird life. Some species will be at their geographical peak, while others may have faded away to a state of rarity – but all are of equal fascination in giving to the county its unique character. Those of us who spend as much of our time as possible grappling with the questions of what? where? and how many? are as much intrigued with the constantly changing patterns as with any notion of exactly what ought to be here.

For our purposes a number of sub-regions are identified (Map 4.2). There are two distinct highland areas, three more loosely defined hill areas, three totally different lowlands, one large isolated peninsula which is mostly hilly (Inishowen), and one complex area which can be identified as the peninsular north. There are any number of variations on the theme, but most people with an eye for landscape and habitats should find this sub-division useful.

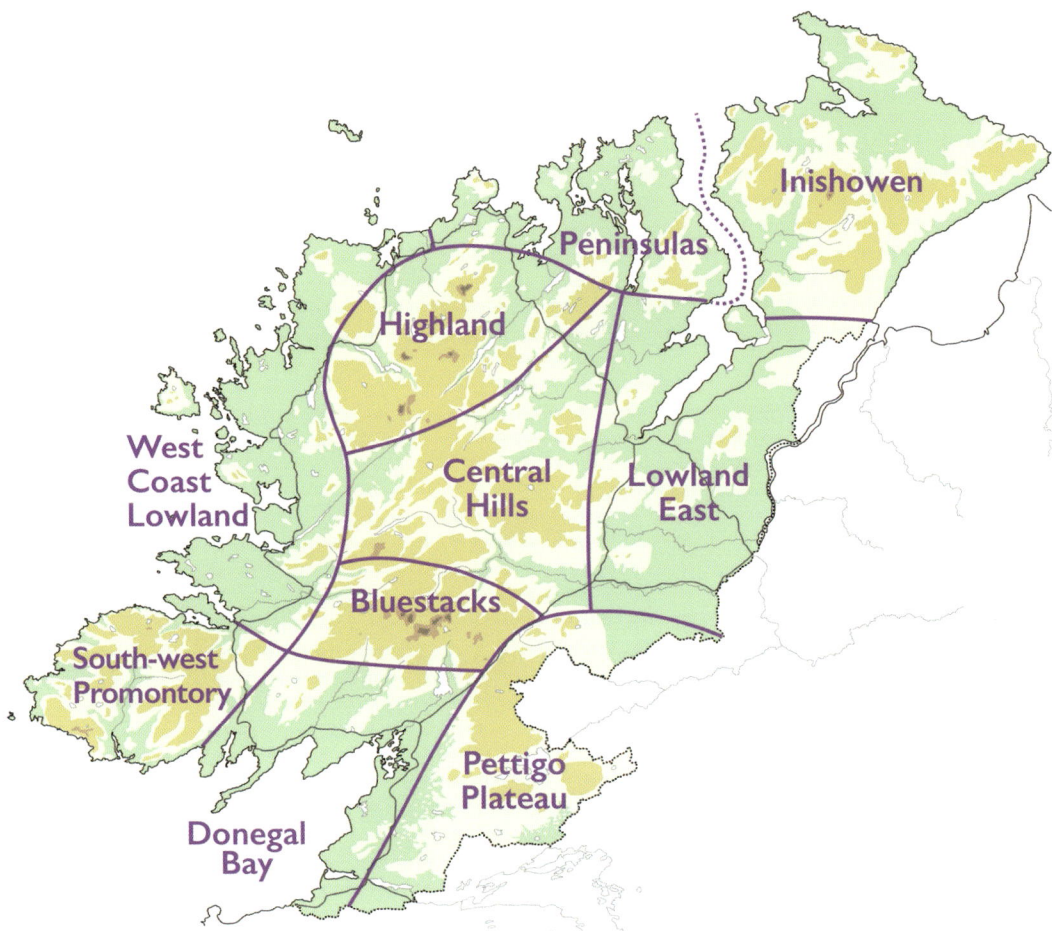

Map 4.2: Donegal's physical regions

Weather and climate

Ireland is a small, rather homogeneous island. So it doesn't have the noticeable contrast in the climate of our neighbouring island, between the north of Scotland and the south of England. The climate of Donegal is, like the rest of the island, equable. In other words, the temperate is rarely too hot or too cold, and every month generally gets enough rainfall, and on the whole we have not yet experienced the extremes of flooding or drought that are becoming familiar throughout Europe. Where we do stand out rather more is in the windiness of this north-west corner.

Our weather is usually delivered by a succession of Atlantic depressions moving east across the country, or towards the north of Scotland. The ocean keeps the climate mild, and the low-pressure depressions keep it wet and windy. But it must not be forgotten that between the showers we have the cleanest air in Europe, and brilliant light that colours the land and the seascapes. These conditions occur throughout the year, but are more frequent in winter. The high-pressure anticyclones centred north of the Azores or on continental Europe, which deliver the warm, settled summer weather we all describe as 'good', trap atmospheric dust, which reduces light quality.

Summer temperatures usually peak in the low twenties (degrees Celsius), and winter lows remain above freezing. Snow, which used to be present in the lowlands for one or two weeks each year, is now expected for only a few days. In the hills of course there is more snow, but even the high peaks are white for only a couple of weeks in total.

Wildfowl and waders, nesting in northern latitudes that freeze up in winter, find Donegal an ideal base for the winter months. Migrating seabirds use the strong north-westerlies that are frequent in autumn to propel them southwards, closer to the shore than normally. Summer land migrants are mostly more concerned about the abundance of our insects than the weather, although there is obviously a connection between the two.

How a changing climate will affect all of this is something that was traditionally deemed to be in the lap of the gods. We now know more about what to expect, and it is mostly not good news. Precisely how it will work out in this small corner of our small island is not yet totally clear. The indications are for increasing storms, with increasing severity – but even here, east Donegal is often in the becalmed eye of the typical named storms making landfall in south-west Ireland and moving north. The way this is already impacting our different bird species, and the prospects for the future, is dealt with in chapters 16 and 18.

One thing is certain – keeping a close watch on our bird populations is the essential first step in planning any evasive actions that might help them cope with whatever the future brings.

The north-west highlands

Can there be any finer mountains in Ireland than these? They are not the highest, but they more than make up for that in grandeur. There are

Muckish Mountain, north-west highlands.

three parallel ranges picking out the north-east to south-west trend which the rocks of Donegal share with those of the Scottish Highlands. The north-westerly ridge is made of quartzite. Its peaks are individually sculpted monoliths rising stark and splendid from dark peat bogs. These are spectacular peaks – Errigal at one end, a white cone, and Muckish at the other end, an immense flat-topped block. The other two ridges, the Derryveagh and Glendowan Mountains, are granite – solid, well-rounded mountains, polished smooth by the movement of ice. What sets them apart is the great gash of the Gweebarra fault which divides them, and is now at the heart of the Glenveagh National Park.

Virtually all of Ireland's mountain birds can be found here. Some, like the Golden Plover and the Ring Ouzel, have retreated from many of their other haunts over the last few decades. The reasons are still unclear, but almost certainly habitat deterioration and climate change are both involved. Their presence here is something we must savour while it lasts, as their eventual disappearance from Ireland seems inevitable.

Banagher Mountain, Blue Stack Mountains.

The Blue Stack Mountains

These are typical granite mountains, high and bare – a desolate landscape with little animal life and less human. There are few roads apart from the Barnesmore Gap, which only cuts off the edge, albeit in some style. They may not have the charisma of the north-west highlands peaks, but the Blue Stack Mountains should not be overlooked. Empty wilderness has an enormous value. Those few birds that can be found are, as often as not, the very scarce species which need such conditions. Peregrine and Merlin haunt the open landscape in summer. Golden Eagles and Golden Plover are in their element, and can be seen at all times of the year. And the discerning birdwatcher wandering those bare high tops in winter can also be rewarded with small flocks of Snow Buntings roaming the summits.

Finn Valley, central hills.

The central hills

The heartland of Donegal is to many people nothing more than a space between the more well-defined areas – a harsh land, where people struggle to win a living from thin peaty soils and dwindling bogs. But for the ornithologist it has its own identity. Many of the county's largest rivers have their upper reaches here, so it should be no surprise that this is prime Dipper country.

Unfortunately, the broad low hills, too poor for profitable farming, yet still with a covering of peat, and soil of a sort, have attracted extensive afforestation. As well as being a major visual intrusion on the landscape, such a dramatic change in the ecology of the area is bound to have a profound impact on its birds. Golden Plover, Red Grouse, Red-throated Divers, Hen Harrier, Merlin, Peregrine, Curlew and Dunlin are all species that we could easily lose if planting on the hills continues without restraint. White-fronted Geese have already almost abandoned the habitat.

These scarce species, and the commoner ones like Meadow Pipit and Skylark, have been replaced by a much smaller group of species well adapted to the coniferous habitat. Siskin, Goldcrest, Coal Tit and Crossbill are the most dedicated, and there are other opportunistic species like Chaffinch and Sparrowhawk.

However, the Merlin, a dedicated open-country specialist, is learning to adapt, and now nests exclusively in the forest canopy. And in the early stage of planting, or re-planting, when grass is freed from grazing pressures, the young trees make convenient song-posts – ideal for scrub and rough grassland species like Whinchat, Whitethroat, Grasshopper Warbler, Willow Warbler and Blackcap. And even Hen Harriers can find secure areas of rough grassland for nesting.

So replacing the open hills with conifer plantations is not an unmitigated disaster, but is certainly a very poor trade-off. Happily, the losses should now be capped with the removal of grant aid for planting on peat soils. And the broadening of the remit for plantations to include reversing biodiversity losses and mitigating climate change has to be a good thing. How effectively it will be implemented remains to be seen.

The Pettigo plateau and Donegal Bay

Donegal Bay's extensive and diverse estuarine and shore habitats are a south-facing counterpart to Lough Swilly. Otherwise, it has close affinities to its neighbouring regions.

The south-east is similar in character to the central hills, but less populated and with its habitats on a larger scale. At its core is one of the largest and most unspoiled bogland wildernesses in the county, the Pettigo plateau. As in the highlands, the number and variety of the birds is very poor, but the quality could hardly be better. Apart from Meadow Pipit, this great expanse of uncut blanket bog is inhabited by Red Grouse, White-fronted Geese and Hen Harriers, of which the latter two are certainly threatened.

Lakes are everywhere, and although mostly empty of bird life, there are enough to ensure that breeding gulls and waterfowl can be found, and there are always small numbers of wintering birds on the richer lakes, Goosander being the most significant. The only inland record in the county of a Little Auk was on Trumman Lough. As with the central hills, there are extensive conifer plantations, but here they merge to form one of the largest areas

Afforested bog and new birchwood, Lough Derg.

in the country. Given good management, there has to be potential here for some of the more demanding species that can utilise extensive forest areas. Great Spotted Woodpeckers have already discovered that potential.

The south-west promontory

The limited area of lowland in south-west Donegal is mainly inland, with the mountains – Slieve League and Slievetooey – on the outer fringe. The result is a cliff-bound coast of unparalleled magnificence. Seabirds of course are no more impressed by 600 m cliffs than they are by the more usual 100 m examples. But the lower sections of these high cliffs, the sea stacks and islands had a number of good seabird colonies in the past, with some of them still present. In winter too this corner of the county has its attractions. There are Barnacle Geese in suitable locations, and a bewildering number and variety of gulls can be seen in Killybegs Harbour.

Less well known is the importance of this south-west promontory to migrating birds. Seabirds passing Rocky Point do so in numbers as good as anywhere in the county, and the relatively sheltered west-facing settlements from Malin Beg to Glencolumbkille draw in land migrants that rival Tory in their diversity and rarity. Better coverage of this remote area would undoubtedly raise its profile for visiting birdwatchers.

Malin Beg to Slieve League.

The lowland east

This sub-region is the only part of Donegal in which good agricultural land predominates. There are hedges here, thick and tall, whereas in the rest of the county, fields are bounded by stone walls or earth banks with scattered bushes. The landscape is mostly rolling, and the tops of the low hills are capped with heather and gorse, which ensure that the scenery is never boring. But oddly enough, the chief glory from the birdwatchers' point of view is the most highly developed land – the slobs along Lough Swilly. These huge arable fields claimed from the estuarine mud are one of the best places in Ireland to see wild swans and geese, and along with the rest of Lough Swilly this is Donegal's primary bird wetland. Lough Fern is one of the better inland lakes in the county for both breeding and wintering waterfowl. The eastern fringe of the area is bounded by the River Foyle and is also of value to waterfowl.

Intensive mixed farmland in the lowland east (the windfarm is in County Tyrone).

The west coast lowland

From the foot of Slievetooey in south-west Donegal, a broad coastal plain stretches all the way to the north coast. It is a zone of estuaries, lakes and bogs protected from the Atlantic (to a degree) by a maze of islands. Against all logic, it has a dense human population. Towns are few and small, but everywhere there is a scatter of houses. From Bloody Foreland to the Rosses it is hard to tell where the villages stop and the countryside begins. Why this remotest corner of Ireland should be so densely populated is beyond the scope of a book on birds, but it is at least a measure of the strength of community life here that it can continue to flourish in this beautiful but barren land.

For better or worse, it is inevitable that the fortunes of birdlife here are intimately linked to human settlement and activity. The limited agricultural land has always supported many interesting birds. But where agriculture survives at all, the old labour-intensive practices of crop rotation, fertilising by seaweed or manure, and mowing by scythe have been replaced by

Continuous rural settlement south from Bloody Foreland.

heavy machines and chemicals. The change has been catastrophic for some birds – it is already too late to save the Corn Bunting, and the Corncrake is on life-support. Neither are among the most musical of birds, but the jangle of the former and the rasp of the latter were, until a generation ago, among the most evocative summer sounds of the western seaboard. The offshore islands of Inishbofin and Tory are two of the last strongholds of the Corncrake in Ireland. As the survival of these birds depends on the nature of the land management, without environmentally driven incentives their future is very uncertain.

Rock Dove just about survives, but interbreeding with its feral descendants is a continuing threat. Cuckoo and Stonechat are more abundant here than elsewhere. So despite the transformation of the landscape in recent decades, the west is still a varied, fascinating and very beautiful area to explore, never knowing what bird waits around the next corner.

Woodland never had much chance in west Donegal – lack of soil, exposure to wind and removal by humans were all against it. The fragments surviving in stream gullies too dangerous to allow access for sheep or cattle were not enough to retain many of the woodland-edge birds that can still thrive in the eastern lowlands. Perversely, the abandonment of agriculture

Machair and cottage farming (the old west), Inishfree Island.

in some of the more densely inhabited areas has allowed patches of scrubby woodland to emerge again. So birds like Treecreeper, Long-tailed Tit, Coal Tit and Bullfinch have been able to move into the west in recent decades.

Inishowen

Inishowen is the most clearly defined of our sub-regions on account of its physical isolation – surrounded on three sides by water. The inland core of the peninsula is very similar to the central hills in its landforms, its settlement pattern and its land use. Its birds too are no different, except that the White-fronted Goose has been totally banished from its bogs by the ever-advancing tide of conifers and peat cutting. The north coast of Inishowen has much in common with the other north-facing areas.

At all times of the year there is good birdwatching in Inishowen, especially in the north, where the mix of sea coast, estuary and farmland maintains the variety. Malin Head is one of the most exciting locations in the county to watch birds. As well as its diversity of habitats for resident birds and regular visitors, its position is particularly useful for studying seabird

Loughros Bay area.

Donegal Bay farmland, Frosses.

movements. By projecting far into the northern seas beyond the general line of the north coast of Ireland, it can on occasion block the passage of birds moving west along the coast. This seems to happen particularly in hard northerly weather, when other seawatching sites tend to be less productive. At such times, skuas can be encountered taking the overland route from Lough Foyle to Donegal Bay.

The northern peninsulas

The block of land between Lough Swilly and the north-west highlands is broken by two other large inlets, Mulroy Bay and Sheep Haven, creating three large peninsulas: Fanad, Rossguill and Horn Head. Mulroy Bay also penetrates Fanad from the inland side, making north-west Fanad effectively a fourth small peninsula. Horn Head is famous for its seabird colonies, and Rossguill for its scenic vistas and golf courses. Fanad is less well known, which may help to explain the survival of its intimate landscape, although this is gradually being modernised, to the detriment of its wildlife.

Both Fanad and Rossguill are topped with good seawatching headlands, and the Fanad Head area in general is of interest for migration and for its

Lowland blanket bog at Malin, Inishowen.

Faugher and Sheep Haven Bay.

non-estuarine wintering wetland birds, like the sea ducks.

The two sea inlets are very different – Sheep Haven is sandy, with a variety of coastal species like Common Scoter and divers, while Mulroy Bay is a deep, almost land-locked fjord, with many small uninhabited islets, used historically by terns and other nesting seabirds.

The most distinctive feature of this area is its semi-natural woodland. A high proportion of all the deciduous wood in the county is within this region – rivalled only by the Donegal Bay area. The total quantity is not large, and the woods themselves are mostly small and degraded, but they are good enough to have held a tiny population of Wood Warblers in recent years, although their survival is in doubt.

Chapter 5

Farmland

Yellowhammer, Tree Sparrow, Corn Bunting, Redwing, Skylark, Fieldfare, gull species, Grey Partridge, Lapwing

VIRTUALLY ALL LAND in Donegal that can support agriculture does just that. Our farmland is long removed from any appearance of being natural or organic, but even so, the processes of the natural world are still at work. Although modern agriculture is focused on food production, wildlife of all sorts is always doing what it can to carve out a place among the fields.

Most of our farmland is devoted to grass, parcelled up as small fields bound by hedges or dry-stone walls, with small woodland copses remaining in awkward corners or on the less productive soils. Globally, this is a very rare and distinctive landscape but has only been specifically named by the

French, as 'bocage'. The hedges around the farms are traditionally managed by layering, on a fairly long cycle, and as far as wildlife is concerned, they have to stand in for much of our missing woodland. A good hedge has the height necessary for trees and bushes to fruit, and is broader at the base so there is sufficient light for leaf growth on the lower bushes and space for a layer of woodland plants. A wide variety of species, from mammals to insects and even plants, can travel along these 'wildlife corridors' between areas of woodland. Now, of course, with tractor-mounted flails and saws, most hedges don't get the chance to retain their height, and they lose the sheltered bottom layer to barbed and sheep wire.

The birds

Birds like Redwing, Jay and Yellowhammer utilise the taller bushes and trees, while Wren, Dunnock and many others feel more comfortable at lower levels. Despite the current degraded state of most hedges, there is still scope for the familiar woodland-edge species to survive in this landscape, especially as our gardens, in both town and country, supplement not only winter feeding but also breeding opportunities in the unkempt corners. You will readily see several species of thrush, tit, finch, crow, dove and raptor, and in summer the hirundines (swallows and martins) and warblers. And there are other species away from the hedges and gardens, like Pied Wagtail.

'Bocage', Finn Valley.

These successful widespread species are mostly generalists which can adapt more easily to changing circumstances.

Field drainage has always been necessary in our wet climate, but mechanisation led to a renewed drive to drain all wet farmland in the latter decades of the last century, which undoubtedly reduced the value of much farmland to birds. So small flocks of wintering Curlew and Golden Plover have largely disappeared, or retreated to the estuaries.

On the better lowland soils in the Foyle basin there is a sizeable area of arable or mixed farming. At the same time as the wet grassland fields were being drained, management of arable crops, the potato in particular, became so efficient that the wild plants long adapted to live alongside farm crops (ruderals) were largely eliminated. This has led to the loss of most of the flocks of wintering songbirds. Brambling has all but disappeared, and abundant species like Chaffinch and Linnet have been greatly reduced in number. Greenfinches, like Linnets, can now rarely muster a decent flock. But at least Goldfinch is still doing well.

There are also scarce and specialised species for which traditional farmland offered the closest approximation to their natural habitats, but sadly the intensified agriculture that replaced it doesn't measure up. The Grey Partridge and Corn Bunting have gone and Corncrake is on the brink, hanging on in response to a succession of management schemes. Quail has gone as a regular summer visitor, although individual birds will probably continue to reach us on very rare occasions. Skylarks are much fewer than in the past, and missing from many areas.

Intensive grassland, Deele Valley.

Wildlife corridor – an expanded hedgerow, Carnowen, near Convoy.

Recognising these trends can be difficult, as our judgement is often clouded by a few experiences, be they good or bad. However, we now have long-term surveillance from the Countryside Bird Survey[1] and analyses like the regular 'Birds of Conservation Concern' reviews[2] which reveal what might otherwise not be detectable. Combining these studies and the evidence from the bird atlases has shown that of eighteen species of seed-eating birds in Ireland, all were decreasing between the first two breeding atlases, of 1968–72[3] and 1988–9[4] – apart from the Tree Sparrow which was then reclaiming much lost habitat. The most steeply declining species were the Corn Bunting at minus 84 per cent and Grey Partridge at minus 86 per cent.[5]

Happily, realisation is dawning that our current farming model cannot be expected to survive without radical change. It has to be future-proofed with sustainable practices that can only come from a recognition that farming must both support nature and emulate it. An expansion of organic farming would undoubtedly help, but Donegal and Ireland in general are far behind the rest of Europe in adopting it. The newer concept of regenerative farming now has more chance of eventually becoming standard, as part of the national strategy to reduce CO_2 emissions and increase biodiversity. The use of cover crops in the arable areas is already bringing hope of recovery for some of the lost wintering passerine flocks. And there are many other 'greener' initiatives being tried out by farmers that should help our declining, but still common, bird species. Those already lost or rare will need more targeted efforts.

Chapter 6

Woods and Forests

Wood Warbler, Woodcock, Great-spotted Woodpecker, Jay, Treecreeper

Woodland

THE CLIMAX VEGETATION in Ireland is generally oak woodland. Yet it is surprisingly rare, having been replaced not only by farmland but also by assorted woods of non-native species like sycamore, beech and various conifers. Remaining fragments of the real thing are to be found scattered throughout Donegal, with a definite concentration north of Letterkenny, between Lough Swilly and the mountains to the west. The most important area is on the western fringe of this zone, in Glenveagh National Park.

There is an obvious link between Donegal and Scotland when you look at a geological map. A north-east to south-west trend in the geological faults is at its most obvious in the Great Glen which divides north-west

Scotland from the rest of the country, and is also evident in north Donegal, in particular at Glenveagh. The passage of a glacier has further gouged that valley into its present form – not by any means unique in Ireland, but there are few better examples. It is a landform much more familiar in Scotland. Add to that Glenveagh's extensive areas of oak woodland, smaller stands of Scots pine in some prominent locations, and a classic Scottish-style baronial castle – and the connection becomes striking.

There is historical evidence of a core of oak woodland in Glenveagh National Park going back to the seventeenth century, which by Irish standards qualifies it as ancient woodland, of which we have precious little.[1]

The older practice of grazing by a mixture of cattle, sheep, goats and red deer would not have been as destructive as modern sheep farming, but it nonetheless kept the woodlands in check. Red deer, which had survived as a native species in Glenveagh until about 1845, were re-introduced from the 1890s onwards. In the absence of predators such as wolves, their numbers can grow until they have destroyed their habitat, and not until starvation kicks in will the population decline. So left to their own devices, red deer can be as destructive as sheep. Since re-introduction, they have been controlled by shooting, initially for sport but now as part of the overall management of the park.

Destructive though deer have been, there is another introduction that has had an even greater impact, and is far more difficult to control. When flowering in May, it is hard to imagine that rhododendron has not always been a part of the Glenveagh landscape. But this beautiful shrub, native to the Caucasus Mountains, is lethal to woodland habitat in Ireland. Virtually

Ancient oakwood, Binnakilty, Glenveagh.

nothing will grow under its dense shade, and that includes ground flora and tree seedlings, both essential for the long-term survival of the habitat. Rhododendron removal is an ongoing battle that still hasn't been won.

Woodland birds

Despite the problems posed by red deer and rhododendron, the woodland at Glenveagh survives, and the birds demonstrate both the potential of restored native habitat, and how much has already been lost. There is a community of migrant songbirds that populate the wet western oakwoods of Great Britain, mainly in western Scotland and Wales. In addition to common Irish species like Chiffchaff and Willow Warbler, it includes a number that are largely absent from Ireland. They are the Wood Warbler, Common Redstart, Pied Flycatcher and Tree Pipit. All of these have been seen at Glenveagh. Wood Warbler has been frequently recorded and almost certainly breeds occasionally at Glenveagh, and in other native woods around the county. The Redstart has bred in Glenveagh on a number of occasions, but is only rarely recorded. Pied Flycatcher and Tree Pipit remain extremely rare and with no evidence yet that they might have bred. However, all four species are scarce migrants which can be seen most years on Tory and at other migration outposts. If greeted by more suitable habitat when passing over Donegal, who's to say they would not take up the offer to stay and breed? The key is surely the management of the habitat.

Native oakwood, Ardnamona.

Most of the remaining fragments of oakwood in the county are all too small to support more than a single territory, and hardly any are managed in a way that would encourage these specialist birds. State-owned woodland nature reserves include Ardnamona and Ballyarr, both of which have high-quality oakwood habitat. Ards is owned by Coillte, the state forestry company, and is largely managed now for its non-timber resources, which includes biodiversity.

Forestry

Without an indigenous timber resource that could sustain a modern nation, Ireland has, since the 1950s, developed a forest industry based on exotic conifers, mainly the Sitka spruce and lodgepole pine from the north-west of North America. Initially, the peaty soils of upland areas were highly favoured, not least because the land was cheap but also because these species could grow well on it, once drained. The peat soils at the middle altitudes are where most of the area under exotic conifers can be found. The outcome is a very large patchwork of forested sites throughout the county, apart from the storm-prone western fringe and the higher mountain summits. The most extensive blocks occur along the southern border with County Tyrone, in southern Inishowen, and on the slopes around the edges of the north-west highland mountain ranges.

Forest birds

We might expect our native woodland species to spread into this new habitat, but very few do this to any significant extent. Ireland mostly lacks the scarce conifer specialists that find a home in the native pine forests of Scotland, and even in the non-native commercial plantations throughout Great Britain. Ireland simply lacks the diversity of suitable woodland and forest species. The only ones we have are the Siskin and Crossbill, and they have done well. Some generalist species like Coal Tit and Goldcrest have also made good use of this new habitat.

What we *do* have is a suite of songbirds that colonise the young forests while they are in the grass and scrub stage of growth – both new forests, and second-generation re-stocked sites. These include Whinchat, Grasshopper,

Conifer
afforestation,
Aghla Beg.

Conifer
re-stock site,
Drummonaghan
Nature Trail,
Ramelton.

Sedge and Willow Warblers, Whitethroat and Blackcap. Hen Harriers also use this habitat for breeding, although they, and the Merlins which nest in the taller trees, both hunt mostly in the remaining unplanted moorland.

Against those limited gains, there were significant losses as upland habitat specialists dwindled through the breakup of large expanses of semi-natural bog and heath. Hen Harriers are the most high-profile casualty, while Merlin, Red Grouse, Golden Plover, Red-throated Diver, Peregrine, Dunlin and Ring Ouzel have all suffered.

However, although still concentrating on the non-native conifers for commercial timber, forestry is changing rapidly in Ireland. When Coillte was established as the state forestry company, its brief was strictly commercial.

Mature conifer forest, Murvagh.

Then the world markets started to demand that timber be certified as from sustainable sources, under such organisations as the Forest Stewardship Council. To qualify for that, Coillte designated a sizeable proportion of its landholdings to be managed primarily for the protection of biodiversity. It has also stopped the practice of planting on peatland, and is restoring some of the more important bogs which had been planted. Private forestry has been slower to change, but the incentive of EU grants, and the requirement to reduce our net output of CO_2 emissions, is having much the same effect. For both state and private planting, the EU has been instrumental in dramatically increasing the proportion of broad-leaved trees.

What still remains to be done is for the management of commercial forestry, both conifer and broad-leaved, to be 'near to nature', as some Europeans describe what is generally know here as 'continuous cover'. Only then will we see the return of woodland ecosystems capable of supporting the full range of species that are on the brink.

We must also bear in mind that Irish woodland birds are primarily adapted to deciduous habitats, and we have a long way to go before the tiny residue of good-quality native woodland is significantly increased. Our breeding population of Wood Warblers is hanging on by a thread, but the impending colonisation by Great Spotted Woodpeckers shows that we have much to gain by continuing the drive to restore and expand our native broadleaved forests – and to maintain the progress in transforming the forest industry.

Chapter 7

Uplands

Golden Eagle and Red Grouse

Habitats

DONEGAL HAS TWO important mountainous areas, in the north-west highlands and the Blue Stack Mountains, and two other significant areas of high ground – central Inishowen in the north-east, and the south-west promontory. These are all based on either granite or quartzite, so in theory the acidic peat soils derived from both rock types should be supporting vegetation dominated by heather on better-drained land, and by sedges and Sphagnum mosses on the wetter bog surfaces. In practice, overgrazing by sheep in recent times has greatly reduced the density of heather. Sheltered gullies in the mountain zone can often retain fragments of woodland, and crags that are inaccessible to sheep and deer, even if treeless, can give refuge to elements of woodland flora, or at least scrubby heath.

Dunlin, on eroding wet heath to Moylenanav (539 m).

JOHN CROMIE

These habitats, along with streams and lakes and larger remnants of oak woodland, can all be found at Glenveagh, which highlights the best that can be found throughout the Donegal uplands. From the birdwatchers' point of view, Glenveagh is a hint of what is typical of the Scottish Highlands but rare or absent throughout the rest of Ireland, which adds spice to the rich and varied bird communities of the national park.[1]

The dominant vegetation of most of the upland zone is bog. If there is more than a metre of peat blanketing the landscape it is defined, not unreasonably, as blanket bog. Where the peat is shallow it is called wet heath – the variant usually found on the higher slopes and on rocky landscapes. The flatter terrains often have at their centre dense networks of small dark pools. They appear to be random and interlinked, but an aerial view reveals patterns which indicate that the bog is stretching, with separate pools opened up and shaped by gravitation.

The central hills have similar vegetation, but being at lower altitude they are much more altered by human activity – mainly peat cutting and afforestation, and of course more sheep-grazing.

The birds

Meadow Pipits, Skylarks and Wheatears are ubiquitous in the uplands. Their numbers are generally an indication of the abundance of insects and the health of the habitats. If these birds are numerous, so too will be raptors like Peregrine and Merlin. Red Grouse feed on heather although their chicks start off on insects before they can cope with an adult diet. They too will be indicators of the health of the uplands. Surveys of the habitat and of individual species indicate that while the situation has deteriorated, it is not as bad as might have been expected.

But there is no doubt that uplands are under severe pressure. Grazing by sheep could be regulated by national policies on farming. Buildings, roads, power lines and other developments continue to reduce and fragment the areas of unbroken habitat. This leaves ground-nesting birds more vulnerable to general predators like foxes and Hooded Crows, never mind the alien mink – the encroachment of human activity and nest predation by mink are the main problems facing our tiny population of Red-throated Divers.

Ring Ouzel is a declining mountain species. Grazing by sheep or deer in the high, rocky gullies favoured by these birds has probably reduced their suitability. Climate change is undoubtedly involved, although in what way is not yet exactly clear. It can affect plants and animals in many ways, such as the destruction or degradation of habitats by uncontrolled wildfires – which seem to be more frequent and more severe with every passing year. The breeding waders, Golden Plover, Dunlin and Curlew are also affected, but they could hopefully respond to management of the overall upland environment.

With a bit of luck, you might expect to find breeding Dunlin, but they have now almost gone and climate change is likely to see them out. Golden Plover are also declining, but Red Grouse are still quite widespread across the county, albeit at low density. The abundant species are Meadow Pipit and Skylark, which sustain a thin but quite healthy population of Merlin, although almost any of our less (or differently) specialised raptor species can also be found.

The logo of the Glenveagh National Park is a Red-throated Diver. Although they do not breed on Lough Beagh, birds in breeding plumage can be seen there in the summer months. A handful of pairs do breed in our upland lakes, at the southern tip of their global range, and have done so since at least the nineteenth century. Both Great Northern and Black-throated Divers have occurred in the breeding season, but with no prospect of actually finding a mate and settling down.

Conservation

How the breeding Red-throated Divers have managed to hang on for nearly a century and a half is a mystery, but ensuring that they continue is not. It will be down to effective protection measures, consistently applied from year to year.

Much of the better-quality upland habitat has already been designated for protection under one or other of the European Directives on habitats or birds. But that has not been able to bring about suitable management. While this is logically something that should be under the direction of the national agency with responsibility for conservation of biodiversity, the National Parks and Wildlife Service, at the moment it is more likely to happen as the outcome of national farming policies and grant schemes. These influence the choices made by land users, from farmers and forest owners to hill walkers.

The presumption that farming is only about the production of food, and that the only goal has to be the reduction in the cost of food to the consumer, is now being challenged. Farming can equally be seen as land management – for various ends, such as carbon-capture, biodiversity enhancement and recreation, as well as for food production.

One excellent example of what can be done is the Inishowen Upland Farmers Project, where farmers within an area dominated by high nature

Errigal (and Little Errigal).

value farmland (HNV) have got together to develop a sustainable farming system that will not be a drain on family incomes, and will also deliver targets on climate change, biodiversity and water quality. This has so far involved a shift from sheep towards hardy breeds of cattle, planting broadleaved trees and reductions in fertiliser use. This is bound to deliver better results for birds – both in the numbers of the common species and in the survival of rare and threatened species like Curlew or Whinchat.

Given the right incentives, farmers and other land users are perfectly capable of making the necessary changes, and are usually willing. As long as national policy-makers make sure that whatever happens is sustainable socially, economically and environmentally, then the future for our upland birds should be assured. But it will need a re-assessment of the true value of a fully functioning upland ecosystem, which is surely now overdue.

The re-introduction of the Golden Eagle has been hugely successful at highlighting the importance of our wild mountain landscapes. It remains to be seen if it will ultimately succeed in restoring a self-sustaining population of eagles to the mountains of Donegal. If that does eventually happen, it will be due in large part to the habitat management that the national park has been able to provide, and elsewhere to a revision of priorities in agricultural support schemes.

Bog pools, Gannivegal Bog.

Chapter 8

Lakes, Rivers and Streams

Kingfisher, Common Sandpiper, Goosander, Dipper

Lakes

THE IRREGULAR TOPOGRAPHY of our hilly county has given rise to an abundance of lakes. In the mountains they are mostly small, and being extremely acidic (oligotrophic, or soft water), they support a highly specialised flora and fauna of low diversity and abundance. Small- and medium-sized lakes are scattered across the west coastal plain, and on the Pettigo plateau and surrounding area. Like those in the uplands and bogs, these are also highly acidic and tend to be relatively lifeless. Many of our largest lakes are to be found around the fringes of the uplands. They are slightly less acidic and consequently have a richer flora, and rather more birdlife. They also give rise to some of our major rivers – the Owencarrow from Lough Beagh, the Leannan from Gartan Lough, the Finn from Lough Finn, and the Eske from Lough Eske. Only Lough Eske and the smaller lakes in the limestone areas south of the Blue Stack Mountains are alkaline (eutrophic, or hard water). Eutrophic lakes support much more life than the oligotrophic ones, but most of our specialist bird species have other considerations in mind when choosing their breeding territories, such as suitable nesting sites.

Two large freshwater (or very slightly brackish) lagoons were created by the movement of sand by wind (Dunfanaghy New Lake), or by longshore drift in inshore waters (Durnesh Lough), in both cases with the effect of holding back the encroachment of the sea. At Dunfanaghy this was precipitated by human activity, and as might be suspected from the name, it happened fairly recently. Prior to the First World War, Dunfanaghy Harbour was open to sea-going vessels, and the tidal area extended into a large part of what is now the New Lake. During the war, there was a demand for coarse grasses to feed the thousands of horses at work on the battlefields, and as bedding for the soldiers. Marram was ideal, and was cut on the sand dune systems around Sheep Haven Bay and Dunfanaghy. With nothing then to check the erosive power of wind on the bare dunes, the sand took to the air and settled in more sheltered areas. One of these was the outlet between the inner and outer sections of Dunfanaghy Harbour. And with tidal inundations halted in the inner harbour, it was only a matter of time before it filled up with fresh water (despite the absence of any streams of consequence entering it). Some fields were also lost, and an old embankment which had been built to limit the reach of the tide was not able to keep out the fresh water. It remains as a reef between the north and south parts of the lake. The depth of water in the outer harbour was also reduced – terminating Dunfanaghy's history as a seaport.

Durnesh Lough is a natural lagoon that is mostly protected from the sea by a drumlin (a low hill of glacial deposits). The narrow access for the sea to the north of the drumlin was blocked at some time in the past by longshore drift of sand.

Both New Lake and Durnesh Lough have surrounding marshy habitats, with Durnesh in particular having very extensive reedbeds and fens. They are both shallow and have rich aquatic vegetation.

BIRDS

The best of the larger inland lakes are Lough Akibbon and Lough Fern. In winter, Akibbon usually has small numbers of the common diving ducks, and a few Whooper Swans. Bird numbers at Lough Fern can be erratic, but it is nonetheless one of the very best sites for diving ducks. In good years, there are large numbers of Tufted Duck, Pochard and Goldeneye, and it has recently become a regular haunt for a small flock of Scaup, at their only inland site. Two rarities are usually now present – Goosander and Ring-necked Duck. Greylag Geese numbers are growing and several hundred can be expected. An unusual feature is the large Cormorant roost among the trees on a small island.

Durnesh Lough and New Lake have similar collections of wintering birds – Tufted Duck, Pochard, Goldeneye, Wigeon, Teal, Mallard, Mute and Whooper Swans are the standard fare at both sites. At New Lake the large resident flock of Mute Swans usually outnumber the wintering Whooper

A 1,000 Barnacle Geese, Dunfanaghy New Lake.

Fen and reedbed, Durnesh Lough.

Swans. Dabbling ducks have included all the usual common species, and occasional rarities like Blue-winged Teal, Garganey and American Wigeon. Diving ducks are more numerous and the New Lake is one of the better sites for Pochard and Goldeneye – a Ring-necked Duck can often be found among them.

Over recent years the number of Barnacle Geese at New Lake has risen and they now regularly peak at over 1,000 birds. At Durnesh, it is the Whooper Swans that have been expanding, and they have also peaked at over 1,000, in the autumn of 2021. New Lake hangs on to its White-fronted Goose flock, feeding on the richer agricultural fields around the lake, but it has been some time since the south Donegal White-fronts have made use of the Durnesh drumlin fields. A flock of Scaup used to be regular at Durnesh, but has now faded away.

An area of marsh and wet machair grassland at the upper (western) end of the New Lake was drained about thirty years ago, but plenty of good habitat remains, with opportunities for Redshank and Common Sandpiper to breed. Sedge Warblers are usually to be heard wherever there is sufficient cover around the shore. The area retains a wilderness of dunes and large seasonal pools between the lake and the sea at Tramore Strand – a refuge and roost for the small flock of Greenland White-fronted Geese. The blown sand also continues uphill to envelop the Horn Head foothills, supporting more machair, and carpeting a moribund forest of maritime and Corsican pine (tailor-made to welcome Great Spotted Woodpeckers).

Breeding species that have abandoned New Lake and the wider area are Sandwich Tern, Dunlin and Red-necked Phalarope.

Streams and rivers

The multitude of small streams flowing from the wet uplands are usually no more than 20 km from the sea, or an inland lake, and often a lot less. Whether on a bedrock of granite, quartzite or more ancient schist-type rocks, the waters will be strongly acidic. This reduces the abundance of the aquatic life that our riparian birds feed on, so bird territories need to be fairly long to ensure an adequate food supply. There are less acidic streams south of the Blue Stack Mountains, where limestone is the prevalent rock type. These are more likely to support a higher density of breeding birds, but there is little hard evidence to show that this applies in Donegal.

Owengarve River,
Blue Stack foothills.

The middle reaches of these streams swell to become rivers, and as they have also reached farmland, they are much more likely than the upland streams to have limits imposed on the development and extent of their bankside vegetation. Gravel extraction further reduces the abundance of aquatic life.

The main rivers in Donegal all flow to the east. The Leannan and Swilly flow into Lough Swilly, and the Deele and Finn flow into the Foyle system, and ultimately to Lough Foyle. These are the only rivers that have significant stretches of less dynamic slow flow, where rather than eroding the bedrock and gravel, they deposit the sand and silt carried from upstream. Sandy banks can then come and go, as the rivers meander across almost level ground.

Arterial drainage has greatly altered the habitat on the lower River Deele, for example. At its extreme, large-scale dredging along the Foyle has simplified the bank-side habitats, as well as altering conditions on the river bed for aquatic life at the bottom of the food chain. Pollution is of course a universal problem, although the number of inland sources of untreated sewage is slowly decreasing. All these interferences with the natural flow of rivers is to the detriment of breeding birds.

Happily, there is a growing awareness that catchment management, with the involvement of the local communities, can lead to a better outcome for both people and wildlife.[1] The Inishowen Rivers Trust is a good example of what can be done.

Dipper country, Bullaba River.

Lowland limestone river, Eany Water.

Middle reaches, River Deele.

BIRDS

Grey Wagtails are found in almost all our waterways, from the fast upland streams to the broad lowland rivers. On the upland streams they are joined by Dippers and Common Sandpipers. Mac Lochlainn's 1984 survey of Glenveagh found eight Dipper territories in the core area of the national park.

The slow-flowing lowland rivers are where we expect the Kingfisher. Vertical banks meeting all the criteria for nest-building can be hard to find, and that may be the limiting factor in determining territory length, rather than food supply. Sharing the Kingfisher's habitat is the Sand Martin. Its riparian colonies are smaller than at other sites with large expanses of sand cliff, such as quarries, but they are probably more frequent. They may well account for the majority of the breeding pairs in the county. Dippers can also be found in lowland areas, especially if old stone bridges provide them nesting sites. Moorhens have greatly reduced in number since the mink population became established along river banks.

Looking to the future, there are a couple of exciting possibilities. From 1969 and through the 1970s, a pair of Goosanders bred in Glenveagh – the first instance of their breeding in Ireland. The species only colonised Scotland in 1871, and they have been spreading south ever since. They are now frequent enough as winter visitors, and the possibility that they could yet establish a permanent breeding population in the county remains. These fine birds have only a toe-hold in Ireland, but even that is enough to significantly enrich the Irish avifauna. It is the lack of riparian woodland rather than the lack of suitable streams that is most likely slowing their establishment in Donegal. Perhaps some inducement in the form of nest boxes would be enough to bring them back to Donegal as regular breeders.

One other species that can be included in this context is the Osprey. There is an average of about one migrant bird seen somewhere in the county each year. That is hardly enough to project a breeding presence, but we can however look to the continuing rise in the Scottish population. Surely this will eventually bring two birds together at one of the many lakes in the county that would be suitable for natural colonisation – or we could mount a re-introduction project.

Chapter 9

Estuaries

Peregrine, Little Egret, Pintail, Black-tailed Godwits, Wigeon, Teal, Shoveler, American Wigeon, Grey Heron, Whimbrel

Habitats

ESTUARIES ARE TIDAL TRANSITIONAL ZONES between the non-tidal rivers and the open sea. In Donegal they are found right around the coast.

Starting at the north-east and proceeding anti-clockwise, Lough Foyle is a cross-border feature, as is the long tidal reach of the River Foyle inland to Lifford/Strabane. These are both excellent bird wetlands. The River Foyle has some of the best reedbeds in the county. Lough Foyle has excellent muddy habitats in the south, although they are overshadowed by the vast expanses on the flatter shores of the County Derry side, in Northern Ireland.

Gweebarra
River, a sandy
west coast
estuary.

By far our most important estuary is Lough Swilly, which has merited a chapter on its own (Chapter 12). Further details on estuaries in general can be gleaned there, along with the particular topics relating to Lough Swilly itself.

The north coast has a number of large estuaries – becoming increasingly sandy towards the west. Trawbreaga Bay in the east has good muddy habitat, which the birds now have to share with the aquaculture industry. Mulroy Bay may look on the map like a smaller version of Lough Swilly, but it is essentially marine, with only a few small streams draining into it. The inner reaches of Sheep Haven Bay are diverse, with extensive salt marshes and muddy habitats at the upper extremes, and sand flats closer to the marine body of the bay. Dunfanaghy Harbour is a smaller, undivided bay but with similar features. Ballyness Bay in the west is an extensive area of sand flats.

Along the west coast, short, fast rivers descend from the hills and generally empty directly into large, shallow, sandy bays. These run from Gweedore Bay in the north to Loughros Beg Bay in the south. The most useful one for birds is the small estuary at Derrybeg, which supports a sizeable area of salt marsh, and a good stretch of sandy mud. The Gweebarra River is an unusual one, with a long, narrow sandy estuary that reaches inland all the way to the edge of the central mountains at Doochary.

The south-facing coast, from Malin Beg east to Donegal Bay, has a series of small estuaries that are more sheltered and have some muddy habitat, but little salt marsh. Teelin and Killybegs harbours, and the estuaries of the Eany and Eddrim Rivers, are the best. Rather more mud is available at Inner Donegal Bay, where there are also some corners of salt marsh

Salt marsh, Derrybeg.

tucked in among the drumlins which break up the outline of this interesting landscape.

At the southern border of the county is the west-facing estuary of the River Erne. It has the largest discharge of all the Donegal rivers, including even the massive Foyle, but it is packed into a relatively small, dynamic space. The habitats are dominated by sand flats at the expense of any decent expanses of mud.

The birds

Estuaries and their associated habitats hold most of our waterfowl – the group of birds for which Donegal is most highly valued.

Most of the birds on Lough Foyle are to be found on the wide, flat County Derry shore, with its adjacent polders. But the Donegal west bank also has good areas of mud, and shingle. Many of the species on the Derry side are also present here in smaller numbers – Teal, Wigeon and Knot among them. Brent Geese are joined by Eider in the waters off these muddy stretches, especially off the small deltas where the streams flatten out after their downhill drop through a series of wooded glens. In summer, Sandwich Terns are regularly present, and have bred.

Birds are more evenly shared between both sides of the River Foyle, which has a broad estuarine stretch between St Johnstown and Carrigans. This can be used by roosting geese and swans, and to a lesser extent by feeding waders, especially Curlew and Black-tailed Godwit. In the 1990s, a small flock of Goosanders wintered regularly.

The reedbeds of the River Foyle were the home of Bitterns in the early nineteenth century. Bitterns, Marsh Harriers, Reed Warblers and Bearded Tits are all spreading in Great Britain, and Ireland is in their sights or is already part of the spreading range of most of these specialist reedbed species. Further climate-driven changes could bring some of the other herons and heron-allies north through Ireland in the more distant future. But for the moment, we have to be content with our own Water Rails and Reed Buntings, while keeping an eye open for the unexpected.

On the northern side of Inishowen, Trawbreaga Bay is a fine estuarine habitat, although the streams entering it are all small. Unfortunately, the area available to birds is now severely reduced by aquaculture. Time will tell if it is also reducing their numbers. Trawbreaga is most noted for its flock of Brent Geese, and now also for increasing numbers of Barnacle Geese in the surrounding fields.

The other north coast estuaries have small bird populations, but are not without interest – Sheep Haven Bay for its Common Scoters, and Dunfanaghy for its Wigeon and as one of the first sites to regularly attract one or two Little Egrets. Ballyness Bay has good flocks of Sanderling, Golden Plover and Brent Geese. Eider, Cormorant and Shag along with Brent Geese are often attracted to the narrow outlet to the sea where the tidal waters speed up and support a rocky, more marine habitat.

The sandy estuaries of the west generally hold few birds. Derrybeg in the north is the best of them, and its position must explain why over the years several American Golden Plover have joined the usual autumnal parties of European Golden Plover, Sanderling, Dunlin and Ringed Plover. But even the Gweebarra River, with its undifferentiated sand flats, will attract Cormorants and the fish-eating Red-breasted Mergansers, especially when the sea trout and salmon are running upstream to their breeding grounds.

On the south coast, Donegal Bay is by far the best of the sites. A large flock of Common Scoter occupies the marine area, but also extends to the upper reaches at Mountcharles which are very definitely estuarine. Other sea ducks, divers and grebes are usually also present.

On the Erne estuary, most birds will be seen roosting on the sliver of salt marsh, or outside the narrow mouth of the river, in or near the churning waters where the river contacts the sea. Cormorants usually gather here in good numbers, and a large moulting roost of Red-breasted Mergansers has favoured the calmer waters among the stony reefs on the north side of the outflow.

Chapter 10

The Coast

Puffin, Chough, Razorbill, Rock Dove, Guillemot, Fulmar

Habitats

HARD COASTS

HARD AND SOFT COASTS are easy enough to interpret, and Donegal has more than its fair share of both. The hard, rocky and cliff-bound coasts are resistant to erosion, and so project further into the ocean, as headlands and promontories. The main stretches of hard coast are (anti-clockwise from the north-east):

1. Inishowen Head and the Glengad cliffs of north-east Inishowen

2. Dunaff Head to the Urris Hills on the Inishowen west coast

3. Saldana Head to Fanad Head on the opposite side of outer Lough Swilly

4. Rossguill and Melmore Head

5. Horn Head

6. The north coast of Tory

7. The north and west coasts of Arranmore

8. Crohy Head and the coast between Dungloe and Trawenagh bays

9. The south-west promontory including Slievetooey and Slieve League

10. Muckross Head

11. St John's Point

12. Doorin Point

13. Rossnowlagh to the Erne estuary.

SOFT COASTS

Soft coasts are usually recognised as sandy or muddy, and can be either eroding or accreting, but are rarely stable for long. They are the stretches between the rocky cliffs and headlands. Stony beaches, with anything from shingle to boulders, are also soft, in the sense that the material can be moved around by the sea, but they can also resist erosion to a degree, as the higher barriers of rounded stones (storm beaches) sap the energy of storms and high tides.

Behind the sandy beaches there is often an extensive zone of sand dunes, sandy grassland (which can be dry or marshy) and ephemeral pools. The general name for this suite of related habitats is machair. It is confined to the western isles of Scotland, and the north-west coast of Ireland where Donegal has the greatest area and number of sites. The most extensive machair systems occur at Doagh Isle, Rossguill, Sheep Haven Bay, Dunfanaghy, Gweedore, Gweebarra and Sheskinmore. These big machair systems all lie behind our longest sand beaches.

Also included as soft coasts are salt marshes – terrestrial vegetation that is tolerant of regular inundation by sea water. These are covered in more detail in the previous chapter on estuaries.

Donegal Bay is a very large area covering most of the county's south-eastern shoreline, and is really an association of varied coastal sites, but all are lumped together under the I-WeBS censusing scheme as a single wetland, with sub-sites. That includes estuaries and a lagoon, which are dealt with elsewhere. So the coastal habitats are:

Horn Head –
a hard coast.

Tormore ridge,
Tory Island.

Soft coastline,
Dunfanaghy
Harbour.

1. A rocky limestone platform in the south-west, from Doorin Point to Mountcharles.

2. Three long sandy beaches at Murvagh, Rossnowlagh and Bundoran. The last two are facing the open waters and have a broad surf zone. Murvagh is different and the extra area exposed at low tide is slightly more muddy than at the other two beaches.

3. Two more varied soft coastlines, from Mountcharles and Eddrim to Murvagh, and at Inishfad, between Murvagh and Rossnowlagh. There are stony, sandy and muddy areas on both of these stretches.

4. A cliff coastline from Rossnowlagh to the Erne estuary, and from Bundoran to the county boundary with Leitrim. This is largely inaccessible to birdwatchers.

Birds

The sea cliffs that can justly claim to host 'seabird cities' are Horn Head and Tory. Those on Stookaruddan off Glengad and Tormore off Slievetooey have been abandoned since the 1960s. Smaller colonies of only one or two species each are present on most of the other stretches of hard coastline.

Most of these colonies have been counted a number of times over the years, including three complete surveys of all colonies in Ireland and Britain, and one national survey in Ireland. These were initially carried out mainly by teams of volunteers, but the later surveys had increasing involvement of NPWS staff and professional contractors. The first was Operation Seafarer, which operated from 1969 to 1970.[1] Then from 1985 to 1987, the Seabird Colony Register repeated the exercise.[2] Seabird 2000 ran from 1998 to 2002.[3] Finally, the National Seabird Monitoring Programme which NPWS conducted from 2015 to 2018 has given a comprehensive verdict on the state of play for all species, although more work needs to be done on three burrow-nesting species – Puffin, Manx Shearwater and Storm Petrel.[4]

The birds that choose to nest in these mixed colonies are Fulmars, Razorbills, Guillemots, Puffins, Kittiwakes, Herring Gulls, Great Black-backed Gulls and Shags. Cormorant colonies are usually shared with few birds of other species – mostly large gulls. Some Black Guillemots nest in holes and crevices at the base of the cliffs, but most of them are spread more widely on rocky shores. Common Gulls mostly nest inland, and also some Lesser Black-backed Gulls, but small numbers of both of these species are usually to be found in the big seabird cities. One or two predators can be expected to breed close to the colonies – Peregrine and Raven do so regularly, and now there is the prospect of Great Skua and White-tailed Eagle joining them in the not too distant future.

While it is the migrants that nowadays attract more birdwatchers to Tory, the cliffs have long been known for their colonies of seabirds. Early observers didn't often count the birds in a colony – their observations on Tory were typically more comparative, for example the third largest Puffin colony in Ireland, or descriptive, for example the Guillemot colony was described as 'a lofty perpendicular rock at the east end of Tory Island ... covered with these birds'.[5]

Seabirds still breed on the northern cliffs, and especially those at the dramatic promontory at Tormore and its associated stacks at the eastern

Seabird city: 90 Kittiwakes and 200 Guillemots, Tormore (east face), Tory Island.

end of the island. The most important species is probably the Puffin, which in Donegal had three historical colonies. Tormore in south-west Donegal has faded out, and the Horn Head and Tory colonies in the north have both declined. At their peak the three colonies were among the most populous in Ireland. Tory was probably the largest, but numbers at both surviving colonies are now at a very low ebb, and although uncounted recently, are likely to be in the low hundreds of pairs.

There are two other notable cliff breeders – Chough and Rock Dove. While the Rock Dove's genes live on in the Feral Pigeon, in its original wild form it is now very rare in Ireland. So Rock Dove and Chough are among our most valued coastal breeding birds. Both can also be found, especially in winter, feeding on coastal heath or grassland. If the cliff tops hold at least some earth, Wrens and Rock Pipits are usually present, and Jackdaws can make use of Puffin or rabbit holes.

It was not until the stay of Philip Redman in 1954 that any serious attempt was made to quantify Tory's breeding birds. This he did for all species, and not just for the colonial seabirds. His account of that expedition was written up as an unpublished typescript at the time, 'Birds of Tory Island, County Donegal in 1954', which was finally published in the *Tory Island Bird Report 2015*.[6] His counts and estimates, along with any comparative figures from the years since then, are presented in an earlier paper in the same series.[7] Table 10.1 presents a selection of this data for the most important species. The major changes are the arrival of Little Terns and Tree Sparrows as breeding species, and the demise of Twite. Other blank entries just indicate a lack of data. The recent wader figures are from detailed surveying by Michael Bell (averages of seven or eight years, depending on species), showing stable trends for all species apart from Lapwing, which is declining.[8]

Inishtrahull has had little attention since 1965, until now with the establishment of Inishtrahull Bird Observatory in 2020.[9] The species breeding then were fewer, but of considerable importance. The Herring Gull colony, at 3,000 pairs, and the 300 pairs of Shag, were the most important colonies in the county for both species. The island remains important for Shags, but the loss of the Herring Gulls, down to a token few pairs, has had a major impact on the county population as a whole. Other breeding species of note were, and still are, Eider and Arctic Tern.

	1954	1961–3	2012–13	2014–24
Fulmar	178		250	
Corncrake	4	8	11	
Oystercatcher	1	6	30	55
Lapwing	20	40	30	22
Ringed Plover	40		25	26
Dunlin				5
Common Snipe				8
Redshank		10	6–8	19
Kittiwake	508			
Common Gull	2	8	50–60	
Herring Gull	30		10	
Little Tern		20	10	2
Guillemot	165			
Razorbill	754			
Puffin	1,700			
Skylark	150		20	
Tree Sparrow		3–4	15	
Pied Wagtail	3		10	
Meadow Pipit	10		10	
Rock Pipit	150		50	
Twite	15–20	2		

Table 10.1: Estimates of breeding pairs on Tory (most passerine figures from 2012–13 refer to birds rather than pairs)

Soft coasts have a different set of species. Ringed Plovers are almost ubiquitous on the sand and shingle beaches. All species of tern breed on sand or shingle, although adjacent rocky islets or low headlands are often used, where the terns are usually joined by a few gulls and Oystercatchers. These colonies are subject to disturbance, and predation by terrestrial predators. Whether for those or other reasons, most of our terns and Oystercatchers now breed on the offshore islands. One of those, Inishkeeragh, has had all five species of tern breeding, but Roseate and Sandwich never managed to establish colonies.

Machair systems are where we have traditionally expected to find a diversity of breeding waders – especially Lapwing, Dunlin, Redshank and Common Snipe. The importance of machair for these species has led to a long series of surveys documenting the population changes over the years, and the changes in the habitat.[10] In recent decades, changes in land management have combined with other pressures to reduce the numbers, but some of the better sites are now being managed for wildlife. Predator-proof fences have been particularly successful at Rinmore and Magheragallan, and, with qualifications, also at Inch Lough and Sheskinmore. Sheskinmore is one of the largest and most diverse machair sites and has, or had, most of the species we could expect – with Chough also usually present all year round. It is now managed on a landscape scale as a nature reserve. So hope remains that the bird losses on our machairs can be halted.

Although not a coastal species, it is on the coastal farmland that Corncrakes are now concentrated, and most especially on the offshore islands of Inishbofin and Tory. Tory must also be mentioned as the most important site in the county, by quite a margin, for breeding waders. Six species are regularly present – Oystercatcher, Lapwing, Ringed Plover, Dunlin, Common Snipe and Redshank. These are all relatively stable in number, but the recent cessation of grazing by sheep may be starting to make breeding difficult, for Lapwing in particular. Golden Plover and Common Sandpiper have bred, but are not regular.

Breeding birds on Malin Head in the 1960s included Grey Partridge, Tree Sparrow, Twite and Corn Bunting, all of them now long gone. Corncrake and Whinchat have both maintained a breeding presence, with Whinchats reaching a peak in the 1990s of about ten pairs each year, but both are now at a precarious level. These are not coastal species, but it is in coastal districts that suitable rough agricultural land is most easily found, then and now.

In winter, most of the seabird species that breed on the cliff colonies can still be found all around the coast, although Kittiwakes and the auks will have departed for a life on the ocean wave. Cormorants and Shags are joined from northern latitudes by Great Northern and Red-throated Divers, and resident Eider numbers are supplemented by winter visitors. These birds are easily found, especially in the slightly sheltered bays with a good broad surf zone of shallow water. Wandering individuals of Long-tailed Duck and Common Scoter can turn up anywhere, but are reliably found at a few favoured localities. For Long-tailed Duck, the north coast of Fanad and Inishfree Bay currently hold flocks of between thirty and sixty birds, and smaller numbers are reliably present in Donegal Bay. Common Scoter flocks are found at Donegal Bay and Sheep Haven Bay, and they have

Donegal Bay (with Brent Geese), Mountcharles.

an irregular but growing presence in Lough Swilly. Loughros More Bay has had occasional flocks.

Donegal Bay is the best coastline in the county for a number of species. Most obviously this is the case for the large flock of Common Scoter which moves around the waters off the three long sandy beaches, sometimes largely concentrated at one beach, and sometimes dispersed among all of them. If you are lucky, and patient, you might spot one or two Velvet or Surf Scoters among them. This is also the best site for Great Northern and Red-throated Divers, and one of the best in Ireland for seeing Black-throated Divers. Grey Plovers are very scarce in Donegal, and the shore between Inishfad and Rossnowlagh is the only place where a few are reliably present in winter. Purple Sandpipers can be found in the middle of Bundoran town, and at Doorin Point. Mute Swans on the shore at Tullaghan look out of place, but that is only our pre-conceived notion of where Mute Swans should be found.

Small parties of waders can be found almost anywhere in winter or on passage, and among the best coastal sites for seeing a good range of species are the Malin Head area, the coast around Tullagh Point, and the whole north Fanad coast. Otherwise birds are dispersed in small numbers around all the soft coasts. There are two wader species that are coastal specialists – Sanderling on most of the sandy beaches, and Purple Sandpiper on a few of the wave-washed rocky shores, the best of which are north Fanad, Inishfree Bay, Muckross Head and Donegal Bay.

Chapter 11

Top Twenty Sites

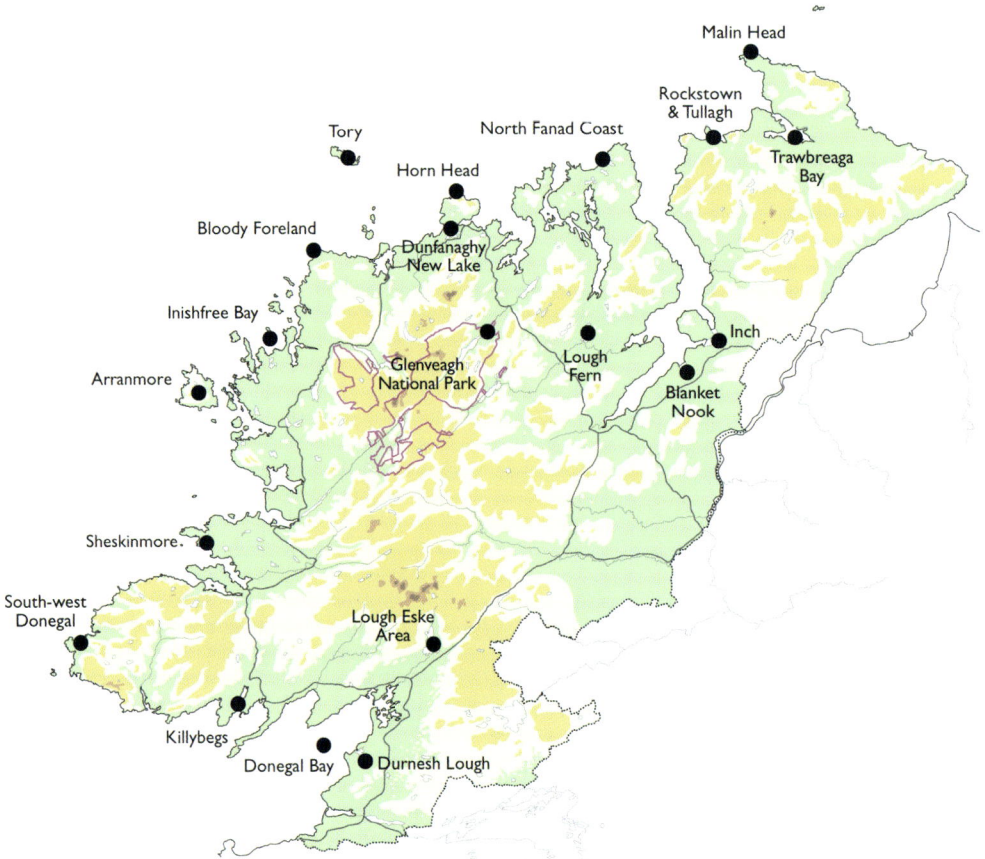

Map 11.1: Top twenty sites for birdwatching

THERE IS NO LEAGUE TABLE of sites, so this is just an alphabetical selection of a few places you might want to prioritise on a short visit. Some are seasonal specialities, and others are worth visiting at any time. You will see their names cropping up frequently in the rest of the book. Places like Inch and Tory are well-known targets for birdwatchers, and rarely disappoint. But for people wanting to explore further, there is no shortage of good habitats and rarely visited hidden corners. Donegal is still a county with scope for discovery.

ARRANMORE

The largest and most populous offshore island in Donegal has an impressive cliff-bound west coast, and a mostly sheltered and settled east coast. Breeding seabirds are concentrated at the north and south-west, and some of the best seawatching in the county can be had at Rinrawros Point in the north-west. Arranmore has had its share of rarities too, and in all seasons. There is a frequent car-ferry shuttle from Burtonport.

BLANKET NOOK

Blanket Nook is like Inch in miniature – a small lagoon, with good viewing from along the disused railway embankment, and parking for a few cars at either end. Regardless of its size, it often has many more waders than Inch – ducks are more or less in proportion. Diversity of waders is as much a feature as the total numbers. Roosting geese can occasionally dominate the lake. Tidal changes on the other side of the railway embankment, and the drainage regime that pumps water into the lagoon from the polder fields beyond its inner embankment, keep the number and variety of birds present within the lagoon constantly in flux.

BLOODY FORELAND

This north-west corner of Ireland is, like Rocky Point at the county's south-west corner, a barren heath. But the seabird migration across the view from the light beacon is a major attraction for birdwatchers. The shearwaters, skuas and other seabirds are often complemented by at least one raptor sighting – usually a Peregrine or a Merlin. In autumn, it is a place where, if you are lucky, you might also find Snow and Lapland Buntings.

DONEGAL BAY

This large and complex area can take a long time to explore fully. It has large sandy bays, like Rossnowlagh and Murvagh, which are prime sites for sea ducks. In between are low limestone cliffs and stony beaches. Two of our rarest wintering waders, the Purple Sandpiper and the Grey Plover, are regularly present in the winter. Inner Donegal Bay is a peaceful sanctuary where drumlin peninsulas and islands hide Kingfishers, Red-breasted Mergansers and Donegal's largest population of Little Egrets.

DUNFANAGHY NEW LAKE

Although a bit smaller than Inch Lough, the New Lake in winter is an excellent site for diving ducks and swans. The surrounding fields can hold Barnacle Geese in large numbers, and a smaller flock of White-fronted Geese. In summer, most of the resident waterfowl species breed here. The New Lake is along the main north coast road just west of Dunfanaghy, but for the slightly more energetic, the sand dunes along its north shore offer even better views.

DURNESH LOUGH

Although treated as part of Donegal Bay for censusing purposes, this is a large stand-alone lagoon, with extensive areas of adjacent fen. Many Whooper Swans join the resident Mute Swans in winter. Diving and dabbling ducks use the open water, and the reedbeds are home to hidden treasures, like the Water Rail. Access is a problem, but the surrounding drumlin hills give good telescope views.

GLENVEAGH NATIONAL PARK

The stunning glaciated mountain valley which is at the centre of Ireland's largest national park has native woodland, bogs and an exceptionally beautiful lake. The woods, from time to time, hold rare songbirds, like the Wood Warbler. It has been the main release site for the re-introduced Golden Eagles, and you have a reasonable chance of seeing one if you spend some time exploring the glen. Beyond the main valley there is a vast area of bogs, mountain heath and lakes, including several of Donegal's highest peaks. If you go for a good hike, nearly all of Ireland's mountain birds are possible.

HORN HEAD

Horn Head has one of Ireland's great seabird colonies along its 8 km of cliffs. Fulmars, Razorbills and Kittiwakes are the most numerous species. Guillemots and Puffins are more localised. To see them, you need to take great care to stay away from steep slopes and cliff tops. The best viewing is from gullies, where the opposite sides are at close range. A boat trip from Dunfanaghy is a good alternative – on a calm day. The Marble Arch area at the western end is a great place to relax on the soft heath and wait to be distracted from the seabirds by Choughs, a Peregrine or maybe a Merlin.

INCH

Inch Lough is the most accessible birdwatching site in the county. It is halfway between Letterkenny and Derry, with an 8 km walkway around the lake, three car parks, four birdwatching hides and a viewing platform, all with disabled access. At any time of the year there will be plenty of birds on the lake – various species of diving and dabbling ducks, geese and swans. Wader numbers are relatively low, but many of them can be conveniently viewed from the hides, including the scarce migrants.

INISHFREE BAY

All of the north-west coast, in the districts of the Rosses and Gweedore, is of interest. It is an intricate mix of rock and sand which, despite the huge human population dispersed throughout, still retains much of its natural features. At Inishfree, there is a large sand dune complex, with wintering Chough, but it is the sea that is the main attraction. It holds one of the best flocks of Long-tailed Duck in Ireland, which will be accompanied by divers, Eider and some waders around the shore.

KILLYBEGS

This busy fishing port is usually attended by hundreds of gulls – thousands in winter. Most of the rare species that have occurred in Ireland have turned up here, and there would normally be small numbers of Glaucous and Iceland Gulls in all their varied plumages. Closest views are to be had among the quays with unloading ships. More leisured inspection of roosting birds at the head of the bay is also recommended.

LOUGH ESKE AREA

Lough Eske is a large limestone lake, with small rock outcrops where Common Gulls, Great Crested Grebes and other species can breed in safety. The surrounding woodlands are among the county's most extensive, and perhaps the finest, where all our woodland species have bred at one time or another. Ardnamona is a national nature reserve, and the grounds of Lough Eske Castle Hotel have extensive woodland, both native and ornamental.

LOUGH FERN

This is the best of the inland lakes in Donegal. Flocks of the commoner diving ducks can sometimes be very large, but numbers are erratic. Among them, scarcer species are often present, like Scaup and Goosander. There is a big Cormorant roost on a small island, and flocks of Greylag Geese in the surrounding fields.

MALIN HEAD

Fine scenery and birds all year round distinguish this popular spot on the tourist trails – the most northerly point on the mainland of Ireland. Malin Head was one of the pioneering sites in Donegal, and Ireland too, for both the study of land bird migration, and seawatching. In winter, Barnacle Geese and Snow Buntings can be expected, and Twite if you are lucky. Summer birds include many of the species of rough coastal habitats, from Stonechats and Whinchats to Linnets and Choughs.

NORTH FANAD COAST

This is a fine succession of sandy beaches, low grassy headlands and wave-pounded rocks. Eiders, Long-tailed Ducks and all three species of divers are usually present in winter, along with small flocks of mixed waders. Choughs often turn up, and there is good seawatching from Fanad Head.

ROCKSTOWN AND TULLAGH

Rockstown Harbour, which features a magnificent shingle storm beach, is divided from the sandy Tullagh Bay by Tullagh Point, where shingle beaches are backed by wet heath with small boggy pools. Most of the wintering coastal birds can be expected here, on land or in the churning waters of the two bays. Several tern species have bred on the small islets that are accessible at low tides, but are unlikely to do so again without some conservation measures to attract them back. Coastal waders, in winter and in the breeding season, are still plentiful.

SHESKINMORE

Sheskinmore Nature Reserve, and the wider area, is one of the most diverse expanses of natural habitat in the county. For birds, there are lake areas where Marsh and Hen Harrier, Glossy Ibis and Large White Egret have all been seen. Twite and Whinchat can be found, and there is even a small flock of Eider and Common Scoter lying offshore in winter. Sheskinmore is not a place to expect large numbers or diversity of birds, but the quality is high, and there is no finer place to spend some time tracking down exactly what is on offer at the time. If the weather or the birds are not up to expectations, there will usually be the call of the Chough to raise a smile.

SOUTH-WEST DONEGAL

The isolated settlements at Malin Beg, Malin More and Glencolumbkille are places to explore at leisure in autumn and spring, when rare migrants can turn up in the cover of gardens, ditches and patches of scrub. A flock of Barnacle Geese is a regular feature in winter, as are small flocks of Chough. Rocky Point is the place to go for seawatching, and the tramp across the bog and heath to reach it is less of a trudge in the knowledge that rarities like Dotterel and Buff-breasted Sandpiper have trodden the ground before you.

TORY ISLAND

Next stop Iceland! There is spectacular cliff scenery and several small wetlands, but it is often in and around the two villages that you can find the rare migrants in spring or autumn. Tory is one of the best places in Ireland to find these birds. In summer, the seabird cliffs are among the most accessible in the county – at a safe distance. There is a quick ferry service from Magheraroarty (about an hour, usually twice a day). If you are a brisk walker, you can cover the whole island between ferries. And there are good accommodation options for those taking a more relaxed approach.

TRAWBREAGA BAY

Trawbreaga is a large, almost land-locked bay with an estuarine character. In winter, it is home to Brent and Barnacle Geese in increasing numbers. The narrow exit lies between sandy beaches, rocky outcrops and machair grassland on the Isle of Doagh (a large sandy peninsula), and large dunes and more beaches at Lagg, forcing water to enter and leave at speed. This is no problem for Cormorants and Shags, or for the Eiders that dive for shellfish. The Doagh and Lagg area is also prime country for Chough at all seasons, with large flocks often present. On the Isle of Doagh the machair attracts seed-eating songbirds, and in the damp areas, breeding waders.

Further details are dispersed elsewhere in these pages, and information on access, etc. can be found for most of them in one or other of the Irish 'Where to Watch Birds' guides.[1]

Chapter 12

Lough Swilly

Teal, Whooper Swan, White-fronted Goose, Lapwing, Pink-footed Goose, Redshank, Oystercatcher, Red-breasted Merganser, Shelduck, Greenshank – and a harbour seal

LOUGH SWILLY is one of Ireland's three genuine fjords, and is very much larger than any of the other estuarine sites in the county. It has many claims to fame, but it holds a unique place in Donegal for its diversity of birds and their habitats, and is one of the great bird wetlands of Ireland.[1] This is largely a measure of its value for wintering birds, but a national analysis, at 10 km² resolution, of breeding waterfowl diversity across the entire country identified the twenty-five most important 'hotspots'. Two sites within Lough Swilly were on the list – Inch Lough at No. 7 and Blanket Nook at No. 25. Also in Donegal was Dunfanaghy New Lake/Magheraroarty at No. 9, Tory Island at No. 11 and Trawbreaga/Malin at No. 15.[2]

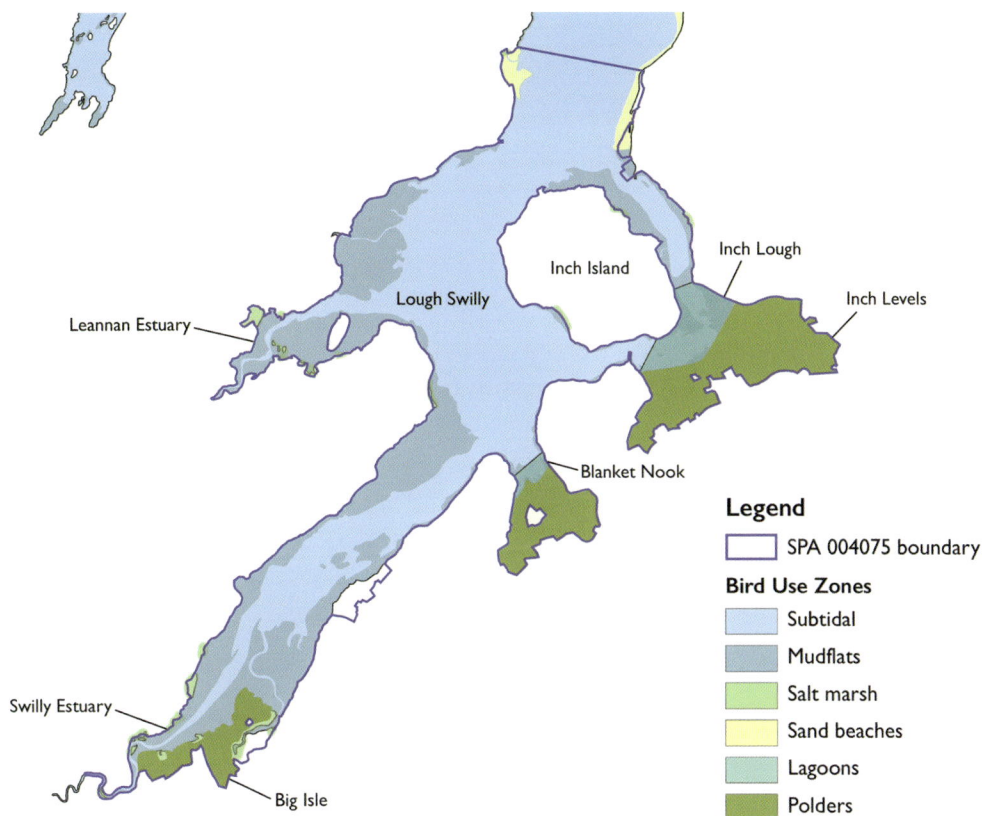

Map 12.1: The habitats of Lough Swilly

The habitats and their birds

The lough is long and narrow – 40 km by about 4 km in the outer half and 2 km in the inner half. The outer half is marine and best treated as part of the coast. The inner half is estuarine, and is the section dealt with here. For the purposes of monitoring by I-WeBS and earlier schemes, and for conservation, it is treated as a single large and varied waterfowl haunt (Map 12.1).

MUD

The agricultural farmland of east Donegal supplies nutrient-rich sediment, which sets Lough Swilly apart from all Donegal's other tidal habitats and sites. This mud and silt, so beloved by waders and wildfowl, is delivered by two main rivers – the Leannan in the mid-west and the Swilly in the south-west. Their estuaries are among the most densely utilised parts of Lough Swilly. The upper reaches are especially valuable as they are the last to be covered by high tides, so birds like Black-tailed Godwits and Redshank can continue feeding there when all other options have gone. As well as the waders, the mud supports dabbling ducks. At low tide they can disappear into the depths of the creeks which are a feature of the two main estuaries.

The vast expanse of mud around Big Isle is the base for huge flocks of Dunlin and Knot. Also in abundance are Shelduck, roosting Cormorants, Redshank, both godwits, Oystercatchers, Wigeon and Teal. Smaller numbers of Shoveler are just as important, and a good flock of Red-breasted Merganser can usually be seen hunting along the shallow water's edge. Most of the flocks of geese, Golden Plover, Lapwing and Curlew feeding on the Big Isle polders also take time out on the mud.

NARROW SHORES

In the middle sections of the lough, the mud zones exposed at low tide are relatively narrow, and attract proportionately fewer birds than the large mud flats. They are broken up with mussel beds where a good variety of birds will feed, especially Curlews, Oystercatchers and Turnstones.

The mud zone usually backs onto salt marsh where Wigeon and Common Snipe feed. They are joined at high tide by roosting Redshank and Curlew, among other species.

The salt marsh in turn gives way to a stony upper zone, also used for roosting, or to a narrow beach of coarse shell fragments. Oystercatcher, Curlew, Lapwing, Dunlin and Knot roost in tight-packed flocks. A few Greenshanks are usually in the mix, and rather more Redshanks. These roosts should only be viewed from a distant vantage point, to avoid disturbance.

These open shorelines are also the principal habitat for Brent Geese, which feed in the rising waters, attended by Wigeon picking up the leftovers.

A small area on the north shore of the Leannan estuary, which had been converted to an agricultural polder, has been reclaimed by the sea following a breach in the sea wall. Over the last couple of decades it has successfully reverted to mud and salt marsh and is now well used by Teal, Shelduck, Curlew and Redshank.

Lough Swilly
west bank, Castle
Shanaghan.

Polders and mud
flats, Big Isle,
Lough Swilly.

POLDERS

Polder is a Dutch term, now used internationally, for flat agricultural land that has been claimed from the sea – the local name is slobs, or levels. Lough Swilly has three large polders – Big Isle, Blanket Nook and Inch Levels. The fields are arable, and for many years were managed in a rotation of grain, potatoes and grass. In recent years the land use is dictated more by whatever the market is favouring at the time. So, many fields are now in permanent grass, fewer in grain or maize, and a very few in potatoes.

At Big Isle, Lapwing, Golden Plover and occasional flocks of Black-tailed Godwit winter in the wet fields, along with Greylag Geese, and it is the only polder area secluded enough to suit the White-fronted Geese. Brent have recently added a string to their bow, by adopting the Big Isle polders – perhaps they learnt the value of the rich grass on the playing fields of Dublin, where for a long time it has been enough to justify invasion of the pitches.

Blanket Nook is less well used generally than Big Isle or Inch Levels, although you can still find the occasional flock of Golden Plover. Greylag Geese use the fields closest to the lagoon. When there is spilled grain on offer, or potatoes, Whooper Swans will also visit for a week or so.

The Inch Levels polder area is the principal base for Whooper Swans, and Greylag, Pink-footed and Canada Geese. In flooded conditions ducks, waders and gulls will move in to join them. In the later years of the last century Whooper Swans used all three of the polder areas through most of the winter, but sadly they now use only Inch Levels.

SANDY SHORES

The mud in the upper estuary gradually gives way downstream to sandy shores at the seaward end of the inner lough. The beaches at Kinnegar and Lisfannan are used by Sanderling when the human presence is not too high. Offshore, a few Cormorants, grebes and Red-breasted Mergansers are usually present.

DEEP WATER

This is one of the minor habitats in terms of the number of birds it supports, but it is still of much interest. There is a core of deep water along the centre of the lough as far as Castle Shanaghan, but the best area for birds lies west of Ballymoney and Inch Island. Here Red-throated and Great Northern Divers can usually be found in small numbers of a dozen or so. Great Crested Grebes are more numerous, and exceed 100, as can Common Scoter. The rarer species like Velvet Scoter, Long-tailed Duck and Slavonian Grebe add icing to the cake. Slavonian Grebe is the most important of these, as our small population of a few dozen birds, at most, could be a sizeable proportion of the remaining Scottish breeding population.

BLANKET NOOK LAGOON

Blanket Nook is a 40 ha lagoon, penned in from Lough Swilly by a disused railway line, the Grange Embankment, which is a continuation of the Trady Embankment at Inch (see under Inch Lough). There is a pumping station which empties water into the lagoon from the network of polder field drains. A sluice gate then lets it flow out to sea at low tide, and closes when the tide starts to come in. But the water levels remain unpredictable – at least to those not conversant with the finer details of the system. Birds favour a moderate level of water – enough to retain mud around the lagoon margins, and to keep the small stony reefs high enough to be available for roosting. When the water is too high there will often be relatively few birds.

The wintering community of birds on the lagoon is similar to that on Inch, but with a greater diversity of waders. Many of these come in when the water level is suitable for feeding, for example Black-tailed Godwits. But many are responding to what is happening on the rest of the lough – so Oystercatchers will come in to roost at high tide, while Cormorants will roost at low tide. In autumn and winter, Blanket Nook holds the main concentration of Greenshank on Lough Swilly.

Where Blanket Nook scores most highly is in the frequency and variety of scarce passage migrants and rare vagrants dropping in for a day or two at any season, but mainly in the autumn. These include Ruff, Spotted Redshank, Curlew Sandpiper and Little Stint. Among the many vagrants, some of the most noteworthy have been American Black Duck, Temminck's Stint, Lesser Yellowlegs, Collared and Black-winged Pratincoles and Gull-billed Tern.

Inch Lough

ORIGINS

The jewel in the crown of Lough Swilly is undoubtedly Inch Lough and Levels. Big Isle and Swilly estuary will often have more birds, and Blanket Nook has great importance despite its small size, but for diversity of birds, species with Internationally Important numbers, all-year interest and accessibility, Inch is one of the finest bird sites in Ireland.

Its origins and history are unusual. Until the mid-nineteenth century, a large arm of Lough Swilly enveloped Inch Island around the east and south, reaching the mainland at the foot of Greenan Mountain (the site of the famous Iron Age hill fort, Grianán of Aileach). This shallow tidal expanse of salt water and mud was claimed from the lough between 1840 and 1859. The first stage was the construction of the Trady Embankment east–west across the centre of the area. As well as its role in draining the waters to the south, the embankment also served as the route of the narrow-gauge Londonderry and Lough Swilly Railway (which operated from 1863 until 1953). Two more embankments were then built north from this to Inch Island, on one of which a road link was established. Between these three embankments and the marshes on the south shore of Inch Island there was a 160 ha section of water cut off from Lough Swilly and also from the mud flats to be drained – Inch Lough. This was kept as a holding tank to receive the waters drained from the south, and also to keep out Lough Swilly's high tides. The salt contained in the mud and enclosed waters, and the seepage of Swilly water in through the sluice and under the causeways, initially determined that it would be a saline lagoon. It has remained salty enough to retain a specialist flora and fauna, even though the sluice gates have been progressively removing the salt over the years.

All that area south of the Trady Embankment, and to the foot of Greenan Mountain, had become cut off from the tides and could therefore be drained, through a complicated system of large and small drains. Known locally as the slobs, or more formally as Inch Levels, this huge expanse of flat land would usually be described as polders, after the original template in Holland. The project was successful – but not initially. Until the late 1950s the patchwork of small fields that emerged were very marshy, and included some areas of unambiguous marshland – great for breeding wetland birds such as Shoveler, but not so good for farming. Since then, the drainage has been upgraded and fields consolidated. Although no longer so attractive to marshland breeding birds, it has been just what the swans and geese needed, and they have thrived.

CURRENT STATUS

Inch Lough and Levels is impacted by a large number of organisations, designations, activities and developments:

- The lagoon and polders are owned by Glenmore Estates.

- The lagoon is leased to the National Parks and Wildlife Service (NPWS),

Inch Lough – with Greylag and Canada Geese, and Mute Swans.

which is currently within the Department of Housing, Local Government and Heritage, and is managed as a Wildfowl Reserve.

- BirdWatch Ireland monitors the bird populations, through the Irish Wetland Bird Survey (I-WeBS). Inch Lough and Levels are counted as one of fourteen sub-sites of Lough Swilly south of a line from north of Rathmullan to Buncrana.

- There is a management agreement between NPWS and Glenmore for the farmland on the levels, which takes account of the needs of wintering swans and geese, and breeding waders, especially the Lapwing.

- Both lagoon and levels are covered by the Special Protection Area (SPA) designation under the EU Birds Directive (Map 12.1).

- The lagoon itself, and its surrounding marshes, are also protected by the Special Area of Conservation (SAC) designation, under the EU Habitats Directive.

- The Office of Public Works (OPW) has responsibility for maintaining drainage throughout the area.

- In 2006, NPWS opened up the Trady Embankment which had previously been overgrown and inaccessible. They also erected the Trady Hide.

These developments made Inch much more attractive to birdwatchers, and improved the detection of rare birds.

- Donegal County Council has constructed tourist infrastructure, funded by InterReg, in a cross-border programme which has done complementary work on the eastern shore of Lough Foyle in Northern Ireland. In 2012, they extended the footpath that NPWS had opened in 2006, to complete the 8 km circuit of the lake and marshes, using all the embankments. They have also built three car-parking areas, and with NPWS have increased the bird hide total to four.

- Donegal County Council is now seeking to upgrade the Trady Embankment path again, as a greenway for cyclists. This would involve expanding the footpath to a 3 m paved roadway, with 1 m gravel shoulders on either side.

- There are already over 100,000 recreational walkers per year on the embankment circuit.

- Three local gun clubs help NPWS achieve its conservation objectives for the site, in return for very limited and controlled shooting.

- The Inch Wildfowl Reserve Trust exists to represent the various stakeholder groups with interests in wildlife, local history, archaeology and community, and to work in partnership with NPWS in delivering conservation of the site.

BIRDS IN WINTER

Inch is justly celebrated for the arrival each autumn of Whooper Swans from Iceland. Seeing them swoop down for their first landfall, and seeing and hearing their greetings when they land, it is impossible not to interpret this as pure joy shared with friends and family at having survived the crossing.

It had long been the availability of freshly harvested grain or potato fields at Lough Swilly that made it worth their while to ignore the landfall opportunities on the north coast. On arrival, their choice was to start with the spilt grain, proceed to the potatoes, and end the winter on emerging cereals or new grass. Their grazing didn't affect the final crop at all, except in very wet conditions when puddling could kill off some patches of grass. What awaits them now is a much less certain welcome, so it is perhaps not surprising that the build-up is smaller, and the stay shorter than at its peak when at least 3,000 would be present shortly after arrival and more than 1,000 would be present throughout the winter. The 500 to 1,000 birds that continue to stay through the winter months can still provide a memorable spectacle. Management agreements and control of human disturbance should be enough to ensure their continuing presence, but it would be nice to see the site again living up to its full potential.

When regular monitoring started, there were two species of wintering wild swans at Inch – Whoopers and Bewick's. Now there are only the Whoopers. Against that, geese numbers and diversity have greatly expanded. In the 1980s, less than 200 Greylags and White-fronts moved around all three of the polder areas. White-fronts have increased in number and have continued to use Inch Lough for roosting, but have abandoned the levels for feeding. They have been replaced by a hugely increased population of Greylags, usually numbering close to 3,000, and by newly established flocks of up to 600 Canada and 300 Pink-footed Geese. Lough Swilly in general, and Inch in particular, is second only to the famed Wexford Slobs in its importance for the *Anserinae* (geese and swans) in Ireland.

BIRDS IN SUMMER

Unlike many other waterfowl haunts, Inch in summer is humming. Mute Swans and Tufted Duck are in large numbers, with enough of them breeding to maintain stable resident flocks. Lesser numbers of other ducks breed in the cover of marsh vegetation, along with grebes, rails and very small numbers of Lapwing, Redshank, Ringed Plover and Dunlin. Breeding waders on the levels have not responded well to intensive agriculture, but

Inch Wildfowl Reserve, Inch Lough.

BirdWatch Ireland and the local gun clubs are doing their best to increase wader populations with improved management of the marshes, and NPWS is working with the farm management on agreements relating to parts of the levels.

The sand island on Inch Lough is home to huge numbers of breeding birds. As well as most of the Mute Swans and Tufted Duck, there are thousands of Black-headed Gulls and hundreds of Sandwich and Common Terns. The site is managed to prevent flooding, and has been monitored for many years by Ken Perry and a succession of NPWS staff.

Monitoring the birds

Lough Swilly's value for wintering waterfowl was recognised by John Robinson Leebody as far back as the nineteenth century.[3] He noted that Brent Goose flocks covering several acres of water could frequently be seen between Inch and the mouth of the River Leannan. At that time the beds of eelgrass would have been healthy, and Leebody's observation chimes well with reports from punt gunners on Lough Foyle who depended on the Brent Goose flocks for their livelihood. Bewick's Swans also visited every winter, but numbers seem to have been relatively small, as Leebody singled out a flock of forty for special mention. Whoopers on the other hand had only been seen once, when five birds visited the inland water (Inch Lough) in 1889.

Leebody's paper was more a selective list of observations than a planned survey. It was not until the 1970s that any attempt was made to enumerate the species on the site as a whole – in the context of an all-Ireland survey, the Wetlands Enquiry 1971/2 to 1974/5.[4] That survey showed Lough Swilly's Greylag Geese, at up to 350, to be of National Importance. Pintail (155) and Shoveler (183) were as numerous as they have ever been since then, but other wildfowl numbers were generally lower than at present. This can be partly attributed to the impact of shooting, both in terms of birds killed and the disturbance caused, although the incomplete coverage of this large and intricate site would have been just as significant. In *Ireland's Wetlands and their Birds*, Hutchinson concluded that the abundance and diversity of waders was poor when compared with east and south coast wetlands – a reasonable assumption from the information he had, but not one that has stood up to the more detailed scrutiny of subsequent years.

Map 12.2: Aerial view of Inch lagoon and marshes (yellow), which is the area of wildfowl reserve, and the polders (red)

The follow-up to the Wetlands Enquiry was the Winter Wetlands Survey, when for the first time there was monthly coverage of the whole lough south of a line from the Mill River outlet at Buncrana, to the north end of Kinnegar Strand (north of Rathmullan), and including the lagoons and polders. It provided three-year mean peaks for each species for the period of the survey 1984/5 to 1986/7. This survey established the fourteen sub-divisions which have been used for monitoring ever since (Map 12.3). The book reporting on the Winter Wetlands Survey also included five-year means for the following period, from 1987/8 to 1991/2, when there were only January counts.[5]

From 1994/5, the Irish Wetland Bird Survey (I-WeBS) has been monitoring sites throughout the country in parallel with WeBS in Great Britain, and feeding into the international monitoring scheme which establishes migratory flyway populations and 1 per cent threshold levels for sites to qualify as of International Importance. I-WeBS is operated by BirdWatch Ireland on behalf of the government, through its National Parks and Wildlife Service.

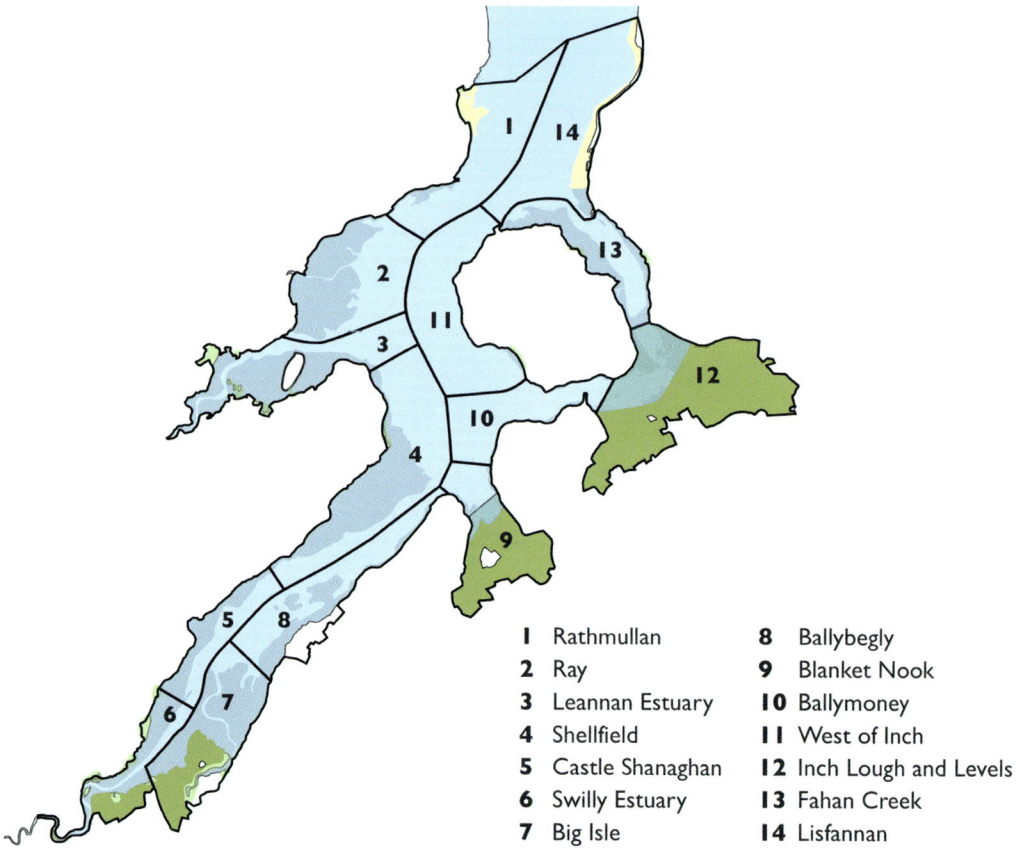

1	Rathmullan	8	Ballybegly
2	Ray	9	Blanket Nook
3	Leannan Estuary	10	Ballymoney
4	Shellfield	11	West of Inch
5	Castle Shanaghan	12	Inch Lough and Levels
6	Swilly Estuary	13	Fahan Creek
7	Big Isle	14	Lisfannan

Map 12.3: I-WeBS monitoring sub-divisions of Lough Swilly

The upshot is that Lough Swilly is now established as one of the most important wetland sites in the country. We have reliable trends for each species, and any changes at sub-site level can be detected. Subsequent reviews have been provided covering 1994/5 to 2000/1 and 2009/10 to 2015/16.[6]

Figure 12.1 shows the comparable sizes of the main species groups, and the changes in each between 2000 and 2020. Most species are relatively stable, but there are quite a few that have increased, and some that have decreased. The two largest groups (Wildfowl and Waders) are generally higher in the second period, Gulls have risen steadily, and the small Others group has increased, but not enough to alter the overall trend. The net effect is that the number of birds using the lough has gone up by about 10,000 over the period. This is in sharp contrast to the national trend

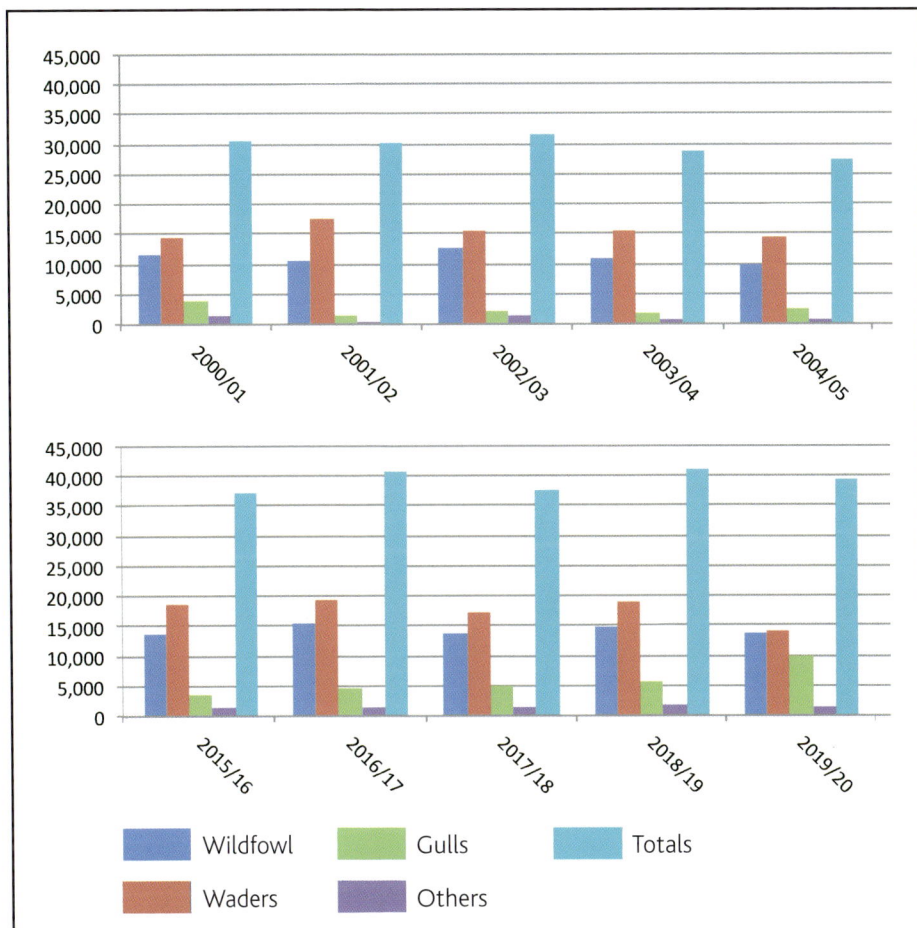

Fig. 12.1: Waterfowl totals on Lough Swilly 2000/1 to 2004/5 and 2015/16 to 2019/20.

for waterfowl, as revealed in a report from the I-WeBS head office team.[7] This shows serious declines for most species – both the short-term five-year mean peaks between 2006/7 and 2010/11, and 2011/12 and 2015/16, and the long-term between 1994/5 and 1998/9, and 2011/12 and 2015/16. Although the national and Donegal time periods differ, the contrast is clear.

On Lough Swilly, the species that have made the greatest contributions to the overall change are given in Table 12.1. Worth pointing out in particular is the meteoric rise, starting in 2014, in the numbers of Cormorants and Great Black-backed Gulls. This is assumed to be related to the development of a fishery dredging for the recently introduced Pacific oysters. Lesser

Direction of change	Species	Mean of 2000/1 to 2004/5 peaks	Mean of 2015/16 to 2019/20 peaks	Change	% Change
Increasing	Black-headed Gull	883	2,390	1,507	171%
	Wigeon	1,198	2,531	1,333	111%
	Black-tailed Godwit	66	1,109	1,043	1,580%
	Herring Gull	254	1,224	970	382%
	Lapwing	1,796	2,710	914	51%
	Knot	321	863	543	169%
	Oystercatcher	1,455	1,934	479	33%
	Canada Goose	48	509	461	956%
	Cormorant	68	525	457	672%
	Common Gull	1,127	1,502	376	33%
	Bar-tailed Godwit	153	520	367	239%
	Brent Goose	318	676	358	113%
	Tufted Duck	231	589	358	155%
	Mallard	871	1,223	352	40%
	Great Black-backed Gull	43	393	350	818%
Decreasing	Dunlin	6,042	4,960	-1,082	-18%
	Redshank	1,954	1,742	-212	-11%
	Whooper Swan	1,425	1,147	-278	-19%
	White-fronted Goose	932	759	-173	-19%
	Great Crested Grebe	194	128	-66	-34%

Table 12.1: Species which have contributed most to change in the total numbers on Lough Swilly (in order of their numerical contribution to the overall change)

Black-backed Gulls show the same trend, but that species has also been responding to other factors. More recent counts suggest that these three species have not sustained their increases.

On Lough Swilly, the species which have made the greatest contributions to the overall change are given in Table 12.1. Worth pointing out in particular is the meteoric rise, starting in 2014, in the numbers of Cormorants and Great Black-backed Gulls. This is assumed to be related to the development of a fishery dredging for the recently introduced Pacific oysters. Lesser Black-backed Gulls show the same trend, but that species has also been responding to other factors. More recent counts suggest that these three species have not sustained their increases.

The experience

Lough Swilly caters for all levels of bird interest, with the footpath and hides around Inch Lough giving many people their first experience of what birdwatching has to offer. Counting or estimating the numbers of each species in mixed duck and wader flocks, at up to a kilometre distance, can keep the more experienced birdwatchers fully occupied. Managing the habitat conditions for breeding waders is perhaps the most challenging task for the ornithological community, and is usually undertaken with the help of professionals. For enthusiasts of all kinds, waterfowl and wetlands offer one of the easiest routes to engaging with the natural world. Thanks to its hilly surroundings and indented shoreline, Lough Swilly provides a multitude of viewing points, where the ever-changing landscape and the wonder of its birds never fails to stimulate and inspire.

Chapter 13

Migrant Hotspots

Lesser Whitethroat, Yellow-browed Warbler, Corncrake, Snow Bunting and Greenland Redpolls on Tory Island

THERE ARE FOUR AREAS in the county that have a particular attraction for migrant songbirds and other terrestrial species. Tory is the outstanding site for most birdwatchers, but each has its own reasons for being of interest.

Tory

THE ISLAND AND THE PEOPLE

If Donegal is special because it is at the far north-west corner of Ireland, then Tory, 12 km closer to Iceland, has to be extra special. The 200 residents don't need to be told that. They resisted government efforts to move them to the mainland in the 1980s, and now have much-improved facilities, including a secondary school, to secure their future. The value of birding tourism to

Loch Aher,
Tory Island.

the local economy is small but significant, not least because it brings people to the island at times when conventional tourism is at a low ebb. Although Tory is more exposed to Atlantic storms than anywhere else in Ireland, its southern aspect gives shelter from the north, and on a good summer's day, with the Donegal highlands on the horizon, it can be delightful.

The islanders were largely self-sufficient when fishing and farming provided most of their needs. Until the twentieth century, farming was managed in the old two-field rundale system. Close around each village was the infield, within which plots for cultivation were allocated each year to different families. Beyond that, on the less favourable soil, was the outfield, on which everyone had the right to run their livestock. Although self-sufficiency had worked successfully for so long, it ceased to be an option. It didn't fit a more modern approach to farming or land tenure, and each family now owns their own plot. Only one fence was needed in the rundale system, to protect the infield from the livestock in the outfield, but now each family plot has to be fenced, and probably fenced internally as well. But even with modernisation, it was very hard for somewhere like Tory to compete in the modern world, and only one or two families now do any farming at all. So most of the new fences are already redundant. Fortunately, their continuing presence doesn't deter the island's most celebrated visitor – the Corncrake.

THE HABITATS

The island is 4 km from west to east, by about 1 km south to north. The land rises gently from the south shore to some of the finest cliff scenery in Donegal. The highest point (83 m) is at the eastern end, on the promontory fort of Dún Bhaloir (Doon Balor), from which projects a magnificent knife-edge ridge, An Tor Mór (Tormore). The peat which covered most of the island has long ago been cut away to the bedrock, and what remains is a mosaic of dry heath, marshy ground with shallow seasonal pools, and bare sandy or stony ground. In early summer this supports a colourful display of wildflowers, dominated by the yellow Bird's-foot Trefoil.

There are three small lakes and a good wet marsh, all very different, but all supporting breeding birds. In the north-west is An Loch Ó Thuaidh (Loch Ahooey), the largest lake, although the water area is broken up with a number of constructed islands of spoil – ideal for nesting Black-headed Gulls and other waterfowl species. An Loch Ó Dheas (Lough Ayes) in the south-west is a brackish lagoon lying inside a high boulder storm beach, which protects the island from the sea but allows seepage of salt water into the lagoon. This is where one expects most of the scarce or rare waders and waterfowl. The south-east corner has another brackish lagoon, An Loch Thoir (Lough Aher), lying just inside a shingle barrier. The final area of open water is usually called the Boating Lake. It is a small ephemeral pond or wet marsh on the inland side of West Town, which has a few small islets on which breeding birds can find sanctuary.

Of much interest to migrant birds are the two village areas of West Town and East Town, where there are a few gardens, planted shrubberies (there are no native trees on the island) and small agricultural plots.

BIRD RECORDING

Migrant birds have been recorded from Tory since Barrington gathered information from lighthouse keepers for his book *Migration of Birds at Irish Light Stations*, published in 1900. One result of note from that period was Ireland's third Red-breasted Flycatcher, which was killed at the light in 1894.[1] There followed a half century with few bird records. A Black-winged Stilt shot in 1916 was a very notable exception – still the only county record.

A fine all-round naturalist called D.J. (Danny) O'Sullivan worked as the lightkeeper in the 1950s. He made note of the birds attracted to the light. He also sent specimens of insects killed by the light to experts for identification, with one outcome being a published list of the *Lepidoptera*, for which he had provided most of the data.[2] For many years he wrote a daily nature column in the most popular national newspaper of that time *The Irish Press*, which naturally enough drew on his experiences of both Tory and Inishtrahull.

Also in the 1950s the need was identified for a network of stations to cover each corner of Ireland, complementing the work of the Saltee Bird Observatory off County Wexford, which had been established in 1950. As a result, bird observatories were established on the Copeland Islands off County Down in 1959, and in 1961 on Cape Clear Island off the south-west tip of County Cork.

At the same time, an expedition was organised to establish the potential of north Donegal for migration study (see below under Malin Head and Inishtrahull).[3] That led firstly to an extensive visit to Tory by one of the members of the expedition, Philip Redman. He stayed on the island through spring, summer and autumn of 1954 and ended the long dearth of systematic bird observations on the island.[4] His visit also led to the formal establishment of the Tory Bird Observatory (TBO) in 1959, which functioned most actively from 1959 to 1964, and to a lesser extent in 1958 and 1965.[5] But for this north-west corner of Ireland, the observatory would be a joint operation at two locations – Tory and Malin Head (see below). On Tory there were many participants, largely under the leadership of R.G. Pettitt.[6] As well as continuing the work started by Philip Redman, the observatory also did a lot of trapping and ringing. The set of thirty-one first records for Donegal (in Tables 13.1–3) show that twelve of those records came within the lifetime of the observatory.

Redman's visit in 1954 had been highly productive. His analysis of seabird movements was starting with a blank slate, but it remains as useful as anything from the seventy years that have followed. He also recorded two first records for Ireland,[7] the first in an impressive series of Irish firsts for the island (Table 13.1). But perhaps his most important work was in relating the arrival of migrants to the meteorological situation, and making comparisons with what was happening concurrently at the Fair Isle Bird Observatory between the Orkney and Shetland islands. He concluded that Lapland Buntings, Snow Buntings and *rostrata* Redpolls were each coming here from Greenland rather than from arctic Eurasia – Iceland (Lapland Buntings and *rostrata* Redpolls), and either Iceland or Eurasia (Snow Buntings). The Lapland Bunting and the redpoll would ordinarily have been regarded as drift migrants (i.e. vagrants), but Redman's work showed that they are annual migrants – in other words, coming here deliberately rather than having been blown off-course.

Blue-headed Wagtail	22 May 1954	Philip S. Redman
Common Rosefinch	8 September 1954	Philip S. Redman
Yellow-breasted Bunting	18 September 1959	Henrietta Cooke
Caspian Tern	30 September 1959	R.G. Pettitt
Arctic Warbler	1 September 1960	Tory Bird Observatory
Rock Pipit (Scandinavian)	4 April 1961	R. Moss
Booted Warbler	27 September 2003	William McDowell, Dennis Weir et al.
Bullfinch (Northern)	25 October 2004	A.A. and P. Kelly
Collared Flycatcher	29 May 2012	Robert Vaughan
Eastern Yellow Wagtail	12 October 2013	Victor Caschera, Jim Dowdall, Jim Fitzharris

Table 13.1: Firsts for Ireland

Black-winged Stilt	1916
Long-billed Dowitcher	1962
Greater Yellowlegs	1964
Yellow-billed Cuckoo	1989
Paddyfield Warbler	1998
Pechora Pipit	2001
Hooded Merganser	2002
Red-rumped Swallow	2007
Spotted Sandpiper	2013
Radde's Warbler	2016

Table 13.2: Tory firsts for Donegal (single records not repeated)

Tree Pipit	1952
Little Bunting	1952
Pectoral Sandpiper	1960
Arctic Warbler	1960
Barred Warbler	1960
Balearic Shearwater	1963
Yellow-browed Warbler	1964
Greenish Warbler	1965
Hobby*	2015
Red-throated Pipit	2015
Marsh Warbler	2020

Table 13.3: Tory firsts for Donegal (other records have followed) * yet to be assessed by IRBC

The most significant records are the firsts for Ireland. The most surprising of these are probably the two wagtails. The Blue-headed Wagtail is a close continental neighbour which one would think might be fairly frequent, whereas the Eastern Yellow Wagtail is the most remote of all the many forms of the former 'Yellow Wagtail', and just about as unlikely a visitor to Tory as they come.

There has been no formal or organised monitoring on Tory since 1964. During the lean years that followed, one of the island's priests, Fr O'Neill, did have an interest in birdwatching. He most famously recorded the Yellow-billed Cuckoo in 1989, still the only one ever recorded in the county. And since the end of the observatory period there has been a slowly growing band of enthusiasts visiting each year, initially from outside the county, but over the last ten or so years including a number of Donegal residents. Their visits have inevitably been concentrated in the autumn migration period, and to a lesser extent, the spring. The final hurdle was the recruiting of a Tory resident, Anton Meehan, whose full-time presence has led to a number of rare migrants and vagrants being recorded that would have otherwise been missed.

With the level of coverage once again being significant, an annual *Tory Island Bird Report* was produced for the six years 2011 to 2016 by Peter Phillips, Chris Ingram and Robert Salter, with an accompanying online resource in the form of the *Tory Island Bird Blog*.[8] These reports may have had a limited run, but the reputation and popularity of Tory as a hotspot for birds continues to rise. With a total bird list of 244 full species it is now almost easier to check what hasn't occurred there than what has!

Malin Head and Inishtrahull

Small though it is, at 46 ha, Inishtrahull was inhabited until 1927 by a fishing community which also farmed under the rundale system, as on Tory. They cured the fish, and also traded with passing ships, so their lifestyle was well above the basic subsistence level. The island had a manned lighthouse, and at the end of the nineteenth century the keepers would send migrant birds that were killed by the light to R.M. Barrington. These demonstrated that Inishtrahull was of interest for rare birds – he received, for example, the first Lesser Whitethroat for the county.

Danny O'Sullivan, who had also been a lighthouse keeper on Tory, had productive postings at Inishtrahull, from 1937 to 1942 and from 1956 to 1964. He recorded a number of Turtle Doves and Garden Warblers and, most significantly, Donegal's first Collared Dove in 1961. The lighthouse was finally automated in 1987, after which the island was completely uninhabited – apart from occasional visits for maintenance of the light.

Inishtrahull.

However, the first focused activity at both Malin Head and Inishtrahull in modern times was that same expedition by Redman, Gibbs and Nisbet which had inspired the setting up of the Tory Bird Observatory.[9] It seems amazing now to recall that the first Great Skuas for Donegal were those seen by that expedition as recently as 1954. Lapland Buntings were also a revelation. Kennedy et al. knew of only seven birds recorded in Ireland[10] – the expedition's total at Malin Head and Inishtrahull was about 300. Inishtrahull had a total of about sixty, and on the mainland, a half-hour count on 28 September revealed sixty-one birds.

The Malin Head Bird Observatory (MHBO), which ran from 1961 to 1965, was another outcome from those 1950s expeditions, and operated over the same period as an arm of the TBO, and also overlapped with Danny O'Sullivan's time on Inishtrahull. It was decided that the MHBO would include Inishtrahull in its remit, although its coverage there was mostly limited to 1965.[11]

The MHBO found the diurnal migration of common species to be more concentrated on the mainland, but the vagrants to be almost all on Inishtrahull. These included Redstart, Garden Warbler and Ortolan Bunting. Although the observatory had full coverage only from 1962 to 1964, it was no less valuable for that.

The area that needs to be covered at Malin Head is larger and more varied than Tory. It fades into the rest of the Malin peninsula, so the boundary drawn by MHBO was somewhat arbitrary, and was chosen on practicality as well as on natural features.

Most of the observatory work was done by Oscar Merne and Dick Devlin at Malin Head. It was expanded in the autumn of 1965, when Brendan Doherty and Joe Donaldson stayed on Inishtrahull for more than a month.[12] Seawatching was their main task. Unlike the unfavourable contrast between Tory and its nearest mainland equivalent at Bloody Foreland, Inishtrahull generally fared better than Malin Head, except for Gannets which were consistently more abundant passing the mainland. Cory's and Great Shearwaters were recorded with a frequency that has not been matched anywhere else in Donegal. Sooty Shearwaters were present in large numbers, but they were 'pottering rather than passing', and were attracted to fishing boats operating just offshore between 16 September and 5 October.

Covering a wider area, but also tapping into the output from the obser-vatory, *The Birds of the Inishowen Peninsula* is a useful summary of the 1960s and early 1970s.[13] Following that, there was a lull in recording at Malin Head which was spasmodically filled from the 1990s onward, mostly by Rónán McLaughlin. He has noted a continuation of the substantial falls of many common species which can, when the conditions are right, arrive or depart in large numbers – Skylark, Mistle Thrush, Redwing, Wheatear, Meadow Pipit and Linnet are some of the main species involved. Lapland Buntings continue to feature at Malin during autumn migration. Rarities have included Hoopoe and Red-footed Falcon. Recent seawatching has been only occasional, but it has, for example, produced the county's second gadfly petrel (2014), as well as impressive totals of skuas and large shearwaters. The re-establishment of a bird observatory on Inishtrahull bodes well for the future, and the island's status as a nature reserve should ensure that both breeding and migrant birds will be well catered for over the whole island.

The far south-west

The three settlements of Glencolumbkille, Malin More and Malin Beg in the far south-west of the county have all proved to be worthy of more attention. They each have had more than their share of exciting finds, given the very small level of coverage. Attention was drawn to the area by John O'Boyle, who found Pied-billed Grebe and two Yellow-browed Warblers in

1988, and a couple of Pied Flycatchers in 1989. This inspired Ian Wallace, Anthony McGeehan and Dave Allen to give the area some concentrated attention in the autumns from 1995 to 2000, leading to a stimulating analysis which complemented the publications of the 1950s relating to the north and north-west.[14]

Migration highlights include two of the county's three records of Firecrest (1997 and 2017), Donegal's only American Bittern (1984) and the first of only three Pallas's Warblers (2016), two of them at Malin Beg. Near Glencolumbkille, that Pied-billed Grebe of 1988 was the only one ever found in Donegal, and Ireland's second.

Other birds which have been seen less than ten times in the county are evenly shared among the three main centres in the area. The total number of records for each species is as follows (south-west/county):

Malin Beg – Reed Warbler 2/6, Siberian Chiffchaff 1/5, Ortolan Bunting 1/9. Malin More – Grey-headed Wagtail 1/2, Citrine Wagtail 1/3, Melodious Warbler 1/6, Short-toed Lark 1/6.

Glencolumbkille – Red-footed Falcon 1/2, Hoopoe 1/9, Red-breasted Flycatcher 1/9.

Malin Beg.

Arranmore

The only Donegal islands with a resident human population of some permanence are Tory and Arranmore (a few others have been recently re-colonised by small, mainly seasonal, populations). Being much the larger island, Arranmore's character is more akin to that of the mainland, with a more diverse settled area on the east and south coastal zone and, of particular value, more trees than other islands. Cut-over blanket bog covers most of the land surface, although peppered with a number of small lakes. Those areas, and the cliff-bound west and north coasts, are not very different from Tory.

Many birdwatchers have of course visited the island, and there is a growing interest in birds among the resident population, but the bulk of recording over several decades has been done by Andrew McMillan.

Although obvious wader habitats are lacking, Arranmore has turned up Ireland's first Semipalmated Plover in 2003, as well as Dotterel, and Buff-breasted, Baird's and Green Sandpipers. It has had more than its share of Snowy Owls and Wrynecks. Some of the rare passerines seen have included Short-toed Lark, Little Bunting, Donegal's only Shore Lark (2009), its first Bluethroat (1996) and Citrine Wagtail (1993).

For beginners there is a useful and attractive bi-lingual book on the birds of Arranmore.[15]

An overview

The emphasis on rarities at these locations begs the question of just how predominant these species are. Out of 376 species and races to have occurred in the county, 173 of them have the status of rarity – almost half of the total (see Chapter 17 and Appendix 5). Breaking these down by the biogeographical regions from which they have come, 115 (66 per cent) are from our own western Palearctic (Europe, the Middle East and North Africa), thirty-nine (23 per cent) come from the Nearctic (North America), and fifteen (9 per cent) from the Eastern Palearctic (Asia north of the Himalayas). That leaves two other rarities – albatross sp. and Wilson's Petrel – which breed in the South Atlantic or Antarctic islands, which don't sit easily within either of the nearest biogeographical regions – the Neotropical (Central and South America) and the Ethiopian (sub-Saharan Africa).

Only two American passerines (a Baltimore Oriole and a Bobolink) have reached us, which is remarkably few, considering the annual arrival of a number of these small songbirds, from a wide variety of species, in the south-west of Ireland. Yet the majority of the rarities reaching us from outside our own region are also from North America – the waders and waterfowl – which are collectively dependable in unpredictable combinations.

A fair number of the birdwatchers who make a special trip to Donegal come late in the autumn migration season, in October, in the hope of seeing a few vagrants from the Eastern Palearctic, even though only thirteen such species (and two races) have so far been recorded here – such is the draw of these exceptionally unlikely visitors. Paddyfield Warbler, Pechora Pipit, Eastern Yellow Wagtail and Yellow-breasted Bunting are among those that have turned up on Tory.

The attraction of Tory in particular has been sufficient to keep Donegal on the map, as a county with exciting possibilities for adventurous birdwatchers. But the full potential of the network of good migration outposts has still to be realised. A handful of individuals have shown over the years that there are a number of them that deserve the same level of attention. This culminated in the establishment in 2020 of the Inishtrahull Bird Observatory (IBO).

Increasing numbers of casual day-trippers to that island in recent years have caused some disturbance to breeding species. But with a brief that covers surveying, ringing and recording, habitat conservation, and management of day-trippers to this privately owned island, the future for Inishtrahull looks bright.

Chapter 14

Gulls Galore at Killybegs

Yellow-Legged, Black-headed, Great Black-backed, Common, Thayer's, Kumlien's, Bonaparte's, Iceland, American Herring, Slaty-backed, Lesser Black-backed, Glaucous and Herring – and a Turnstone

Why Killybegs?

KILLYBEGS is Ireland's biggest fishing port. The ships that tie up here are vessels that can roam the North Atlantic, provide all creature comforts for the crew, and stay at sea for as long as it takes to fill their enormous holds with fish. They tend to come in to disgorge their catches when the worst storms are forecast, and those connoisseurs of the best fish waste, the gulls, will either follow them in or be there in advance, waiting.

Fish processing would normally have sent a stream of pink effluent into the harbour, attracting Iceland Gulls in particular, as they like to dip-feed

Gulls and boats at Killybegs.

like Kittiwakes on the smaller morsels. A proper sewage treatment plant has reduced that outlet, but there are still plenty of other opportunities available. When fish or crustaceans are piped into waiting lorries on the piers, gulls will be there to take their cut. If a number find themselves feeding in the bed of a lorry, the mêlée can leave them filthy and without room for take-off, so they have to be tipped out under the tailgate, minus their dignity, but very happy to tumble into the harbour for a long wash. Such hazards are clearly worth it, as thousands of gulls can be present in the winter months. The majority are Herring Gulls, with large numbers of Great Black-backed, Black-headed and Common Gulls, and a few Kittiwakes and Lesser Black-backed Gulls to complete the home-grown set. But there are more.

The gulls

The situation of Killybegs at the north-west corner of Ireland makes it the port of choice for gulls from the north – Iceland, Greenland and beyond. Add to these the occasional bird from almost any compass point, and the 'gull capital of Europe' epithet begins to make some sense. This is of course subjective, and not to be taken too seriously, as it depends on which gulls we are talking about. But there is at least no doubt that Killybegs is the best place in Ireland to see the northern gulls – Glaucous, Iceland and Kumlien's. The first two of these are almost invariably present from December to April, with young birds predominating among the full spectrum of ages and plumages. Davy Hunter's count of eighty-five Iceland Gulls on 20 February 2002 is surely phenomenal.

The list of rarer species is also unequalled in Ireland. Table 14.1 gives the cumulative totals so far for these species. The outstanding one has to be Slaty-backed Gull, from the western Pacific (Siberia, south to Japan). It could have travelled west through Eurasia, or east, across two oceans and a continent – the former route is the more likely, in light of one or two recent sightings in eastern Europe. Almost as unexpected when it appeared was the Caspian Gull, but that is a species likely to become more familiar as its breeding range continues to spread westwards. Thayer's and Bonaparte's have opened the door for birdwatchers to anticipate other North American gulls still to be recorded in Donegal – Glaucous-winged and Franklin's are already on the Irish List. The only recorded Donegal species not yet seen at Killybegs are the Sabine's, Ivory and Laughing Gulls. The absence of Sabine's is odd, as Killybegs would be an ideal refuge for storm-blown migrant birds. Ivory Gull has finer tastes than could be satisfied with the Killybegs menu, but if beached whale carcasses are not available for one of these very rare visitors, surely Killybegs will eventually be discovered as a stop-gap. Laughing Gull would seem to fit the profile of the other North American vagrants that usually find their way to Killybegs. Finally, the Baltic Gull, a distinctive race of Lesser Black-backed, has only been recorded once, from the opposite end of the county, but could yet make an appearance at Killybegs.

There is a particular difficulty with the large northern gulls. Although not universally agreed, many authorities recognise a ring (or cline) of related forms around the northern latitudes. What starts as Lesser Black-backed Gull in north-west Europe, and grades eastwards through progressively paler subspecies, eventually reaching the American Herring Gull, which passes the baton across the Atlantic to Herring Gull. But Herring Gull occupies the same ground as the Lesser Black-backed Gull, and rarely hybridises with it. So we now have two distinct species in Britain and Scandinavia, linked by the ring of subspecies. And of course American Herring Gull has also been recently promoted to a full species.

But hybridisation does occur – between the adjacent races, and with adjacent species like Glaucous Gull or any of the races of Iceland Gull. Confusion about the parentage of a hybrid bird often means that it goes unrecorded, but fortunately some gull hybrids are fairly obvious. 'Viking Gulls' (Herring x Glaucous) are the most frequently identified, with Glaucous x Iceland and Herring x Iceland also noted.

Species (subspecies in italics)	First Date	No. Seen
Kittiwake	-	-
Bonaparte's Gull	2018	3
Black-headed Gull	-	-
Little Gull	2014	1
Ross's Gull	1983	1
Mediterranean Gull	1984	21
Common Gull	-	-
Ring-billed Gull	1982	9
Great Black-backed Gull	-	-
Glaucous Gull	1974	517
Iceland Gull	1975	704
Kumlien's	1983	86
Thayer's Gull	1998	3
Herring Gull	-	-
Herring Gull *(Scandinavian)*	1997	61
American Herring Gull	1990	11
Caspian Gull	1998	1
Yellow-legged Gull	1998	18
Slaty-backed Gull	2015	1
Lesser Black-backed Gull	-	-

Table 14.1: All species and races of gulls identified at Killybegs

Herring Gulls (and two Great Black-backed), Killybegs east bank.

When not gorging themselves on fish, the gulls repair at low tide to the head of the bay, where they can scrub up in freshwater at the outlet of the Bungosteen River. The tidal mud here is a good place to scan at leisure for rare species – if you can find a viewing point that suits the light. With all their maintenance duties completed, the gulls chill out on the roofs of the town, showing a preference for large corrugated roofs, which the many businesses supporting the fishing industry have conveniently provided.

Despite the numbers of birds usually present in the port area, it is always worth taking a tour of the whole bay. Birds can be seen at various locations around the shore. However, the main attraction of the east bank is the view across the bay to the edge of the newly developed quay at the southern end of the town. This is where birds feed on effluent from the processing factories, and where there will always be ships to offer perching facilities. Access for birdwatchers to the new quay itself must be arranged in advance with the port authorities.

Other birds

One regrettable aspect of the development of the port has been the land-claims which have engulfed some fine shingle banks used by roosting gulls and waders. One small area remains, where a decent flock of Turnstones is usually present, along with a few Redshanks and Oystercatchers. A few Black Guillemots are usually present somewhere among the flocks of gulls resting in the waters near the shore. At the head of the bay, where the gulls go to wash, there is usually a small number of ducks – Mallard, Wigeon and Teal.

Of the less familiar birds, a Grey Phalarope turned up in 1998, and a Velvet Scoter in three winters during the 1990s and 2000s. While the three years were not consecutive, they were close, so it is most likely that only a single bird was involved.

Killybegs in context

It might seem that the attraction of winter gulls at Killybegs is secure, but nothing is certain in these fast-changing times. The county's landfill sites had been 'go-to' places for rare gulls in the latter decades of the last century, but these have all been closed. The fishing industry is in a state of perpetual crisis – overfishing, international disputes over fishing rights, and climate-induced changes in fish populations and locations all contribute. The warming of the Arctic is quite likely to result in the withdrawal of the northern gulls from southern outposts like Ireland. Already there has been a fall-off in the numbers of Glaucous, Iceland and Kumlien's Gulls. But for the moment, Killybegs still provides a fine spectacle each winter, and the prospect of seeing something really rare continues to tantalise.

Chapter 15

Seawatching

Sooty Shearwater, Leach's Petrel, Sabine's Gull and Pomarine Skua

An overview

G IVEN THAT quite a few birdwatchers spend quite a lot of time engaging in this rather esoteric pursuit, it is worth looking at it in a little detail. Seawatching is the practice of sitting on an exposed headland to watch the flow of migrating and pelagic seabirds, usually at its best in strong onshore winds during late summer and autumn. If you dress up properly, have a comfortable folding seat and a hot flask, you will soon get immersed in the rhythm of the sea and forget about everything other than the flow of seabirds across your telescope's field of vision. It is very satisfying to discover that birds can be confidently identified, sometimes at considerable distances, by a very brief flash of plumage, a hint of their shape and, above all, their wingbeats and style of flight. And you will be seeing exotic species that you will not see otherwise – various shearwaters, petrels and skuas, amongst others.

The objective of the exercise can vary. Some of the most dedicated will be just waiting for a new species – something really rare in Ireland, maybe rare anywhere, but one that could potentially pass across this corner outpost of the European continent from time to time, if only someone is at the right place, at the right moment, to see it. Whether that is our main focus or not, everyone is alert to the possibility, and will happily share in the brief event – when it happens. For others the main focus will be the logging of numbers, matching these with conditions, and working out patterns that will reveal something about the abundance and trends of these otherwise inaccessible birds. Is what we see a reflection of changing conditions on remote breeding cliffs and islands, or the impact of climate change, or over-fishing? Or could it be related to how plastic or sound pollution may be reducing life expectancy or breeding success?

Whether you are a beginner or an old hand, seawatching is time out from everyday life, leaving you refreshed and stimulated by your glimpse into another world.

Seawatching was pioneered in Donegal on Inishtrahull, in 1953 by Philip Redman and in 1965 by Brendan Doherty and Joe Donaldson.[1] Their work inspired much of what has been done in more recent years. So it may come as a surprise that both Inishtrahull and its western counterpart, Tory, have barely featured at all since the 1960s. Tory attracts fairly large numbers of birdwatchers in autumn, but they come during weather conditions more suited to passerine migration, when the passage of seabirds is so thin that seawatching doesn't feature on their agenda. Any time that it is tried, it is the south-east corner of the island that is favoured, where birds can be seen travelling through Tory Sound, between the island and Inishbeg. Inishtrahull has been ignored largely because it is so inaccessible and without accommodation – but that is now beginning to change.

After that early burst of activity, seawatching in Donegal more or less re-started around 1994. Reported counts have not always been timed, and these could refer to anything from half an hour to all day. So most of the analysis here is based on 160 timed counts reported since 1994. Listing from the north-east, these have been at Malin Head (12), Fanad Head (24), Melmore Head (14), Tory Island (1), Bloody Foreland (101), Arranmore Island (3) and Rocky Point (5).

Bloody Foreland has had a lot more attention than the other sites, although that is not because it is better in every way. Rocky Point is the site with most birds. This is because any westerly wind has the long fetch south from Arranmore to concentrate birds inshore. To regain their southerly route, all these birds have to pass Rocky Point. Malin Head, Rinrawros Point on Arranmore, and possibly Melmore Head can also be better than Bloody Foreland on a good day. But sheltered viewing conditions are not good at Malin Head, Melmore Head lacks good access and natural shelter, Arranmore is very inaccessible and Rocky Point is both inaccessible and without good shelter. However, direct comparisons between sites are rare. The most comprehensive and convincing came from the steady south-west to west winds that blew throughout August 2024. Twenty-four counts revealed very convincing matches in the numbers of most species across all the main sites except Arranmore (which wasn't counted). But this was only true when sites were counted at roughly the same time of day. So the difference between time of day was greater than the difference between sites.

That leaves Fanad and Bloody Foreland which are both reasonably accessible and with some good shelter. The viewing point at Fanad Head is high above sea level, which allows birds to be spotted even in the depths of a deep swell. But it also means that telescopes trained at mid-distance can miss birds passing along close to shore (or vice-versa). Bloody Foreland has a small light beacon which gives some shelter from all useful winds except the northerlies, and has easy access (if you have the patience to drive at the speed dictated by a bog road with rock outcrops and deep pools). The elevation is a better compromise for detecting most of the passing birds than at Fanad. Whatever way you balance these variables, the conclusion will be that all of these headlands are worth a visit. For most of the locals, however, Bloody Foreland has the edge.

Minor sites that have had good seabird passage at times, but for which there are few recent records, are St John's Point and Rossnowlagh Lower. Both are within Donegal Bay and tucked well away from the exposed Atlantic coast, which limits the number of times when conditions will favour them.

A good day

So what can be expected on a good day – let's say Bloody Foreland in late August to early September, with north-west wind at Force 5 or 6 coming around the top of a big anticyclone in the Atlantic to the west of Ireland

Bloody Foreland on a good day.

and France, or at the rear of a depression centred over Scotland? You will have driven in the dark through heavy showers, to emerge into the dawn, and, perhaps surprisingly, few if any showers (the winds don't drop their contents until they have to rise over Bloody Foreland hill). By the time you have moved yourself and your kit ('scope, tripod, provisions and chair) to the beacon, and arranged yourself comfortably on the bed of rocks and sheep guano in its lee, the light will be just about good enough. Occasionally you might see something passing in the gloom that you will not see again all morning – often one of the smaller skuas, or if you are lucky, a large shearwater.

Light changes frequently with the direction and angle of the sun, and with cloud conditions. So what would seem fine to anyone else can often make detecting and identifying birds at sea much more difficult. Mid-morning in our favoured weather conditions usually brings on a bright glare when birds and sea blend into each other, and as birds are often fewer by then anyway, the end of a three-hour morning session usually works out to be nicely timed. Recording birds on half-hour shifts usually requires a fifteen-minute break in the middle of the standard three hours. But if good birds are on tap, you might go on for another hour, and on a really good day you will continue for as long as your eyes can cope with the growing strain.

On our typical good day, almost all birds will be moving south, which at the north-facing Bloody Foreland means passing to the left (west), and going into the wind – no problem to them. The few birds that go east are usually seen in slack winds, or earlier in the summer when they might be local breeders – auks, Kittiwakes and Fulmars commuting between breeding sites and feeding grounds. Gannets consider Bloody Foreland a reasonable commute from Ailsa Craig in Scotland.

Gannets are undoubtedly the most reliable birds. They will almost always be visible somewhere in your view throughout every seawatch, but often have to be ignored when the number and diversity of other species demands all your concentration. So take a few minutes every so often to savour this magnificent bird. Many of them slip past underneath the field of your telescope, which will probably be trained on the mid-distance. For these you will need only binoculars to see the subtle detail in their plumage and anatomy, but if you do pick them up in your 'scope you will be rewarded with a very special close encounter.

Other species have shorter seasons. Manx Shearwater are early migrants, mainly in August, but also July and early September. They will be passing at several hundred per hour, or over a thousand on a particularly good day. Mostly they are close in to the shore, in ones and twos, and in loose batches of a dozen or more birds. They give great views as they scud across the surf, just over the top of your view of the low cliff edge. Sooty Shearwaters always liven up the shearwater passage. A typical day in late August will have reasonable numbers – up to fifty per hour – even though they won't peak until September. Sooties are always enjoyed by everyone. Their long thin wings, powerful flight and classy plumage show up Manx Shearwaters as the poor relations.

August is the time when the large shearwaters might turn up – but usually don't. When they do, there can be good numbers, although if that's what you really want, it happens more often off Arranmore Island. There used to be a lot of birds identified simply as large shearwaters, but the relaxed mastery of the air demonstrated by Cory's is so different from the bursts of hurried wing-beats used by Great Shearwater (and the much smaller Manx Shearwater), that the plumage details now usually follow as secondary evidence of their identity.

Skuas are birds that always give a buzz, but you won't get many in August. You might think the Great Skua is a particularly drab bird, but the rich brown plumage and dazzling wing flashes can light up a dull sea otherwise populated with birds in monochromes, however stylish those might be.

Razorbill passage.

Species	Totals seen in 4 hours	Rate/hour
Eider	9	2
Kittiwake	25	6
Great Skua	177	44
Pomarine Skua	13	3
Arctic Skua	6	2
Puffin	5	1
unidentified auks	143	36
Red-throated Diver	1	
Great Northern Diver	55	14
European Storm-petrel	7	2
Leach's Storm-petrel	4	1
Fulmar	9	2
Sooty Shearwater	465	116
Manx Shearwater	61	15
Gannet	2,312	578
TOTAL	3,292	822

Table 15.1: Birds at Bloody Foreland on 22 October 2015

Fulmar is at its peak in late August, passing mostly as single birds, and is probably the hardest species to pick out against the light. Auks have not yet reached their peak in August and only occasional birds will be noted – by October there will be so many that you might give up trying to count them. In contrast to the Fulmars, auks, when numerous, usually pass in batches or straight lines of twenty or thirty. Kittiwakes in August/September are in a lull between the peak of local birds in July and peak passage in October, but you will see a few. Do not ignore them, or you might miss a Sabine's Gull. It is not the only species you need to be particularly alert for. Storm Petrels are often hidden between wave crests, giving only the briefest flash of something small and slow – patience and concentration are then needed to get a re-sighting. Leach's Petrel will not become the more frequent of the two species until well into September. Since the 1990s, Wilson's Petrels have been spotted from 'pelagic' boat trips, up to 80 km west of land, close to the edge of the Atlantic shelf. Just a few of these Antarctic wanderers are now being spotted from Bloody Foreland. And just as rare at this end of the country are the three virtually indistinguishable gadfly petrels (Fea's, Zino's and Desertas) – always hoped for, but only three have been seen so far in Donegal.

So what does a really good day at Bloody Foreland deliver? They are of course all different, but one that sticks in my memory was 22 October 2015. Chris Ingram, Theo Campbell and myself started at 8.15 a.m. and record-ed 3,292 birds over four hours (Table 15.1). The wind was from the west, starting at Force 7, which is probably as much as we could have handled, but dropping to Force 5 by the end. We've seen more birds on other days, and we had no rarities. What we did have was a good spread of species, and record rates of passage for both Great Skuas and Great Northern Divers. With three of us on the job, we managed to count everything, including the Gannets, which would have been impossible for one person on their own.

For a greater diversity of rare or uncommon species (ignoring Fulmar, Manx Shearwater and Gannet) the count in Figure 15.2 is as good as it gets.

SeaTrack – the not so good days

We naturally choose to go seawatching on days that we think will be good. Fortunately, between 2010 and 2014 BirdWatch Ireland persuaded us to take our chances on set dates. This was for the SeaTrack project, designed

Species	Totals seen in 5.5 hours	Rate/hour
Common Scoter	3	1
Kittiwake	17	3
Sabine's Gull	2	
Sandwich Tern	1	
Arctic Tern	14	3
Great Skua	8	1
Pomarine Skua	5	1
Arctic Skua	11	2
Long-tailed Skua	8	1
Guillemot	1	
Red-throated Diver	3	1
Great Northern Diver	1	
Wilson's Petrel	1	
European Storm-petrel	3	1
Fulmar	NC	
Cory's Shearwater	40	7
Sooty Shearwater	215	39
Great Shearwater	231	42
Manx Shearwater	NC	
Balearic Shearwater	1	
Gannet	NC	
TOTAL	565	102

Table 15.2: Birds at Bloody Foreland on 24 August 2024
NC = not counted.

primarily to determine the true status in Irish waters of the critically endangered Balearic Shearwater. The project was a success, and did establish the southern Irish waters to be of great importance. Regrettably, we found only a few Balearic Shearwaters this far north, but what we also discovered was that you could sometimes have a good day in conditions that we would have previously thought not worthwhile.

As predicted, the best winds were found to be nearly always from the north-west. Westerly winds can also be good, but are less reliable. South-west early in the season is good for the large shearwaters, coming from the south. We were surprised to find that even the offshore south-east winds can occasionally have reasonable numbers of birds, particularly Gannets, Kittiwakes and auks. Northerly winds are quite good, particularly late in the season, although the lack of shelter from that direction remains a serious deterrent. North-east winds are not good, and we have yet to experience what the east wind might have to offer.

As with direction, we had our own pre-conceived notions of what should be the best wind strengths – they had to be Beaufort Wind Forces 5, 6 or 7. SeaTrack's requirement to count on set dates led to the realisation that slack winds of less than Force 4 don't affect the routines of Gannets, Kittiwakes or auks, although we had been correct in assuming that other species will be fewer, or not show up at all.

Pelagics

For the dedicated seawatcher still waiting for a new species, or a close-up, leisurely view of ones normally glimpsed briefly at telescope range, there is another option available. Take a 'pelagic' boat trip, out into those summer blooms of plankton that are visible from space as vast green whorls, or to the edge of the Atlantic Shelf where there is an upwelling of nutrients from the oceanic depths.

Charter a boat in advance at one of the north-westerly harbours, where the variety of craft now available has greatly increased with the development of sport angling and scenic trips for tourists. The main harbours are Burtonport, Bunbeg, Magheraroarty and Portnablagh. Try to plan for a calm day in late summer. If all goes well, you should start finding birds from 10 or 20 km offshore, but if you are not having luck, be prepared to go the same again. Luck can be greatly assisted if you bring along fifty litres or more of 'chum' (fish offal from local fishing boats). This lure was first adopted by shark-anglers, but is also the only way to conjure seabirds out of an apparently empty ocean. Within half an hour of emptying a barrel of this foul-smelling liquor, hundreds can surround your boat, and give wonderful views. All species are possible, but the Storm Petrels are

Great Shearwater and Fulmar, Rockall Bank (245 nautical miles west-north-west of Donegal).

NIALL T. KEOGH

usually most reliable in good numbers – with one or two Wilson's Petrels often among them. Fly-pasts by various shearwaters and skuas will always command the attention of the photographers. Sea legs are recommended.

The verdict

It's hard to beat the buzz you can get from seeing birds which breed in the South Atlantic islands (e.g. Sooty Shearwaters) converging in front of your deck chair with others from the high-arctic tundra of Siberia (e.g. Pomarine Skuas), or from Canada (e.g. Sabine's Gull), or en route to winter in the Antarctic summer (Arctic Terns). There are seawatching headlands further south in Ireland which can get more of some interesting or rare species than we can expect. But Donegal has its own particular mix, and on the right day it can be the best place in Ireland to witness a bird spectacle only seen by the few who know it is there to be seen and make the effort to be at the right place, on the right day.

Chapter 16

Looking Ahead

DONEGAL'S DISTINCTIVENESS lies in its position at the north-west corner of the island, its long and varied coastline, and the contrast between its lowland east and mountainous west. But Ireland is small, so there is no great divergence between north and south, unlike for example the north of Scotland compared with the south of England. So the changes in Donegal's avifauna will not be too different from those experienced by the rest of the country.

Global trends

The prospects for birds in all Irish counties are subject to global trends in land use and climate change, both of which have profound impacts on birds and their habitats. There is a rapidly growing series of reports charting the decline in the number of individuals in all classes of vertebrates, including birds. For example, from 1970 to 2019 there was a drop of 29 per cent in the number of birds in North America (USA and Canada) – a loss of 2.9 billion birds in fifty years.[1] An almost identical number has been lost from the EU since 1980 – 600 million lost from an area five times smaller than North America.[2] At the same time, BirdWatch Ireland (the Irish partner of BirdLife International) has calculated that 25 per cent of Ireland's bird species are showing severe decline, and an additional 37 per cent are showing moderate decline.[3] We would usually notice and record local population reduction in our rare species, but we are slower to detect it in the common species. The Countryside Bird Survey is doing great things nationally to address that problem, and in time will probably be able to quantify the thinning out of many of the species that we still regard as common – although any declines prior to the start of the survey in 1998 would remain unquantified.

Of huge significance for our future bird populations is the catastrophic decline of insect numbers throughout the world. There are plenty of national statistics to demonstrate this in the developed countries of Europe, but nothing is more convincing than the disappearance of something people of my generation remember vividly – the experience on warm summer evenings of having to stop and clear dead insects from car windscreens. This loss of insects is bound to be impacting the breeding success and survival of our insectivorous species, most of which are summer migrants from Africa.

Climate change has already been mentioned, and will frequently crop up in the individual species accounts. It is now generally accepted as a reality, and already having an impact – warmer summers and winters, more rainfall, floods and droughts, storms more frequent and more severe, and rising sea levels. But there is another possible outcome that is still being downplayed as much as possible. The weakening of the Atlantic Meridional Overturning Circulation (AMOC) threatens to block the North Atlantic Drift, which is what keeps Ireland much milder than would be normal for our latitude. Without it, we could suffer arctic winters like those experienced by Labrador. This slowing of the ocean's circulation is partly caused by the vast quantities of fresh meltwater flowing from the Greenland ice sheet, which will also raise sea levels by very much more than the 3.6 mm per year we've had so far – it could be 5 m in total. If these changes happen, all bets are off as regards the future of our avifauna.

There are also unpredictable events, of which avian influenza is one. In the winter of 2021/2, some swans fell victim to it at Inch, but nothing compared to what happened the Svalbard population of the Barnacle Goose, all 40,000 of which winter on the Solway Firth in south-west Scotland. One third of that population died. The summer of 2022 had a devastating impact on breeding seabird colonies in Scotland, where thousands of birds died, most conspicuously on the Bass Rock, the world's largest breeding colony of the Northern Gannet.[4] The hope was that winter would kill off the virus, as had always happened with previous outbreaks. That didn't happen, and in February 2024 the Royal Society for the Protection of Birds and the British Trust for Ornithology reported a 75 per cent loss in the Great Skua population of the UK (which is home to most of the world's 16,000 pairs) and a 15 per cent decline in Northern Gannets.

Trends in Donegal

Projecting current trends may give an idea of what the future holds for many species, but they have to be treated with caution. For a start, the amenability of the different species to accurate assessment is hugely variable.

	Decreasing	Stable	Increasing	Mixed	TOTAL
Birds present at all seasons	23	37	19	8	87
Summer visitors	12	10	5	2	29
Winter visitors	10	15	11		36
Passage migrants	1	11	1		13
Rarities	5	6	7		18
Formerly winter visitors	1				1
Formerly summer visitors	1				1
Extinct	3				3
TOTALS	56	79	43	10	188
Percentages	30	42	23	5	100

Table 16.1: Abundance trends for 188 taxa

And without constant county-wide surveillance by birdwatchers and field-workers we are left with insufficient data to detect trends for many species, and have to rely on evidence of varying value, and on indications from beyond the county boundary. So the following statistics should be treated only as pointers (Table 16.1). This table summarises the trends indicated in Donegal over the past twenty years for 188 taxa out of the total of 376, as given in the Check List (Appendix 5). The 188 taxa not included are mostly rarities, a few cryptic species like Jack Snipe, and passage migrants for which the records are erratic, such as Pomarine Skua or White Wagtail.

It is worthwhile spelling out a few of the details in the table. Starting at the lower end of the decreasing column, Nightjar, Grey Partridge and Corn Bunting have lost their own battles to survive in Donegal. Quail and Bewick's Swan could be added to the extinct tally, but for the moment they can still qualify as rarities. Corncrake would have gone extinct had it not been for the national effort to save it, and its recent annual totals (Figure 18.23) puts it into the stable column. But it remains vulnerable.

Sadly, 30 per cent of all our bird taxa are in the decreasing column. This would be reasonable if it was matched by a similar figure in the increasing column, but that is not the case – only 23 per cent are increasing. The national assessments mentioned above are more refined, and even more depressing. They are mostly based on much harder data than we have available for the county, and show 25 per cent of species severely declining, and 37 per cent moderately declining.[5] For wintering waterfowl, the I-WeBS coverage of Lough Swilly provides relatively hard data (Figure 12.1 and Table 12.1) and more encouraging results than the national analysis.[6] Perhaps Donegal's remote location accounts for a delay in the full impact of whatever forces are reducing the numbers of birds in Ireland as a whole.

Four contrasting species out of the twelve declining summer visitors illustrate the magnitude of the challenge we face. Little Tern suffers from human disturbance, Fulmar is probably short of food as marine fauna move north, Ring Ouzel's habitat is deteriorating as well as climate change being certainly involved, and Spotted Flycatcher is a summer migrant dependent on insects everywhere it travels throughout the year. So multiple changes in how humans are affecting the planet, and how we are managing our own share of it in Donegal, are needed if we are to retain these species in the long term.

It is likely that Donegal will soon be the final county in Ireland to lose its breeding Ring Ouzels and Golden Plovers. Climate change is involved with both, but poor management of our uplands and wetlands will speed their demise, and will be the main reason for the probable loss of other breeding waders – Curlew, Dunlin and Redshank. In the uplands, the fate of our breeding Hen Harriers also hangs in the balance. Birds of arable farmland, like Barn Owl, Stock Dove, Yellowhammer and Tree Sparrow, are all decreasing – will they further boost a growing list of former species?

The species in the mixed column include two of the most concerning – Lapwing and Curlew – which are stable as winter visitors but decreasing rapidly as summer breeders. Also among the mixed species, the Greylag Goose is stable at the moment in winter, while the breeding population is increasing. Chaffinch is almost certain to be following the national upward trend as a breeding species, although we still lack firm local evidence, but we can be confident that its winter numbers in Donegal are decreasing.

There is some comfort in that the largest cohort of species in Table 16.1 is the one with stable populations – as we should expect. We might pick out some members of this group as having potential for increase, but without local evidence of any change yet, the grounds for predictions here are shaky.

The cohort of currently increasing species includes some that we might soon be welcoming as additions to our regular breeding species, and some others that are not yet regular visitors. Resident Mediterranean Gulls are now almost guaranteed, and Reed Warbler might eventually advance from its present vagrant status to that of regular summer breeder. Garganey and Little Ringed Plover are showing signs that they might eventually breed, and given a bit more time they could be joined by Cattle Egret. Could Ring-necked Duck be now at the stage the Mediterranean Gull was at a couple of decades ago? If they go on to breed regularly, it would be our first known colonisation by an American species.

The continuing northward spread of Bank Voles will take some years yet to reach Donegal, but it almost certainly will happen. If so, that will be good news for some of our raptors. Hen Harriers, Kestrels, Long-eared and Barn Owls will benefit – and Short-eared Owls could even extend their breeding range to Donegal in response. Raptors in general are already responding well to the lack of persecution, which of course was one of the pre-requisites for the national re-introduction projects for Golden and White-tailed Eagles and Red Kite, and now Osprey. The un-aided return of the Marsh Harrier is not impossible.

Table 16.2 is based on viewing current trends in light of the changing environmental factors relevant to each species. Firmer predictions are almost impossible to derive from known data, so speculation like this is only on the grounds of being better than nothing. What it shows is that the possible gains are a mixed bunch driven by a wide variety of factors. Some are already moving northwards, probably in response to climate change. Other species also on the northward march, like Great White Egret, might not breed due to the lack of suitable

GAINS	LOSSES
Garganey	Stock Dove
Ring-necked Duck	Golden Plover
Goosander	Curlew
Little Ringed Plover	Dunlin
Mediterranean Gull	Woodcock
Cattle Egret	Redshank
Osprey	Little Tern
Marsh Harrier	Red-throated Diver
Red Kite	Golden Eagle
White-tailed Eagle	Barn Owl
Reed Warbler	Wood Warbler
	Ring Ouzel
	Redstart
	Tree Sparrow
	Twite
	Yellowhammer

Table 16.2: Potential gains and losses in our breeding birds over the next twenty years (in systematic order)

habitat. The losses will include northern species retreating in response to the same climatic forces that are driving southern species towards us. There is also a significant proportion that are suffering from factors like habitat deterioration, or disturbance, that are not related to climate change and could still be fixed if we had the will to do it.

There is much that can be done to lay out a welcome mat for new arrivals, and to slow, or even reverse, the decline in others. Among the key changes needed are the following:

- Limiting grazing on uplands, and where possible replacing sheep with lightweight, winter-hardy breeds of cattle – Galloway, Dexter and Kerry are all able to look after themselves on hill pastures, and encourage biodiversity.

- Restoring appropriate management of our native grasslands, particularly the machair systems, where winter grazing should keep scrub-invasion at bay, and limited cultivation on a long rotation would be possible, as in the Outer Hebrides.

- Restoring our native woodlands, and creating new deciduous woodlands with appropriate management, as, for example, with the 'continuous cover' forestry system.

- Eliminating the pollution of our existing wetlands.

- Controlling development in the countryside to retain open spaces.

- Creating wildlife corridors, at every scale, whereby species populations can avoid being isolated in areas that are too small to guarantee their survival.

- Increasing the number of large areas where different habitats can determine their own boundaries and grade into each other naturally – Sheskinmore Nature Reserve and Glenveagh National Park are two of the few examples that we already have.

As indicated in the final bullet point above, there are many advantages in letting nature decide how to proceed. Re-wilding is still controversial, and not appropriate in many situations. But working towards it as the ultimate management prescription has to be kept in mind.

Many of these changes are already happening through local initiatives. For example, the Inishowen Uplands Farmers Project is proving that low-impact farming can be profitable.[7] And the Inishowen Rivers Trust is bringing communities together to restore rivers to a more healthy state.[8] The Inch Wildlife Reserve is a co-operative effort between national agencies and local stakeholder groups.[9]

But of course, changes can only be made on the necessary scale if they are fully accepted as national priorities, and support given to those who will have to implement them – farmers, conservation agencies and voluntary organisations, and local authorities. It will need considerable investment, but that is not to say that it will be economically un-viable. We just have to learn to recognise and reward benefits that are not yet part of the economy. Re-thinking what we mean by 'the economy' will be to the benefit of birds and other wildlife, and also to ourselves. Preserving and encouraging biodiversity in our own small corner is an obligation that we share with every other small corner – if abundant life, human and other, is to remain sustainable.

Chapter 17

Introduction to Species Accounts

Species entries

SEQUENCE AND NAMES

S EASONED BIRDWATCHERS expect their reference books to list species in a conventional order – from the most ancient lineages to the most recent. The trouble is that science constantly produces more evidence on the relationships of species, and also on higher-level groups like families. This means that changes in both the sequence and the scientific names are more frequent than most of us would like. Obscure subspecies are sometimes elevated to full species status (these changes are usually welcomed), or deleted as being just evidence of the natural variability of the common form. The most widely accepted reference on these matters, and the one followed by the Irish Rare Birds Committee (IRBC), is the International Ornithological Community list (IOC World Bird List) which at the time of writing stood at version 11.1.[1]

Approved vernacular names also change, to suit international needs. We have only one common species in Ireland called Heron, but for the benefit of people from countries with greater diversity, we have to refer to our bird as the Grey Heron. That change is one of the few that might be adopted locally, but does Arctic Skua really have to become Parasitic Jaeger? The approach taken here is to retain the names used by our resident birdwatchers. Where a different name is being promoted by IOC, it is placed, in brackets, alongside the name still used locally. For a very few birds, like Grey Heron and Rosy Starling, the new names have already been adopted, so the old vernacular is dropped.

Irish names are also included. These have been culled from two main sources – the *Dictionary of Bird Names in Irish*[2] and the BirdWatch Ireland website, and with valuable help from Micheál Mac Gloinn.

STATUS

A thumbnail statement is given on the status of each species. The definition of each term is given below (status summaries, abundance and trend), and the details for each species will be elaborated in the full entry.

OBSERVER NAMES

Where known, the name of the first finder of a species is given. The first finder for each subsequent record is given, where the number of records is five or less. For records that were rejected, but for whatever reason are still worthy of inclusion, the observers' names are not given – Albatross is a good example. For those that are not yet verified, names where available are given.

RANGE AND MIGRATIONS

At some point in the text a thumbnail statement of the normal winter and summer range of each non-resident species is given.

REFERENCES

Literature references can at times clutter up the text, so frequently cited texts are occasionally omitted – it should be fairly clear what these are, with reference to other similar species.

SEASONS

For each species entry, breeding, wintering and migration details are always given, where applicable – but not necessarily in discrete packages.

STANDARDS

Where applicable, the standards for designating a population of waterfowl as of National or International Importance are given, usually at or near the end of the entry.

Also, threatened species that are Red-listed (the highest of the three categories – Green, Amber and Red) have a similar entry stating why they are Red-listed.

Status summaries

These are conventional terms which can explain a lot about the *when*, *where* and *why* that define the presence of any bird. They are given at the start of each species account.

RESIDENT

Birds present throughout the year, and breeding.

SUMMER VISITOR

Species that are present here only in the summer breeding season, migrating to warmer climes for the winter, although food availability is often a more important driver than temperature. The most significant group of summer visitors are the songbirds that winter in Africa.

NATURALISED / FERAL BREEDER

Species that have been introduced by humans, deliberately or as escapes from captivity, but are now surviving in the wild on their own terms. These are mostly wildfowl or game species. We have only three in Donegal – the Mute Swan, Pheasant and Canada Goose. Our breeding Greylag Geese were thought to have been feral but have now been re-assessed as native (see under the species entry).

WINTER VISITOR

Species that come mostly from northern latitudes to spend the winter months with us. As with summer visitors, they are here as much for the feeding opportunities as for the climate, but there is no doubt that climate change is now diverting some species. Many of our winter visitors are wildfowl and waders.

PARTIAL MIGRANT

Part of a local population which moves seasonally, usually within the country/county. Not many species are obviously in this category, but many can make short movements unnoticed. If not all of the local population moves, those that do might not be noticed. They include thrushes, finches, etc.

PASSAGE MIGRANT

Species that only stop off (or fly past) Donegal en route to or from their summer or winter homes – typically in autumn and spring. They are mostly seen on the coast and offshore islands, and include seabirds that spend the non-breeding parts of their year on the move, and will only be seen at sea or from prominent headlands. Cohorts of some species that are also present in the county as summer visitors, residents or winter visitors can also be passage migrants. The sojourn in Donegal of some passage migrants may be brief, and the numbers small, but it is not accidental as with the species in the next category.

VAGRANT

Accidental arrivals in Donegal. These tend to be birds that were migrating between summer and winter territories elsewhere but had lost their way, most likely having been driven off-course by adverse weather conditions. Following an unplanned ocean crossing, most of them will make landfall at the first opportunity, so will be found only at extremities like Tory Island or Malin Head. Another supply of vagrants comes from continental birds on their return migration from Africa. A few will sometimes continue past their intended destination (when Tory could be their last chance to stop and re-direct themselves southward). These overshooting birds tend to appear in spring and early summer rather than in autumn. Vagrant wading birds and waterfowl will usually make their way to a suitable wetland where they can feed – the Lough Swilly lagoons at Inch and Blanket Nook have been the most attractive.

Abundance

Terms like 'rare' or 'scarce' can be allocated to a range of numbers, but as these will vary with the size of the population or the lifestyle of the bird, they cannot be regarded as definitive. However, for non-breeding migrants and vagrants, bands indicating bird numbers (or number of records), as suggested below, are useful. For breeding birds, the use of 'rare', 'scarce' or 'uncommon' would have to be related to the density requirements of the species – so if we had fifty pairs of Hen Harriers we might regard them as common, but fifty pairs of a species capable of existing at higher density, such as Yellowhammer, would place it in the uncommon bracket. The best approach is to take abundance bands in the context of what you read in the text. In any case, where their use would not be very helpful, the temptation to allocate one has been resisted.

ABUNDANT

Reserved for the exceptionally common and numerous species, like Blackbird or Meadow Pipit.

COMMON

This term is useful for birds found throughout the countryside. Those that most people are familiar with can usually be described as common.

UNCOMMON

Between 201 and 500 records. Counting becomes progressively pointless above 500.

SCARCE

Between 11 and 200 records.

RARE

For vagrants this means ten records or less, with all of them listed, and is indicated in the status summary. For residents or regular visitors, there are no firm boundaries.

FIVE-YEAR RUNNING MEAN

Five-year running means are mostly used to produce graphs showing the population trends of waterfowl, whose annual peak counts are recorded. The average of the highest count in each year from one to five is the five-year mean. In year six it will be the average from years two to six, in year seven it will be years three to seven … and so on. The result is a smoother graph, ironing out the highs and lows and making the trend easier to detect. However, the highs and lows often tell us more, so running means are used only where it is useful.

Trends

For almost exactly half of the species, it has been possible to give an indication of how well their populations are faring. This is summarised in the status line in the species accounts (Chapter 18), and in the checklist (Appendix 5).

AERC category system

The AERC (Association of European Records and Rarities Committees) has a system which defines the degree of 'wildness' of each species. This is intended for countries, but still has some value at county level. Almost all species recorded in Donegal belong to the first category (A) – species that have occurred naturally in the wild since 1 January 1950. The exceptions are listed in Appendix 6.

Records

IRBC decisions on rare birds have reached 2021, so all rarities after that date, and some before it, await verification. Otherwise, the coverage is complete to the end of 2023, with notable rarities in early 2024 squeezed in.

Sources and monitoring

In the early days, there was a succession of authoritative books on all the birds of Ireland written by the experts of the day, starting with Thompson (1849–51),[3] followed by Ussher and Warren (1900),[4] Kennedy, Ruttledge and Scroope (1954),[5] Ruttledge (1966),[6] Ruttledge (1975)[7] and finally Hutchinson (1989).[8]

THE GLOBAL STATUS

For all but the resident species, it has been necessary to give some details of where Donegal's birds come from, or go to, and if relevant, their international status. For this there are three main references – *An Atlas of Wader Populations in Africa and Western Eurasia,*[9] *Atlas of Anatidae Populations in Africa and Western Eurasia*[10] and *Handbook of the Birds of the World.*[11]

IRELAND AND BRITAIN

Starting with *The Atlas of Breeding Birds in Britain and Ireland*,[12] there is a series of mass-participation surveys (now treated as citizen science), each resulting from a number of years' fieldwork and published as a series of atlases which map all species at 10 km² resolution.

BIRDWATCH IRELAND

The primary voluntary bird organisation in Ireland, promoting bird conservation, field research and membership activity.

NPWS

The National Parks and Wildlife Service is the government agency responsible for implementing national and international wildlife legislation. It conducts or commissions many reports on flora and fauna, and the ongoing programmes of other organisations, such as the I-WeBS surveys by Bird-Watch Ireland.

BTO

British Trust for Ornithology. One of the main drivers of field research on bird populations and conservation. Does not operate in Ireland, but often collaborates with Irish organisations on projects covering Britain and Ireland.

BOU

British Ornithologists' Union. Despite its name, this venerable organisation has a global brief on all matters ornithological.

CBS

The Countryside Bird Survey. The common breeding species throughout the countryside, especially the territorial ones that are well dispersed, are not so easily surveyed. So a network of 1 km² plots have been surveyed since 1998. A standard methodology is used to gather data on the numbers of all species regularly encountered. This allows national populations to be extrapolated, and trends calculated, but for many species a longer time-run is needed before these will be reliable. Details are given only where useful.

I-WeBS

The Irish Wetland Bird Survey and its predecessor the Winter Wetland Survey have been counting the waterfowl of estuaries, lakes, etc. since the 1980s – usually by teams of local volunteers. The results are published in the *Irish Birds* journal and elsewhere. The scheme is run jointly by BirdWatch Ireland and the National Parks and Wildlife Service. It enables trends in species numbers and site usage to be continuously under review. The 1 per cent thresholds for International and National Importance are calculated every few years. These are specified here only for those species for which they are of relevance. Whooper Swans, Barnacle Geese, White-fronted Geese and Greylag Geese are each monitored separately, in addition to I-WeBS. Graphs of peak counts on Lough Swilly show blanks for 2020/21, when there was no counting due to the Covid-19 pandemic.

NEWS

Non-Estuarine Waterbird Survey. As I-WeBS surveying mostly doesn't include the open coast, NEWS does it every five years.

RARE BIRDS

Following on from the production of *Birds of Ireland*,[13] an annual *Irish Bird Report* (IBR) commenced in 1953, edited by Ruttledge up to 1971 and then by various editors on behalf of a succession of committees. It dealt mainly, but not exclusively, with rare birds. It was incorporated into the journal *Irish Birds* from 1977, and narrowed its focus to become the *Irish Rare Bird Report* from 2005 onwards.

The Irish Birding website has been publishing daily sightings of rare and scarce birds since 2000. Long-staying birds can be reported many times. Unlike the other resources, it doesn't critically assess its records – that task is the responsibility of the Irish Rare Birds Committee.

RARE BREEDING

Birds records of these are handled differently from 'rare birds', as confidentiality is often necessary. Preceded by a summary paper on the previous decade, the Irish Rare Breeding Birds Panel has been the repository of all such records since 2002 and it publishes an annual report.[14] This was edited by Paul Hillis from 2002 to 2012.[15] Since then there has been an almost annual change in authorship, but always published in the annual journal *Irish Birds*.[16]

BREEDING BIRD SPECIES

Upland birds, waders on machair, inland gulls, terns, Hen Harrier, Peregrine, Merlin, Corncrake, Chough, Ring Ouzel and Twite have all been surveyed separately, often in the context of research into their conservation needs, and most of them more than once.

SEABIRDS

The breeding seabirds of Britain and Ireland have been surveyed a number of times, starting in 1969 to 1970 with Operation Seafarer, published as *The Seabirds of Britain and Ireland*.[17] This was followed by a national inventory[18] and another Britain and Ireland census,[19] before ongoing national monitoring followed.[20] These surveys have provided most of the data on breeding seabirds, but the difficulty of such undertakings means that the results are not as neatly comparable as for wintering wetland birds, for example. So they are supplemented with spot counts wherever these are available. Where this requires extra clarity here, the comparative results are presented in tables.

Conservation

The main designations and evaluations affording legal protection for the conservation of species and their habitats are defined here. The National Parks and Wildlife Service (NPWS) has responsibility for implementing these, but only the state-owned areas are also managed by them. Maps for SPAs and other designations can be found on the NPWS website.[21]

IBA Important Bird Areas have been identified by BirdWatch Ireland, as the Irish partner of BirdWatch International. Their only formal status is that they provide the basis for the areas chosen as SPAs.

SPA

Special Protection Area, under the EU Birds Directive. Most often these are wetland sites, and sites for colonial breeding species like seabirds. A few dispersed species like Chough and Corncrake have recently been added. In general it can be assumed that almost any area that is noted for its birds will have an SPA designation – with understandable exceptions, like Killybegs.

SAC

Special Area of Conservation, under the EU Habitats Directive. These are essentially for the protection of native vegetation types (bogs, woodland, etc.), but often overlap with SPAs.

NNR

National Nature Reserve. Usually, these are under state ownership, and are already either an SPA, an SAC, or both. In Donegal we have Inch Levels Wildfowl Reserve, identified by NPWS as 'Ireland's premier wetland site'; Sheskinmore, a dune, lake and marsh complex; five areas of native woodland; and three bogland sites. Public access is usually provided.

NATIONAL PARK

These are larger areas owned by the state and managed for conservation. In Donegal we have Glenveagh (170 km²), the management of which tries to maintain a balance between protecting the important habitats and species, and catering for the public in one of the county's most popular tourist destinations.

INTERNATIONAL AND NATIONAL RATINGS

For waterfowl (wildfowl, waders, etc.) the convention is that 1 per cent of a flyway (e.g. East Atlantic and West Africa) population qualifies a wetland site as internationally important for that species. Likewise for Ireland, 1 per cent of the national population estimate confers nationally important status. These ratings inform the designation of legally binding designations like SPAs.

BOCCI

Birds of Conservation Concern in Ireland. Using mainly data from its own surveys, there is a five-yearly assessment of the threat levels for all species, taking into account both population and distributional changes – the most recent being in 2021.[22] The internationally used traffic-light rating of Red,

Amber and Green has been adopted for Ireland. To avoid cluttering the text, and to increase awareness of the most threatened species, these ratings are only mentioned for the Red-listed species.

IUCN

International Union for Conservation of Nature. This is the body that compiles the Red Data lists of threatened species, and sets the standards for lower-level listings e.g. National.

Chapter 18
Species Accounts

Fulvous Whistling Duck

Dendrocygna bicolour | Feadlacha odhar

ESCAPE

This widespread tropical species of tree-duck was a surprising addition to the birds on Lough Swilly in 2017. Two birds moved between Inch and Blanket Nook from at least 6 August to 25 November (C. Ingram et al.). There is no doubt that they were escapes from a zoo or bird collection.

Pale-bellied Brent Goose (Brant Goose)

Branta bernicla hrota | Cadhan

WINTER VISITOR
INCREASING

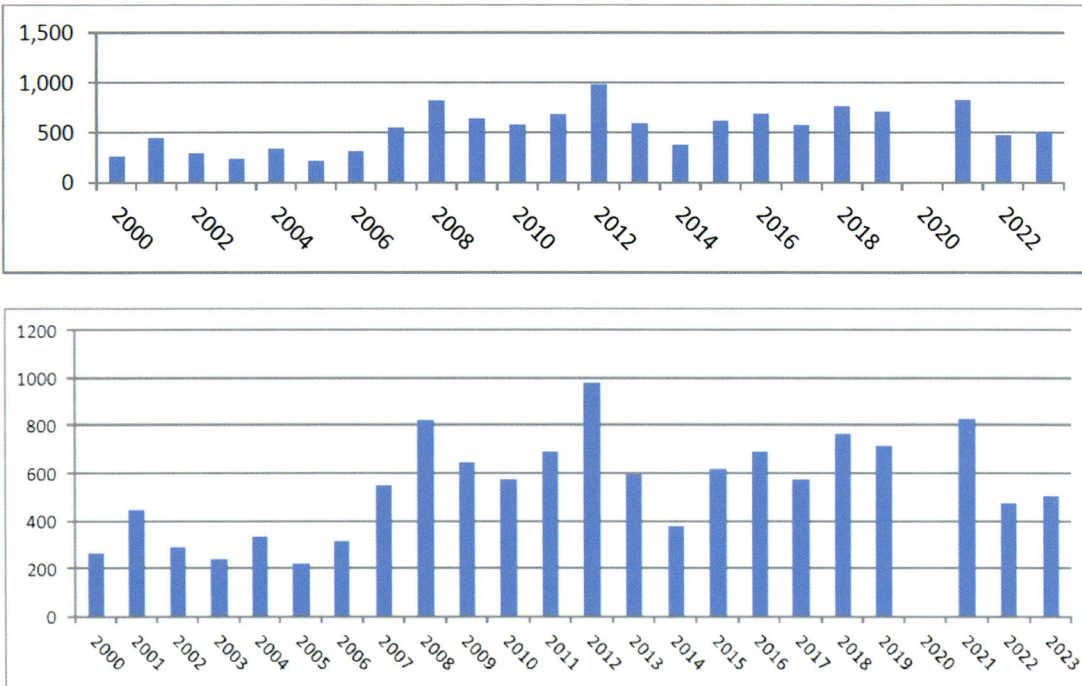

Fig. 18.1: Pale-bellied Brent Goose annual peaks on Lough Swilly from 2000/1 to 2023/4 (counting was abandoned in 2020/1 due to the Covid-19 pandemic)

Brent Geese in Ireland comprise almost the total population that breeds in the Canadian arctic islands beyond Greenland, and winters on this side of the Atlantic. The numbers on our estuaries were much higher than now in the time prior to a disease that almost exterminated their main food supply (eelgrass) in the first half of the twentieth century.

Small parties can be seen almost anywhere around the Donegal coast between October and March, but they favour the sheltered inlets – Trawbreaga Bay, Lough Swilly, Dunfanaghy Harbour, Ballyness Bay, Trawenagh Bay and Donegal Bay. Numbers at these sites are usually in the low hundreds. The highest count was of 984 birds in Lough Swilly from 10 to 11 November 2012, and that number was itself the sum of five sub-flocks and a number of small groups. The number using Lough Swilly is creeping steadily upwards, in keeping with the national trend (Figure 18.1). The most recent five-year mean peak, to 2023/4, is 661. Donegal Bay had a mean peak to 2015/16 of 375, and a peak count of 573.

The Internationally Important figure (1% of the flyway population) is 400.

Qualifying Sites: Lough Swilly

The Nationally Important figure (1% of the Irish total) is 350.

Qualifying Sites: Donegal Bay

Subspecies
Dark-bellied Brent Goose
Branta bernicla barnacle | Cadhan dhubh
RARE WINTER VISITOR

Nine Dark-bellied Brent have been seen in Donegal. This race is from breeding grounds in arctic Russia, and winters around the North Sea. The first record was on the late date of 6 May 1979 at Trawbreaga Bay (D.J. Radford et al.). There was one at Donegal Bay in 1993. Lough Swilly had a pair in 2003 and a single in 2004. In Inishowen there was one on the west bank of Lough Foyle in 2009, one at Malin Head in 2011, and one at Culdaff in 2016. Finally, another late bird was present on the west coast at Carrickfinn on 16 May 2021.

Canada Goose

Branta canadensis Carrick Finn | Gé Cheanadach

NATURALISED WINTER VISITOR, AND RARE BREEDER
INCREASING

Canada Geese were introduced into Britain, from where they spread to Ireland at least 100 years ago. They remained very local until recently, with none being seen in Donegal until 1987, when fourteen birds appeared on Lough Swilly (R. Sheppard). From 1993 a flock started to build up. Their presence is somewhat erratic, with large numbers only reliably found from late summer to early winter. Having reached a peak of 626 on 20 September 2019, there has been a fall-off. Time will tell if this is a genuine reversal of fortunes, or just a temporary blip (Figure 18.2). Only one other site in the Republic has numbers over 100. The first pair to breed locally was at Inch in 2019, and again in 2021, but otherwise our birds are assumed to come from the naturalised population in County Fermanagh. Single birds or small groups appear occasionally around the coast (see Todd's Canada Goose, below).

Canada Geese, Inch Levels.

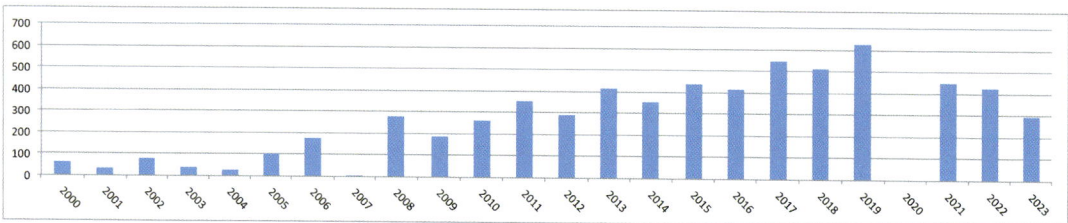

Fig. 18.2: Canada Goose annual peaks on Lough Swilly from 2000/1 to 2023/4

Subspecies

Todd's / Lesser Canada Goose

Branta canadensis interior / parvipes | Gé Cheanadach Todd/beag

RARE WINTER VAGRANT

A lone Canada Goose that looks different could be an escape, but we can usually assume that in the north-western corner of Ireland, and amongst a flock of Barnacle Geese, such birds will be genuine vagrants. Canada Goose is divided into many races (our feral birds are largely derived from only one), and some of these have now been put into a separate species, the Cackling Goose (see below). A review by IRBC summarises the Irish records of both species up to 2013, and accepted a small number of the Irish vagrants as being Todd's Canada Goose (*interior*).[1] The majority were of indeterminate race, including a few suspected of being Lesser Canada Goose (*parvipes*).[2] There is the additional problem that some small Canada Geese can be confused with larger forms of Cackling Geese (see below, and also under Cackling Goose).

For Donegal, only two vagrant birds have so far been accepted by IRBC. The first was from 2 to 12 March 2010, at Trawbreaga Bay, and returning again from 14 February to 8 April 2011 (C. Mellon et al.). It was reported as Lesser, but was only accepted as 'presumed to be of North American origin'. The second was accepted as 'probable Todd's'. It was seen at Dunfanaghy New Lake on 25 February 2017 (A. McMillan et al.).

Reports of nine other Donegal birds are as follows. The first record was of two birds at Inch on 6 March 1993 which just might have been large Cackling Geese, but either way, they were not feral birds (C. Mellon, R. Sheppard). One reported as Lesser was at Trawbreaga Bay on 18 February 2015, with 2,000 Barnacle Geese (R. McLaughlin). A suspected non-feral bird was seen on Tory on 10 September 2015 (R. Sheppard et al.), and three possible Todd's were at Dunfanaghy on 20 January 2016, with 600 Barnacles (C. Ingram). Another probable Todd's was on Tory from 20 to 21 October 2016 (C. Ingram). Finally, a probable Lesser was with Greylag Geese at Big Isle on Lough Swilly from 19 to 20 December 2020 (R. Sheppard).

Barnacle Goose

Branta leucopsis | Gé ghiúrainn

WINTER VISITOR

INCREASING

Ireland's Barnacle Geese are from the population breeding in north-east Greenland, and wintering exclusively along the coasts of north-west Ireland and west Scotland, where they have traditionally favoured the uninhabited, wind-blasted offshore islands.

Barnacles are very mobile, with frequent movements between sites. But there is a lot more movement within groups of sites which are fairly well separated from other groups. This would make censusing relatively easy, were it not for the many sub-sites on remote offshore islands. The solution to this problem has been a five-yearly series of national aerial surveys, now increasing to every three years, with complementary ground counts.[3]

The 1993 census found flocks at Dunfanaghy New Lake, and on eleven offshore islands, and in 1998 at Trawbreaga Bay plus eight islands. Increasingly, they have been lured into the richer grasslands of the mainland, which has allowed more complementary ground monitoring to take place. So by 2020 there were 3,688 on nine mainland sites and 677 on five islands. Figure 18.3 reveals the overall pattern since 1983, which should be viewed against the backdrop of a continuing rise in the Greenland population.

Peak counts (aerial or ground) for the most important sites, with the years in brackets, are Malin Head 1,800 (2013), Trawbreaga Bay 1,775 (2018), Dunfanaghy New Lake 1,300 (2018), Shalwy 868 (2021), Gweebarra Bay 700 (2020), and Rathlin O'Birne 560 (2013).

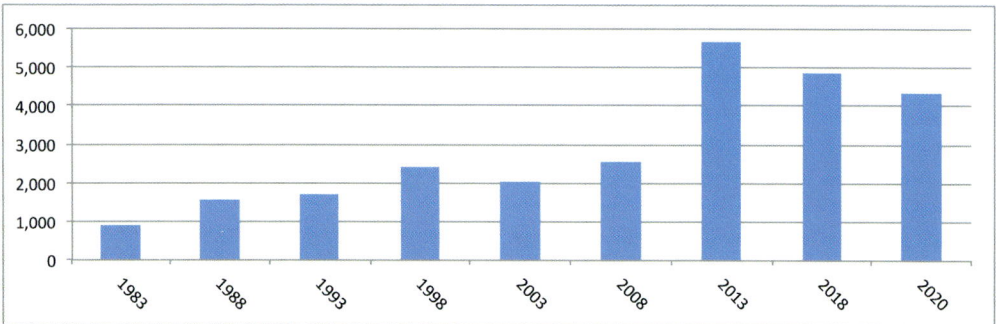

Fig. 18.3: Total numbers of Barnacle Geese in County Donegal at national census dates

Peak counts from other sites which at times exceed the current nationally important threshold are: Loughros Point 230, Drumnatinny 228 and Gola 170. Sheskinmore held a regular flock in the 1980s but this site is now rarely used by Barnacles.

The list of qualifying sites below is only what it would be if the designations could keep pace with the birds.

The Internationally Important figure (1% of the flyway population) is 810.

Qualifying Sites: Malin Head, Trawbreaga Bay, Dunfanaghy New Lake

The Nationally Important figure (1% of the national population) is 160.

Qualifying Sites: Doagh (Melmore), Inishdooey, Dooey (Gweebarra), Rathlin O'Birne, Shalwy

For hybrids, see Cackling Goose.

Cackling Goose

Branta hutchinsii | Gé grágaíleach

RARE WINTER VAGRANT

There have been a number of reports of 'small Canada Geese' over the years, often in the company of Barnacle Geese. Since several of the smaller subspecies of Canada Geese were together promoted to a full species (as Cackling Goose), more attention has been paid to them – but that doesn't mean that any more birds are passing the scrutiny of the Irish Rare Birds Committee.[4] No Donegal records have as yet been assigned to a particular subspecies, but most, if not all, are likely to be Richardson's Cackling Goose *Branta h. hutchinsii*, with a breeding distribution in the arctic Canadian archipelago, well to the east of the three Alaskan races. However, some of the larger individuals may have been Taverner's Cackling Goose *Branta h. taverneri* from arctic Canada to the west of Richardson's.

There have been claimed sightings of eleven birds, of which two had already been reported from County Sligo (and accepted as Cackling Geese of unspecified race). Only three (so far) count as acceptable records of the species for Donegal, none of them assigned to a subspecies.

The first record accepted by the IRBC is of a bird at Dunfanaghy on 3 November 1996 (A. McMillan et al.). There were then the two birds previously seen in County Sligo – one at Dunfanaghy on 25 February 2001 (C. Batty) and one at Trawbreaga Bay on 3 March 2012 (C. Cassidy). The other two acceptable Donegal records were in 2017, the first at Kilcar on 16 January (T. Campbell et al.) and the second in Inishowen on 3 November (C. Cardiff et al.). This one had been initially reported as a Richardson's Goose.

There are a further four birds as yet unverified. The first was on 27 February 1996 at Dunfanaghy (P. Farrelly). Another appeared at Inch on 2 April 2002 (D. Hunter, D. Steele), and there was one with Barnacle Geese on Inishkeel on 7 January 2006. This was thought to be Taverner's (R. Sheppard). And finally one was at Malin Head on 13 October 2013 (R. McLaughlin).A hybrid bird at Inch Levels on 12 December 2015 was probably a cross between Barnacle and Cackling Goose. It was accompanied by a Barnacle Goose, among a large flock of Greylags (R. Sheppard and C. Ingram). (See also Todd's / Lesser Canada Goose, above).

Bar-headed Goose

Anser indicus | Gé stríoc-cheannach

ESCAPE

This central Asian species is not likely to reach Ireland naturally, so the three birds first recorded on Inch Levels on 4 September 1997 (M. and S. Guthrie) are assumed to have been all or part of a single escape. One remained into 1998 and was last seen on 13 September.

Snow Goose

Anser caerulescens Asner | Gé shneachta

RARE WINTER VAGRANT

There are six records of Snow Goose in Donegal (plus records from 1995 and 2021 that have not yet been assessed by the IRBC).

This arctic Canadian species, whether or not arriving with Greenland White-fronts, will usually be found in their company, or with Greylags. There was one on Tory, one at Dunfanaghy, one at Sheskinmore, and the rest were on Lough Swilly.

The first was the Tory bird, on 14 October 1951. Not until December 1990 at the Inch / Blanket Nook area of Lough Swilly was a second bird seen (G. Gordon). Then one bird was seen from 21 March to 7 April 1993 at Dunfanaghy. There was one from 25 to 26 March 1995 at Inch. From 6 December 2009 to 15 February 2010 a bird was at Big Isle (Lough Swilly) and what was presumed to be the same individual returned to Blanket Nook, where it stayed from 17 to 26 November 2010. This was followed by a white phase bird, also at Big Isle and at Inch, from 3 December 2011 to 6 February 2012. The Sheskinmore bird stayed from 7 November 2012 to 29 March 2013. The most recent record was of a dark phase bird at Inch on 22 September 2021.

Only two of these birds were not of the white phase. The 2009–10 bird was 'intermediate', which implies that it was the result of a pairing between a white bird and a blue phase bird. The 2021 bird was identified as 'dark', presumably blue phase. A hybrid with Barnacle Goose was recorded at Inch on 25 October 2013 (B. Robson).

Greylag Goose

Anser anser anser | Gé ghlas

WINTER VISITOR, AND SCARCE BREEDER
STABLE IN WINTER, INCREASING AS BREEDER

The breeding of Greylag Geese was first recorded on 20 June 1998 when a brood was seen at Sheskinmore, followed by a nest at Inch in 1999. There is now a good population on Mulroy Bay, with over 100 birds in post-breeding flocks. Around the coast, isolated pairs are turning up on a number of offshore islands. Inishsirrer had a pair in 2017, and on Tory they had reached four pairs by 2021. These birds have undoubtedly come from the Scottish population which had formerly been regarded as feral but is now recognised as a re-establishment of the native British and Irish population, which had been contracted to a very small remnant in north-west Scotland. The remnant and the re-established populations in Great Britain have each now expanded in both number and range, and are presumed to have merged.[5] They have also spread to Donegal in late summer (see below), but whether our breeding birds are derived from this flock or from a direct colonisation from Scotland of breeding pairs is not known.

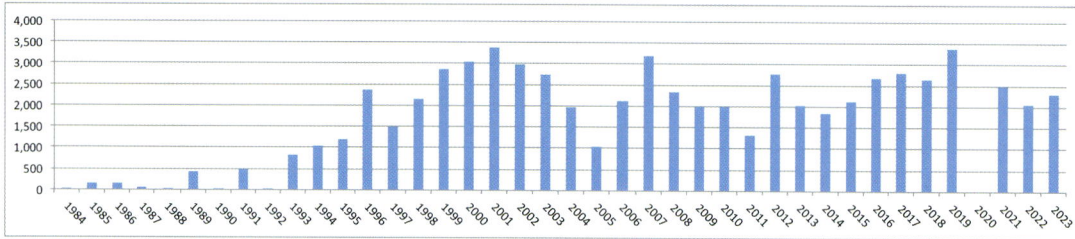

Fig. 18.4: Greylag Goose peaks on Lough Swilly from 1984/5 to 2023/4

Greylag Geese, Inch Levels.

Greylags were only irregular winter visitors in small parties until the second half of the twentieth century. By 1979, 162 were reported as the highest total for some years. The total on Lough Swilly jumped to 435 in 1989/90 and continued to the remarkable ceiling of 3,384 by 2001/2 (Figure 18.4). Although that rise in numbers through the 1990s was spectacular, and the annual totals since then have been erratic, the overall trend since 2000 has been one of stability at a level that makes Lough Swilly easily the Greylag's most important wintering site in Ireland.[6] Closely associated with Lough Swilly is the cross-border site along the River Foyle which held rather more Greylag Geese than Lough Swilly in the 1970s but has not increased since then on anything like the same scale as Lough Swilly. While these birds can roost on the wide open tidal space of the River Foyle at St Johnstown, they are just as likely to merge with the Swilly birds on Inch Lough.

Lough Fern currently supports about 150 birds, and is therefore of National Importance. Other minor sites are Drumnatinny, Dunfanaghy New Lake and Gweebarra Bay.

The Lough Swilly wintering totals are largely composed of immigrants from the Icelandic breeding population. They also include the late summer flock referred to above, which is presumed to be from the expanding native British population in Scotland. These were first noticed in 1992, when 136 birds were at Blanket Nook on 21 August. Numbers are still growing, with a peak of 1,166 birds recorded at Blanket Nook on 10 August 2020. A team from NPWS and the Inch Wildfowling Club put numbered neck collars on about fifty of these early birds. Re-sightings established that they moved freely around Ireland, and beyond, but a direct link with the re-established population in Scotland was still not proven.[7]

The Internationally Important figure (1% of the flyway population) is 980.

Qualifying Sites: Lough Swilly

The Nationally Important figure (1% of the national population) is 35.

Qualifying Sites: Lough Fern

Taiga Bean Goose

Anser fabilis fabilis | Síolghé Taiga

RARE WINTER VISITOR

In the nineteenth century, H.C. Hart wrote that in Donegal, the Bean Goose was the ordinary wild goose found inland in winter. If correct, these birds would have been Taiga Bean Geese, as this species (at that time only recognised as a subspecies) was more numerous in Great Britain than it is now, and occupied habitat in the Scottish Highlands not unlike inland Donegal. Hart did acknowledge the additional presence in the county of both

Greylag and White-fronted Geese. However, he seems to have understated the presence of White-fronts, saying only that the species 'probably occurs every winter'.[8] The conclusion has to be that, despite his general reliability and expertise, Hart was mistaking Greenland White-fronted Geese for Bean Geese. To give him some credit, it is likely that he was taking the word of others rather than reporting his own observations, as he himself was mainly a summer resident in the county. And there is further evidence from Ussher and Warren that Bean Geese were not common in Donegal: 'Sir Victor Brooke stated that [Bean Goose] was the common goose of all the north of Ireland. This certainly does not apply to the Lough Swilly district, so remarkable for wildfowl, where this species is rare.'[9]

All Bean Geese in modern times have been seen around Lough Swilly, and many have stayed for extended periods. Certain identification is frequently not attempted, to some degree on the presumption that any Bean Geese in Ireland are Tundra Beans (from the high arctic), unless identified as Taiga Beans (from the boreal forest zone to the south). While this is certainly reasonable for single birds, as is suggested from ringing returns, there is a case to be made that small parties are more likely to be birds from the flock of wintering Taiga Bean Geese in Scotland. There have been five such records.

1. A flock of twenty-one birds at Castlewray, on 31 January 1982 (R. Sheppard), were confidently identified as Taiga Bean, in direct comparison with one probable Tundra Bean and a Pink-footed Goose. At that time, all Bean Geese records were assessed simply as Bean Geese.

2. Two birds were identified as presumed Taiga on 19 November 2011 (B. Robson and M. Tickner) and peaked at five on 17 March 2012. However, these were accepted by IRBC as Tundra (see also under Tundra Bean Goose, below).

3. A flock of thirteen birds at Inch Levels on 30 October 2015. These birds were presumed at the time to be Tundra, but with reservations (S. Feeny et al.).

4. Two birds on Inch Levels on 4 December 2016 were suspected of being Taiga (R. Sheppard).

5. Three birds at Ardee (Lough Swilly) on 26 November 2017 were larger and with longer necks than accompanying Greylag Geese, and so were suspected of being Taiga (R. Sheppard).

It must be stressed that the IRBC have not yet recognised any of Donegal's Bean Goose records as being of Taiga Bean Goose (see also the entry on Tundra Bean Goose, below).

Pink-footed Goose

Anser brachyrhynchus | Gé ghobghearr

WINTER VISITOR
INCREASING

All our Pink-feet breed in Iceland and eastern Greenland. Most winter in Great Britain and around the North Sea coasts. There have been occasional records of these turning up as vagrants in Donegal since the first one seen in Ireland was shot on 19 October 1891 on Lough Swilly. An immature arrived on Tory on 26 September 1954 and was shot the next day. The third was shot on Inishtrahull on 16 October 1960. Two were shot from a flock of twenty-seven at Trawbreaga Bay on 26 September 1964 – a remarkably early date – and the skins presented to the Ulster Museum.

Beginning in the 1980s, one or two birds were seen most years on Lough Swilly, with a build-up starting around 1993 (Figure 18.5). The remarkable

Pink-footed Geese, Inch Levels.

DEREK BRENNAN

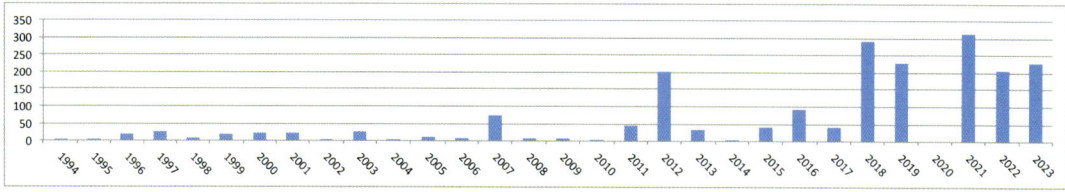

Fig. 18.5: Pink-footed Geese totals on Lough Swilly from 1994/5 to 2023/4

flock of 200 in 2012 was an early indication of what lay ahead. While numbers remained between 50 and 100, parties were dispersed and sometimes hard to find. But since a leap was made in 2018/19 to 293 birds, the flock has tended to settle into a small number of fields, usually on Inch Levels, but also at Blanket Nook. The peak of 314 as far back as 12 March 2022 suggests that this increase could have stalled. If so, it may be related to a sharp rise since the winter of 2021/2 in the number of birds wintering on the east coast, in County Louth.

Tundra Bean Goose

Anser serrirostris rossicus | Síolghé Tundra

RARE WINTER VAGRANT

As mentioned in the entry on Taiga Bean Goose (above), single Bean Geese in Ireland are presumed to be Tundra, unless identified as Taiga. There have been nine such birds either convincingly identified or presumed to be Tundra. The first was seen on 14 April 1976, and reported as probably of the Tundra race, and in the company of Taiga birds (R. Sheppard) – see under that species, but note also a detailed account claiming that four birds in County Louth in the winter of 1997 to 1998 constituted the first record for the race (as it then was) in Ireland.[10]

In Donegal, the 1976 bird was followed by a series of single birds on 7 October 1978 (G. D'Arcy and D. Watts), 12 September 1983 (unattributed), 30 April 1988 (D. Brennan et al.) and 22 October 1989. Five birds in 2011/12 were accepted by IRBC as Tundra, but the initial two were reported as probably Taiga (see also under Taiga Bean Goose, above). Another single bird was seen on 7 September 2014 (K. Bennett). Finally, there were three separate birds for fairly long periods between 8 January and 3 April 2023, all confidently identified as Tundra Bean Geese. Two were on

Lough Swilly (C. Ingram, T Campbell, D. Brennan et al.) and one on Tory (G. Meenan, R. Vaughan et al.). All were well photographed.

White-fronted Goose (Greenland)
(Greater White-fronted Goose)

Anser albifrons flavirostris | Gé bhánéadanach

WINTER VISITOR

STABLE

As the entire population of this distinctive and globally rare subspecies of the Greater White-fronted Goose winters exclusively in Ireland and Scotland, it has long been regarded in Ireland as a top priority for bird conservation, and a very challenging one at that. Its breeding grounds are in western Greenland. Ruttledge and Ogilvie wrote in 1979 that it had traditionally used seven sites (or groups of sites) in Donegal (see also under Taiga Bean Goose, above).[11] The seven sites were:

- Moors west and north-west of Lough Derg, and Brownhall
- Sheskinmore, and other sites in the Rossbeg – Glenties peninsula
- Lough Sallagh and Barnesmore Lakes
- Moors north-east of Fintown
- New Lake Dunfanaghy
- Blanket Nook and Inch Levels were regarded as part of a core flock along the River Foyle in County Tyrone, but were only supporting about sixty birds
- Glenveagh, but this site had already been abandoned by 1979.

Since Glenveagh was abandoned, Barnesmore and Fintown followed suit – all of them bogland sites. There are now four flocks in the county. These all have access to some nitrogen-rich agricultural grazing, but only two have increased through the 1980s and '90s. The four remaining sites are:

- The Lough Derg flock, which now oscillates between the Pettigo plateau and Donegal Bay, normally peaking at about 80 but with a high of 269.

- The Sheskinmore flock survives with twenty to thirty birds.

- The Dunfanaghy flock increased through the 1990s to hold 200 to 300 birds. This peak has declined, with disturbance a particular problem here. Around 60 birds are now normal, but there are occasional peaks of 130 to 160.

- The few birds that Ruttledge and Ogilvie reported for Lough Swilly have abandoned Blanket Nook and Inch, and are now concentrated at Big Isle and another smaller area, where they appear to feel more secure. Their numbers rose to a five-year mean peak of 1,008 in 2003/4, settling back to the 700s for the last decade. The more recent five-year mean to 2019/20 was 759, and a peak of 1,100 in 2022/3 is very encouraging. After the Wexford Slobs, Lough Swilly is the most important site in Ireland for this goose, and one of only three that rate as internationally important (Figure 18.6).

The decline in numbers of the Greenland subspecies was due initially to hunting at their Icelandic stop-over, but since the start of the new millennium has been mainly due to problems on the Greenland breeding grounds deriving from climate change. Here in Ireland, a study by NPWS showed that large flocks with more than ten feeding sites available to them can tolerate disturbance. As this is no longer an option for most of the Donegal flocks, disturbance is a critical factor threatening their survival.[12]

The Internationally Important figure (1% of the flyway population) is 190.

Qualifying Sites: Lough Swilly

The Nationally Important figure (1% of the national population) is 100.

Qualifying Sites: Dunfanaghy New Lake

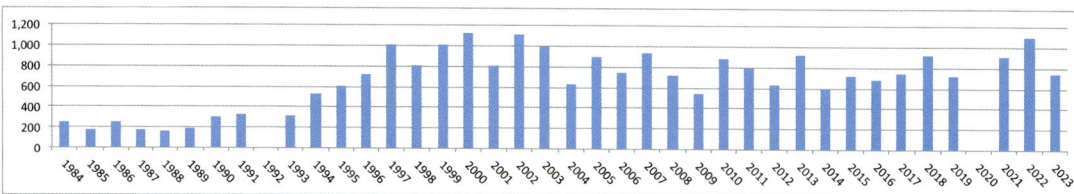

Fig. 18.6: Greenland White-fronted Goose peak counts on Lough Swilly from 1984/5 to 2023/4

Subspecies

White-fronted Goose (Russian)

Anser albifrons albifrons | Gé bhánéadanach (Rúiseach)

RARE WINTER VAGRANT

There are three records on Lough Swilly of this race, which has a breeding range in the Russian tundra and a declining wintering population in Great Britain. The first was a bird at Inch Lough on 20 January 1982 (R. Sheppard). Two birds were seen at Farsetmore on 15 January 2012 (R. Sheppard). A third bird, at Blanket Nook from 26 April to 3 May 2022, has still to be assessed (T. Campbell et al.).

Black Swan

Cygnus atratus | Eala dhubh

ESCAPE

The first of these native Australian birds to be seen in Donegal was at Inch on 1 December 1991 (R. Sheppard). The second was on Mulroy Bay on 1 September 2000. Probably the same bird was at Inch on 23 September, and the species has been a constant presence there ever since. Their origins are not known, but they most likely have escaped from a wildfowl collection somewhere in Ireland – or perhaps Great Britain.

A pair first bred successfully on Inch Lough in 2003, and in most years since then. It is not known if the young birds die or disperse – or indeed if it was the original pair responsible for all the broods. But so far, there are never more than three adults plus a brood of cygnets present, and for the last two seasons there has only been one bird. Although frowned on by conservationists who see them as potentially threatening aliens, there is no escaping the fact that these birds are extremely popular with the large numbers of visitors who now come to Inch, as much for exercise as for the wildlife experience. It could be argued that the Black Swans act as unofficial ambassadors for the legitimate, native species that the Inch Wildfowl Reserve was set up to protect.

Mute Swan

Cygnus olor | Eala bhalbh

RESIDENT

STABLE

There is little, if any, migration of Mute Swans to or from Ireland, so the population has the status of a discrete flyway. As a long-lived species, few pairs need to breed each year to maintain the population. In the breeding season they are scattered in small numbers around the coastal lakes, often only as single pairs. At Inch they breed at unusual density on the small island, but productivity is low due to the frequent flooding. Summer numbers are higher at Inch than in winter. A count on 25 July 1989 of 484 birds included 53 young birds in 21 broods. At Dunfanaghy New Lake a summer flock of ninety-eight birds on 4 August 2022 included twelve young birds in four broods.

The winter population at Inch has been slowly increasing in recent years, with the record total being 430 birds on 15 October 2022 (Figure 18.7), but the higher summer total above prompts caution for any claims of more than stability. For Durnesh Lough (Donegal Bay) the mean peak for 2011/12 to 2015/16 is 203, and for Dunfanaghy New Lake it is 99 (Lewis et al., 2019). Mulroy Bay in the past had a large flock which was fed from a bread

Mute Swans, Rossylongan, Donegal Bay.

JOHN CROMIE

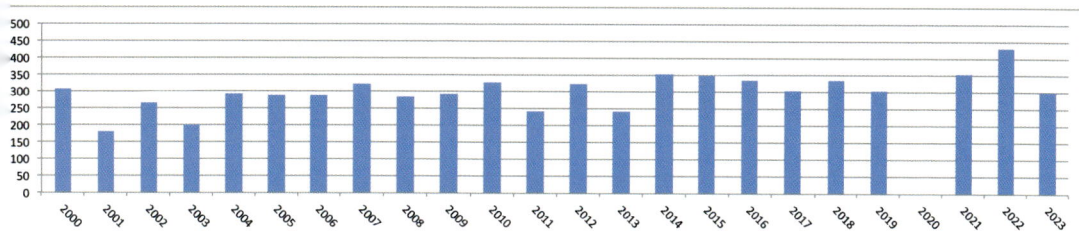

Fig. 18.7: Mute Swan annual winter peaks at Inch Lough from 2000/1 to 2023/4

factory on the shore at Milford. There were 128 birds on 7 February 1982. After the factory closed, the swans faded away through the 1990s and were gone by the turn of the millennium. Inland Mute Swans are scarce.

The Internationally Important figure (1% of the flyway population) is 90.

Qualifying Sites: Lough Swilly, Donegal Bay, Dunfanaghy New Lake

As the Irish birds comprise a discrete population, the national qualifying levels are the same as the international ones, so the sites are the same.

Bewick's Swan (Tundra Swan)

Cygnus columbianus bewickii | Eala Bewick

FORMERLY COMMON WINTER VISITOR, NOW A VAGRANT
RED-LISTED BOCCI-4

In the nineteenth century, Hart recorded wild swans 'probably of this species' wintering on a number of lakes in Fanad. His description of their 'continual yelping' does suggest that his identification was correct. His highest count was eighty-four on Rinboy Lough.[13] Inch Lough was also noted by Ussher and Warren as being visited regularly. More recently, forty-eight were present in 1963.

For Lough Swilly, numbers varied from 50 to over 400 between 1967/8 and 1980/1 (Figure 18.8). The birds used all of the flat agricultural lands around Lough Foyle, Lough Swilly and the River Foyle as a single site, within which movements were frequent – hence the erratic totals at Lough Swilly. The mean total for those sites combined shows a high peak of 693 in 1974/5, at a time when that represented 6.9–7.7 per cent of the total European population of 9–10,000 birds.[14] But it was to be the last time the

Bewick's Swans reached such heights, as a decline set in the following year (Figure 18.9).[15]

The decline was not apparent on Lough Swilly until after its final high peak of 205 birds in January 1989, when numbers there started to collapse (Figure 18.10). Since the last decent flock, of forty-eight birds, was seen in 1995/6, there have been only occasional sightings of no more than five birds, the most recent being two on 21 January 2006. The species is now effectively extinct in north-west Ireland. Three birds at Malin Head on 1 January 2015 are probably best regarded as vagrants.

Bewick's Swan breeds in the tundra zone of arctic Russia, and has been retreating generally from the western outposts of its wintering range as artificial feeding of birds in England, and later the milder winters, have both made the longer journey no longer necessary (short-stopping). The north-west of Ireland was the first outpost to suffer, but the retreat has now spread throughout Ireland and fewer than twenty birds occur in most winters, and are confined to the extreme south-east.

On grounds of a 99 per cent decline in its wintering population, Bewick's Swan is Red-listed in BirdWatch Ireland's 'Birds of Conservation Concern in Ireland 4: 2020–2026'.[16]

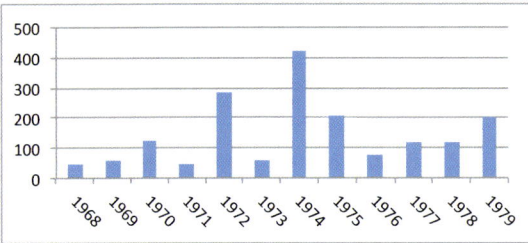

Fig. 18.8: Fig. 18.8 Bewick's Swan peak counts on Lough Swilly for 1968/9 to 1979/80

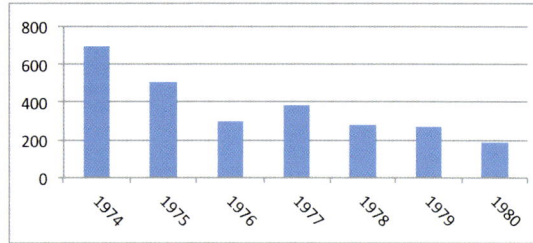

Fig. 18.9: Bewick's Swan peak counts on Lough Swilly, Lough Foyle and River Foyle combined for 1974/5 to 1980/1

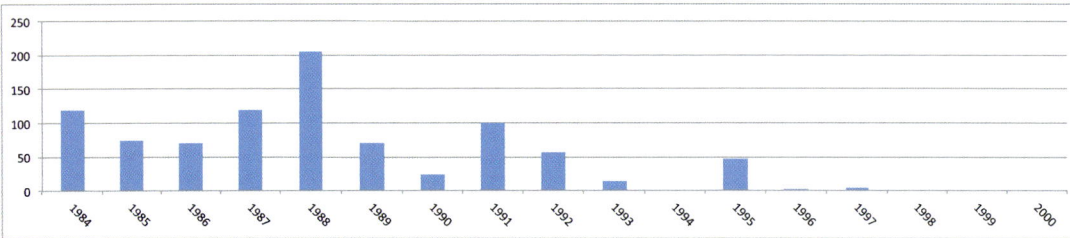

Fig. 18.10: Peak counts of Bewick's Swan on Lough Swilly from the start of the Winter Wetlands Survey in 1984/5, to 2000/1

Whooper Swan

Cygnus cygnus | Eala ghlórach

WINTER VISITOR, PASSAGE MIGRANT, AND RARE BREEDER
STABLE IN WINTER, DECLINING AS MIGRANT

Whooper Swan was only an occasional winter visitor in the nineteenth century. By the mid-twentieth century it was as numerous as the Bewick's Swan on Lough Swilly and the River Foyle. Elsewhere it was the only northern swan species present. Numbers continued to rise through the next half century, peaking each year when immigrants from Iceland made their first landfall in Ireland on Lough Swilly. While many birds remained on Lough Swilly, it is likely that a high percentage of the Icelandic breeding population was feeding up for a few days on the polders before moving on to other wintering sites throughout Ireland. Spilt grain from harvested wheat and barley fields was the target, but the birds that remained for the winter soon moved on to the harvested potato fields. The arrival of spring then gave them a chance to feed on new grass, prior to the return flight to Iceland, which is usually quite late in April.[17]

Since harvesting has become more efficient and the fixed cropping pattern has broken down, numbers have declined, both in autumn and through mid-winter, despite the continuing rise in the national population.[18]

Mid-winter peaks averaged 690 through the 1980s and '90s, but since then have been much lower, and for the last decade, when there were no winters with more than 600 birds, the average is now 422. At least that level seems relatively stable (Figure 18.11). A decline in the much higher autumn peaks started in the second decade of the present century, as shown in Figure 18.12 – although we have failed to count some recent autumn peaks, as a result of birds arriving and moving on very quickly. Either way, it shows that Inch does not appear to offer the very obvious welcome that had been the case in earlier years – given that the national population continues to increase.

Large flocks used to be present in both autumn and winter at the three big polder sites of Big Isle, Blanket Nook and Inch Levels. Big Isle was first abandoned in winter and eventually in autumn. Blanket Nook is now less reliable than before, but retains its value for the autumn arrivals and its relative unimportance in winter. The retreat is apparent on Table 18.1,

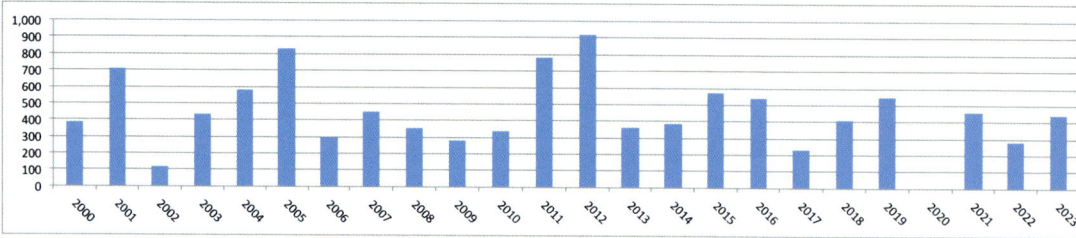

Fig. 18.11: Mid-winter totals of Whooper Swans on Lough Swilly from 2000/1 to 2023/4

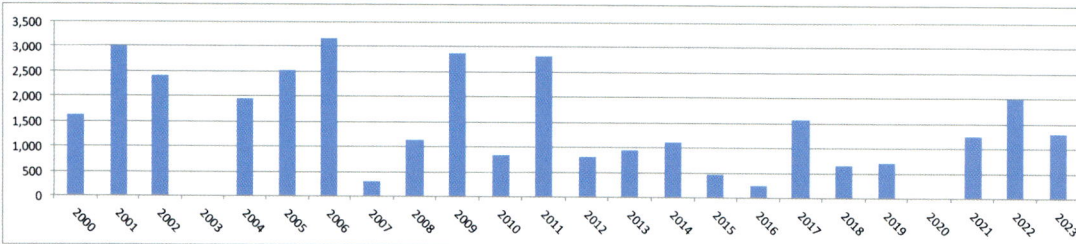

Fig. 18.12: Autumn peaks of Whooper Swans on Lough Swilly from 2000/1 to 2023/4

proceeding to the bottom left corner of the table. In autumn, large flocks can appear anywhere in the area if there is a suitable field of grain or potatoes available. But these don't last long, so birds are largely dependent on Inch Lough and Levels throughout the winter. The same pattern was repeated on the River Foyle, which held 300 to 400 in the 1990s but is now used mainly by flocks of Lough Swilly birds tempted eastwards by the presence of a suitable field in autumn. But Lough Swilly, Lough Foyle and the River Foyle have always been intimately connected for Whooper Swans as a large site complex.[19]

Elsewhere, there are two other outstanding sites. Durnesh Lough (part of the Donegal Bay complex) had a three-year peak to 1986/7 of 103.[20] The five-year peak to 2000/1 was 140.[21] To 2015/16 it was 359, with one particularly high count of 682 in 2012/13.[22] That pattern of increase has continued, with 1,022 birds present on 17 October 2021.

Numbers at Dunfanaghy New Lake come close to the Nationally Important threshold. Clooney Lough and the inland Lough Akibbon support smaller numbers. To these must be added many small flocks and parties, usually a family or two and rarely more than a dozen birds, all around the coast and in the upland lakes – a total of seventy-eight known sites covered in the 2020 national census, although only a small proportion of those are occupied at any one time.

Year	Big Isle Aut.	Big Isle Win.	Big Isle March	Blanket Nook Aut.	Blanket Nook Win.	Blanket Nook March	Inch Aut.	Inch Win.	Inch March	Lough Swilly TOTAL Aut.	Lough Swilly TOTAL Win.	Lough Swilly TOTAL March	Peak Month Aut.	Peak Month Win.
05/06	763	25	X	7		X	1,751	805	X	2,537	830	X	Oct	Dec
06/07	261			1,130			1,695	293	180	3,177	293	197	Oct	Jan
07/08	1		X	143	38	X	160	408		312	451	X	Nov	Dec
08/09		17	X	735	36	10	377	480	359	1,141	357	420	Nov	Dec
09/10	59	22	22	2,270	9	1	551	250	410	2,877	281	433	Oct	Feb
10/11				32	19		764	320	432	855	339	432	Oct	Feb
11/12	1,370	2	7	730	19	6	681	762	1,223	2,810	783	1,271	Oct	Jan
12/13	46		X	610	4	X	48	913	X	804	917	X	Oct	Feb
13/14			12	23	84	14	889	276	614	945	360	640	Oct	Feb
14/15			X	750	3		68	383	618	1,125	389	618	Oct	Dec
15/16				92	24		358	544	618	460	568	618	Nov	Feb
16/17	13	11			2	2	222	524	643	235	537	645	Nov	Dec
17/18					2	72	1,201	225	138	1,566	229	214	Nov	Dec
18/19			7	1	37	36	651	373	529	652	411	575	Nov	Dec
19/20				9	27	5	689	519	714	700	546	751	Oct	Feb
21/22				9	27	5	689	519	714	700	546	751	Oct	Feb
22/23	23		X	848	98	X	1,134	207	X	2,005	279	X	Oct	Feb
23/24				126	25		1,162	422		1,292	439		Nov	Jan

Table 18.1: Seasonal totals of Whooper Swans on Lough Swilly from 2005/6 with sub-totals for the three main sub-sites only. Seasons are autumn (September to November), winter (December to February) and spring (March). Counts of zero are left blank. Counts not done are marked with X. There were no counts in 2020/1.

Whooper Swans, Inch Levels.

Whooper Swans were first recorded in summer in 1937, when four birds were present.[23] Since then summering birds have been seen at a wide range of lakes. But it was in 1992 at Inch Lough, where up to a dozen regularly remain over summer, that they were first proved to have bred in Ireland. Breeding has continued irregularly since then, most recently in 2023. A pair also bred on Tory in 2003. Whoopers breed in the boreal zone of northern Europe, and the birds that visit Ireland are all from the Icelandic breeding population. So anything other than occasional breeding in Donegal is unlikely.

The Internationally Important figure (1% of the flyway population) is 340.

Qualifying Sites: Lough Swilly, Donegal Bay, River Foyle

The Nationally Important figure (1% of the national population) is 150.

Egyptian Goose

Alopochen aegyptiaca | Gé Éigipteach

RARE VAGRANT OF FERAL ORIGIN

The Egyptian Goose is a tropical African species which is often kept in captivity. Escapes have established self-sustaining populations widely across Europe and Great Britain. These feral populations are now more likely than zoos and collections to be the source of any birds seen in Ireland. One bird was seen on Tory on 22 April 2016 (C. Ingram), and four birds were at Dunfanaghy on 23 January 2022 (N. Newell). One bird has remained, with the most recent sighting at the New Lake being on 17 September 2023.

Shelduck (Common Shelduck)

Tadorna tadorna | Seil-lacha

WINTER VISITOR AND UNCOMMON BREEDER

STABLE

Shelduck breed widely around the coast in small numbers. They use sandy areas with nesting opportunities in rabbit holes, but at places like Blanket Nook they are more likely to use cavities in the embankments. They do sometimes nest well inland. I recall watching a brood being led across a busy main road with 1.5 km of well-fenced farmland still to cross.

As a winter visitor from northern Europe, the species is much more abundant. Numbers start low, and build up over the winter to a peak during February or March. Lough Swilly has consistently held 600–800 birds since the 1980s, with an outstanding peak of 1,251 birds on 16 February 2019.

Other notable sites with peak counts for 1994/5 to 2000/1 are Trawbreaga Bay (46), Donegal Bay (38) and Ballyness Bay (23).

The Internationally Important figure (1% of the flyway population) is 2,500.

The Nationally Important figure (1% of the national population) is 100.

Qualifying Sites: Lough Swilly

Ruddy Shelduck

Tadorna ferruginea | Seil-lach rua

RARE VAGRANT

An invasion took place in 1892, reaching Donegal in some numbers – a flock of twenty at Sheskinmore on 25 June, three at Coolmore on 4 August, six at Lough Swilly on 19 August and one on 8 September. These birds were undoubtedly from one of the wild populations, in south-east Europe/Middle East or in north-west Africa.

More recently, two birds were seen at Ramelton on 20 September 1991 (R. Sheppard), and two appeared at Inch Lough on 23 September 2016 with one of them staying to 8 April 2017 (T. Campbell, C. Ingram et al.). For these modern records, an origin in one of western Europe's feral populations is much more likely.

Mandarin Duck

Aix galericulata | Lacha Mhandrach

ESCAPE

Until recently there had only been one record in Donegal of the Mandarin Duck from east Asia. This was on Arranmore on 9 May 2017 (M. Bell). It would not have been a wild vagrant from its homelands, but rather an escape from captivity or a bird from a feral population somewhere in Europe. A small population in County Down appears to have died out, but there are still some birds free around the Dublin parks.[24] It is possible that Dublin is the source of our bird, or Britain. However, following a spate of sightings on Lough Swilly in 2020 and 2021, involving at least two but probably three birds, the most likely source is a local collection.

Garganey

Spatula querquedula | Praslacha shamhraidh

SCARCE MIGRANT

Garganey is a scarce summer visitor to the southern counties of Ireland from its wintering range in West Africa, south of the Sahara. It had apparently not reached Donegal until 1988, when the first was seen at Dunfanaghy on 17 May (P. Mackie). Since then a further thirty-one birds have been noted, mostly as singles, but including six doubles. The seasonal distribution is more or less what might be expected of a non-breeding migrant – three birds in April, eleven in May, eleven in August and five in September. The sightings have been at well-watched sites with suitable habitat, so twenty-one birds have turned up at Inch Lough, five at Blanket Nook, three at Dunfanaghy New Lake and two on Tory.

Donegal is a bit too far north for successful breeding, and there is no evidence of any attempts. Perhaps climate change will rectify this in the near future.

Blue-winged Teal

Spatula discors | Praslacha ghormeiteach

RARE VAGRANT

There are only three records of this North American species in Donegal. The first was shot on Durnesh Lough on 18 October 1991 – a sad end for a bird that had survived an Atlantic crossing, but thankfully no longer typical. One was seen at Dunfanaghy New Lake on 15 October to 17 November 1998 (R. Sheppard et al.), and what was presumed to be the same bird from 23 October 1999 until 23 January 2000, and then also on 3 June and 26 October. Finally, a female was seen at Inch Lough on 24 September 2011 (D. Brennan).

Shoveler (Northern Shoveler)

Anas clypeata | Spadalach

WINTER VISITOR AND RARE BREEDER

INCREASING

RED-LISTED BOCCI-4

There are several records of breeding pairs of Shoveler from before 1877. Ussher and Warren knew of a pair breeding at Port Lough in 1889, and a few pairs near the coast at Coolmore, Rossnowlagh. Leebody said that one or two pairs bred near Inch.[25] The species remained extremely local in the first half of the twentieth century. A pair was noted at a marsh on Inch Levels in 1962, before the site was drained, and on Inch Lough in 1968. Although up to two pairs have summered frequently in recent years on Inch Lough, it was not until 2020 that breeding was again proven.

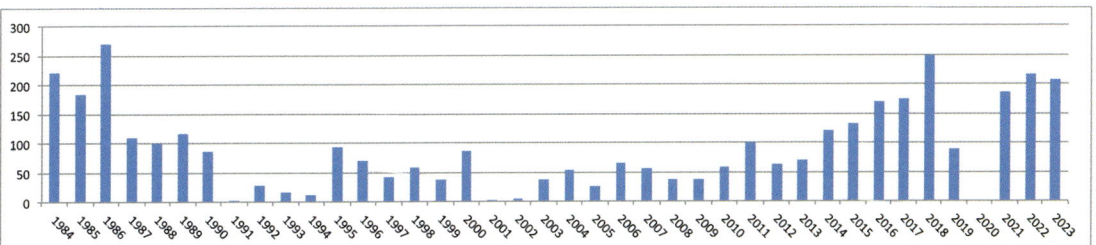

Fig. 18.13: Annual peak counts of Shoveler on Lough Swilly from 1984/5 to 2023/4

The wintering birds that come here from eastern Europe have had mixed fortunes over the years. Between 200 and 300 birds could be found on Lough Swilly in the 1980s, but the numbers dropped to a five-year mean of only 31 in 1995/6. They have crept up steadily since then to the current figure of 163. Annual peaks are shown in Figure 18.13. They are creatures of habit, concentrating on the vast areas of mud at the Swilly estuary and Big Isle, and with a few birds usually also at Inch Lough. Shovelers are very scarce elsewhere in the county. One or two can usually be seen at Dunfanaghy New Lake.

The Internationally Important figure (1% of the flyway population) is 650.

The Nationally Important figure (1% of the national population) is 20.

Qualifying Sites: Lough Swilly

On grounds of a 52 per cent long-term decline in wintering population, the Shoveler is Red-listed in BirdWatch Ireland's 'Birds of Conservation Concern in Ireland 4: 2020–2026'.[26]

Gadwall

Mareca strepera | Gadual

UNCOMMON WINTER VISITOR AND RARE BREEDER
STABLE

There is only one instance of proven breeding of this species. Two females with ducklings were seen on 13 July 1964 (E. and R. Millar) at an undisclosed lake. Three males and another female were nearby. A pair were seen on Roaninish in June 1947, but did not remain. The male of a pair was seen to chase off a Mallard at Blanket Nook on 4 May 2017, an indication at least of breeding intent. The core breeding range of Gadwall is in temperate latitudes, but our birds come from an outpost in Iceland.

Gadwall has always had a modest winter presence in Donegal. Up to fifteen birds were regularly seen at Inch Lough in the 1960s, and less frequently at Blanket Nook. Single figures had been the norm since then, until 2016/17, when fifteen birds were seen at Inch Lake, followed by twenty-six in 2018/19. The only evidence of migration was a flock of fifteen birds seen passing south at Rocky Point on 19 September 1998, and forty birds present at Donegal Bay (presumably Durnesh Lough) in September 2012.

Wigeon (Eurasian Wigeon)

Mareca strepera | Rualacha

WINTER VISITOR

INCREASING

There is no evidence of Wigeon having bred in Donegal, although a few birds are not infrequently seen in summer months.

The wintering birds are mainly from Iceland and northern Europe and arrive early, in September and October. They are found all around the coast in small parties, and especially in sheltered bays and estuaries, where a fringe of salt marsh on which they can graze is an attraction. They sometimes accompany Brent Geese, eating the leftovers of vegetation pulled up by the Brent from depths the Wigeon cannot reach.

The most important site is Lough Swilly where Wigeon are found all around the lough on the tidal mud, and in the two lagoons. Peak counts have hovered around 1,000 to 1,500 since the 1980s, and only rose above 2,000 in 2013/14. It's possible that this increase is not being sustained (Figure 18.14). The phenomenal total of 4,383 was recorded on a day (5 November 2016) when many species at Inch were in high numbers (totalling 13,404) – but could some of the Wigeon flocks have been double-counted? This is always a concern when the counter is faced with many separate, mobile flocks, perhaps of mixed species.

For other sites, the peak counts in the period 1994/5 to 2000/1 were Trawbreaga Bay (386), Donegal Bay (331), Dunfanaghy Harbour (183), Ballyness Bay (174), Dunfanaghy New Lake (121). Smaller flocks can be found in other sites that are not frequently counted, for example Sheep Haven Bay.

The Internationally Important figure (1% of the flyway population) is 14,000.

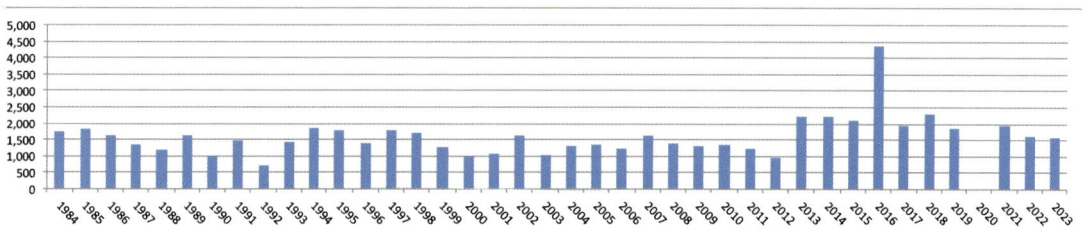

Fig. 18.14: Wigeon numbers on Lough Swilly from 1984/5 to 2023/4

The Nationally Important figure (1% of the national population) is 560.

Qualifying Sites: Lough Swilly

American Wigeon

Mareca americana | Rualacha Mheiriceánach

SCARCE VAGRANT AND RARE WINTER VISITOR

There are twenty-three records of American Wigeon. The first in the county was a bird shot at Trawbreaga Bay on 26 September 1964 (M. Brush et al.). The skin is held in the Ulster Museum. At the time, with only about three other Irish records, it was thought that the possibility of it being an escape could not be ruled out. However, with at least twenty-one undoubtedly genuine vagrants since then, it is likely that this bird was also genuine.

An exact total is problematic. There are a number of sites that have had records in consecutive years. That is not unreasonable, as some sites, like Trawbreaga Bay, are quite likely to attract fresh migrants on more than one occasion. But all repeat records in consecutive years come under suspicion in light of a male which was present at Blanket Nook every winter from 1996/7 to 1999/2000, another one from 2013/14 to 2019/20 which was mainly at Culdaff estuary, and a third one at Malin Town from 2014/15 to 2016/17. These many sightings are presumed to be just three individual

American Wigeon, Culdaff.

DEREK BRENNAN

birds returning each winter to their favourite location, having summered each year somewhere in northern Europe.

Twelve birds have been seen at Lough Swilly (seven at Inch, one at Blanket Nook and four elsewhere), four at Trawbreaga Bay, three at Dunfanaghy and one each at Ballyness Bay, Culdaff, Malin Head and Tullaghan.

Mallard

Anas platyrhynchos platyrhyncos | Lacha Fhiáin

COMMON RESIDENT

STABLE

Other than the atlases, which confirm its presence in most of the 10 km² in the county, there are no surveys that have focused on breeding Mallard. As pointed out by more than one author, they use a bewildering variety of nest sites, so are to be found almost anywhere throughout the county. Numbers are increased locally with the release of hand-reared birds by gun clubs.

The large resident population of Mallard is supplemented in winter by birds from further east in Europe. On Lough Swilly, the most important site in the country, the population has peaked at 1,769, in September 2014,[27] and in Donegal Bay at over 200. River Foyle has a peak of 464 and Trawbreaga Bay 176.[28] Smaller numbers are widespread, with little sign of change, either locally or overall.

The Internationally Important figure (1% of the flyway population) is 53,000.

The Nationally Important figure (1% of the national population) is 280.

Qualifying Sites: Lough Swilly, River Foyle

American Black Duck

Anas rubripes | Lacha chosrua

RARE VAGRANT

There are two records of this close relative of the Mallard. A male at Milford from 22 December 2007 to 5 January 2008 (D. Breen et al.) and a different, long-staying male at Blanket Nook from 29 December 2007 to 31 May

2008 (D. Brennan et al.). Could they have travelled together and parted company on arrival?

Pintail (Northern Pintail)

Anas acuta chorea | Biorearrach

WINTER VISITOR

INCREASING

There has been no evidence of Pintail breeding in County Donegal, and it is one of the least common of the regular winter-visiting ducks. Most of them come from the Icelandic breeding population. Leebody claimed that a few wintered at Inch, and that the species was plentiful during the latter half of February and the beginning of March.[29] Over twenty birds wintered on the mud at Big Isle on Lough Swilly in the late 1980s, with a peak of forty-seven recorded in January 1987. After that, they were very few and erratic until 2014/15, since when twenty to thirty birds have been returning each winter. A few birds also appear at Inch Lough.

Small numbers (single figures) were regular at Dunfanaghy New Lake through the 1990s, but now only the occasional single bird is seen any-where other than Lough Swilly. Sites with occasional records are Mulroy Bay, Ballyness Bay and Sheep Haven Bay.

The only record of migrating birds was of a party of nine that flew past Bloody Foreland on 1 October 2008.

Pintail, Inch Lough.

Teal (Eurasian Teal)

Anas crecca crecca | Praslacha

WINTER VISITOR AND UNCOMMON BREEDER

STABLE

Teal breed mostly in upland lakes, but are few in number. The breeding atlases all show a sparse presence in summer, and very few 10 x 10 km squares with proven cases of breeding.[30] This was not always the case, as Hart thought they were more numerous than Mallard in the mountain lakes, and that 'great flocks of flappers come down in August to some lakes and tidal inlets in the north of Donegal'.[31]

Immigrants from northern Europe come in large numbers to Lough Swilly. There are generally over 2,000, with a peak of 3,711 on 26 October 2013, and have been fairly stable since at least 2007/8 (Figure 18.15). River Foyle had a peak of 1,125 birds in December 1996, but more typically has 200 to 500. The Lough Foyle west bank can also have 300 birds, but usually fewer. They are present in smaller numbers at other wetlands, and along rivers and in marshy ground throughout the countryside.

The Internationally Important figure (1% of the flyway population) is 5,000.

The Nationally Important figure (1% of the national population) is 360.

Qualifying Sites: Lough Swilly, River Foyle

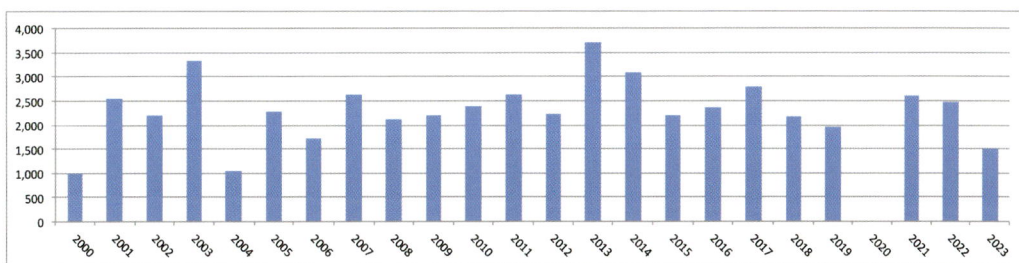

Fig. 18.15: Teal peak numbers on Lough Swilly from 2000/1 to 2023/4

Green-winged Teal

Anas carolinensis | Praslacha ghlaseiteach

WINTER VAGRANT

Green-winged Teal is the North American equivalent of the Eurasian Teal. It is almost an annual visitor, with twenty-four birds seen so far.

This species was long thought to be just another race of the Eurasian Teal, and was first identified here as such on 10 January 1982 at Ballyshannon (S. Fleming). Vagrant birds tend to appear in flocks of Eurasian Teal, and while most records are from Lough Swilly, they have appeared at a scattering of sites. Females are barely distinguishable from their old-world counterparts, with no accepted records for Ireland, so all records have been of males. All records have also been of single birds, apart from three that were seen at Inch on 28 October 2007. Inch Lough has had a total of nine, Blanket Nook three, and three were elsewhere on Lough Swilly. At Dunfanaghy New Lake there have been four, and Tory has had two. There are single records from Ballyshannon, River Foyle and Malin Head. Like most vagrant ducks, quite a few have been long-staying birds, sometimes straddling two years, from autumn arrival to spring departure.

Red-crested Pochard

Netta rufina | Póiseard cíordhearg

RARE VAGRANT

This is a south European species which is often kept in collections, so there is no certainty about the origin of our two records.

The first sighting was of a male at Lough Fern from 11 to 15 November 1990 (R. and L. Sheppard). This bird was with a tight pack of about 700 Pochard, which it dominated from the centre, maintaining a patch of open water to itself. The second one was a female on Inch Lough on 8 September 2012, which stayed in the area until 3 November (B. Robson et al.).

Pochard (Common Pochard)

Aythya farina | Póiseard

ERRATIC WINTER VISITOR AND RARE BREEDER
STABLE
RED-LISTED BOCCI-4

Pochard is a bird of the steppes and eastern Europe which has expanded its breeding range in fairly recent years to western Europe.[32] It remains largely a winter visitor in Ireland, but has given indications of breeding on

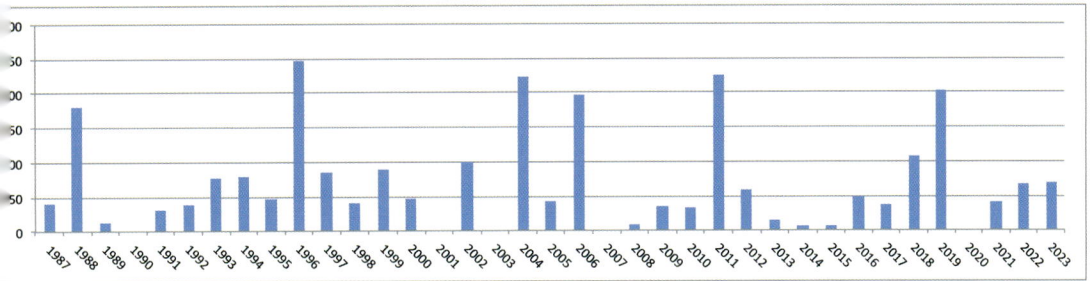

Fig. 18.16: Pochard peak winter counts at Inch Lough from 1987/8 to 2023/4

a number of occasions, the first at Port Lough in 1948. They are sometimes to be seen in summer at Lough Fern. Proof of breeding came in 1997 when there were 'a few broods' recorded at Inch (A. McGeehan).

They are erratic as winter visitors – never predictably present at any one site, and numbers are hugely variable. Large flocks occasionally winter at Dunfanaghy New Lake, Lough Fern and Inch Lough.

The most regular haunt to have consistently good numbers in Donegal is Dunfanaghy New Lake. Anything between 100 and 300 is normal, with a high count of 950 on 8 February 1994 being quite exceptional. Counting has not been as regular as it should be, but from the evidence available there is no hint of any overall trend.

Pochards on Lough Fern are more erratic than at New Lake, and until recently were less frequently counted. But the peaks have been higher. The highest counts were 2,500 on 28 October 1984, 1,208 on 3 March 1996, and 700 on 15 November 1990. A typical flock, when there is one, would be less than 200.

Inch Lough is also a good site, and has the advantage of frequent counts over a long period. Figure 18.16 shows an apparent seven/eight-year cycle between the high peaks, but there is no other supporting evidence to back that up. Background levels between the peaks declined after 2002, although since 2015 it looks as if they might have recovered.

Other sites which occasionally hold Pochard flocks include Lough Akibbon, which had 500 on 23 November 1996, and Durnesh Lough with 165 on 13 February 1983 and 176 on 11 November 1989. Mintiaghs Lough in Inishowen had eighty on 19 February 2000.

The Internationally Important figure (1% of the flyway population) is 2,000.

The Nationally Important figure (1% of the national population) is 110.

No site has been recognised as Nationally Important, but a case could be made for Dunfanaghy New Lake, Lough Fern and Inch Lough.

On grounds of a short-term decline of 77 per cent in its wintering population, Pochard is Red-listed in BirdWatch Ireland's 'Birds of Conservation Concern in Ireland 4: 2020–2026'.[33]

Ferruginous Duck

Aythya nyroca | Póiseard súilbhán

RARE VAGRANT

There is only one verified record of this eastern European species in Donegal. A bird was seen at Inch Lough on 1 January 1990 (T. Campbell). After a long gap, a second one was present at the same site from 24 to 31 October 2022 (T. Campbell), and what seems to be the same bird was seen again the following year, on 28 October 2023 (D. Brennan).

Ring-necked Duck

Aythya collaris | Lacha mhuinceach

SCARCE VAGRANT AND WINTER VISITOR
INCREASING

Although still very scarce, this duck is one of the most frequent of our North American vagrants. The first bird was seen at Dunfanaghy New Lake on 31 March 1984 (R. Sheppard et al.). There have now been ninety-seven records – excluding repeat sightings, but including five hybrids.

Four sites have had most of the records – Inch (32), Dunfanaghy New Lake (21), Blanket Nook (14) and Lough Fern (11). It is rather odd that Durnesh Lough has had only four and Tory only two, both of them seemingly in direct line of fire for an Atlantic crossing. Figure 18.17 shows the increasing number of records in recent years. This could be simply a bias from the increasing number of observers, but even during those years with fewer birdwatchers in the 1980s and '90s, most of the best sites would have been visited. So the increase in birds looks genuine.

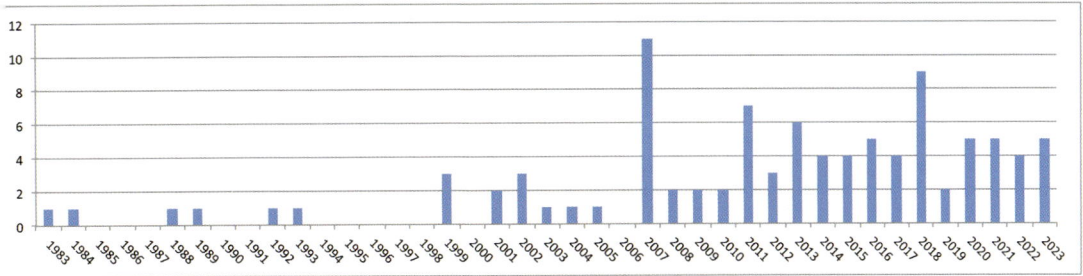

Fig. 18.17: Annual totals of Ring-necked Duck from 1983/4 to 2023/4

Ring-necked Duck, Inch,

DEREK CHARLES

Some birds appearing at the same site in successive years, and wintering there, are likely to be returning visitors. One such was a bird first recorded at a small upland lake, Lough Shivnagh, on 22 November 2011 and every subsequent winter to the last sighting on 22 February 2019. This suggests that at least some of these birds are successfully setting up home on this side of the Atlantic, but whether migrating north–south or summering somewhere more local is not clear. Quite a few of the summer records have probably been of birds first seen in the previous autumn or winter.

Potential breeding is suggested by a few records of birds apparently paired with Tufted Ducks. One hybrid was reported on 14 January 2012 (M. Tickner and D. Brennan), and a male hybrid was well photographed at Inch on 1 May 2020, and another on 9 May 2023 (D. Brennan). A fourth was at New Lake, Dunfanaghy on 17 September 2023 (A. Kelly and J. Dowdall), and probably the same bird in January and February 2024. And there have been others that were not reported. The bright side of this confusing history is that these birds offer clear evidence of individual Ring-necked Ducks breeding on this side of the Atlantic, and we can surely look forward to a pure breeding pair in the future.

However, a presumed hybrid Ring-necked Duck x Greater Scaup was also present at Inch during the autumn of 2022, and well photographed. That bird could not have been bred in Ireland. The two species would most frequently come in contact with each other in Canada, where their breeding zones overlap. But neither species there should feel the need to hybridise (see Chapter 14), and if it did happen, for the exceedingly rare offspring to then cross the Atlantic seems just about impossible. So the origin of our hybrid was probably in the Scaup's breeding haunts in Iceland or Scandinavia, where the Ring-necked Duck's preferred partner would not have been available.

Tufted Duck

Aythya fuligula | Lacha bhadánach

RESIDENT

STABLE

Tufted Duck was rare as a winter visitor in the nineteenth century, and didn't breed until after 1900. They are largely resident throughout most of Europe, and this seems to be the case in Donegal. On 13 June 1997 there were between 100 and 200 pairs nesting on the Inch Lough islet,[34] but the normal total seems to be well under 100 pairs. Other lakes with good breeding populations are Dunfanaghy New Lake and Durnesh Lough, and the inland Loughs Fern and Akibbon. Smaller numbers nest on a number of other lakes, mainly coastal, but also inland in the south-east.

Tufted Duck breed throughout Europe, with the northern birds moving south for the winter. The number of these Scandinavian and Russian birds moving to north-west Europe has been increasing.[35] Winter immigration does boost the resident population in Donegal, but to what extent is not clear. Tufted Ducks are not found on tidal or marine waters, so winter visitors are confined to the lakes where there are also resident populations. Figure 18.18 shows counts starting at 1996/7, after which the brief post-breeding peaks were usually included. The long-term peak at Inch Lough is a count of 1,269 birds on 28 October 2006.

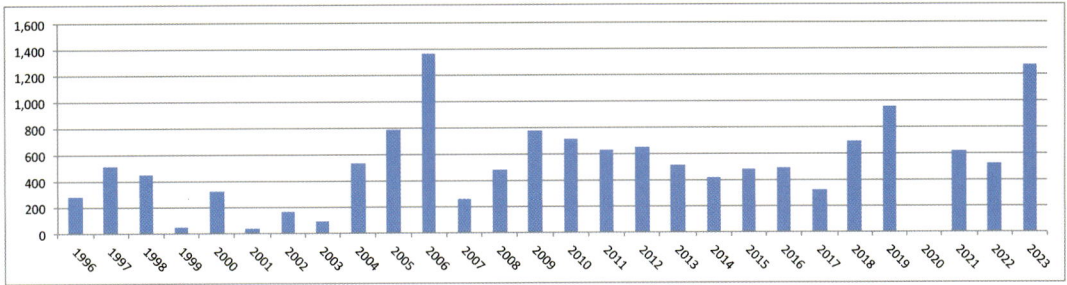

Fig. 18.18: Tufted Duck annual peak counts at Inch Lough from 1996/7 to 2023/4

Other sites with peaks of 200 or more are Lough Fern (1,500 on 28 October 1984, 537 on 26 September 1982 and 305 on 16 December 2018), Durnesh Lough (238 on October 2012), and Dunfanaghy New Lake (200 on 8 February 1984). Apart from these occasional very high numbers, the normal winter complement at most sites could possibly be composed almost entirely of local birds.

The Internationally Important figure (1% of the flyway population) is 8,900.

The Nationally Important figure (1% of the national population) is 270.

Qualifying Sites: Lough Swilly

Scaup (Greater Scaup)

Aythya marila marila | Lacha iascán

SCARCE WINTER VISITOR
DECREASING
RED-LISTED BOCCI-4

Scaup is one of those species with a wide breeding range in the Arctic, although our birds come exclusively from the relatively small population in Iceland. They winter mainly in Scotland and Ireland, but are increasingly choosing to join their much more numerous comrades from their continental breeding range (also declining) around the southern shore of the North Sea and the Baltic.[36]

Leebody wrote in 1892 that Scaup is 'next to Wigeon the most plentiful of the winter visitors to Inch. A few Scaup remain on the inland water all summer, but I believe they are wounded birds'.[37] Scaup on Lough Swilly have always oscillated between Inch Lough and the deeper tidal waters. The

167

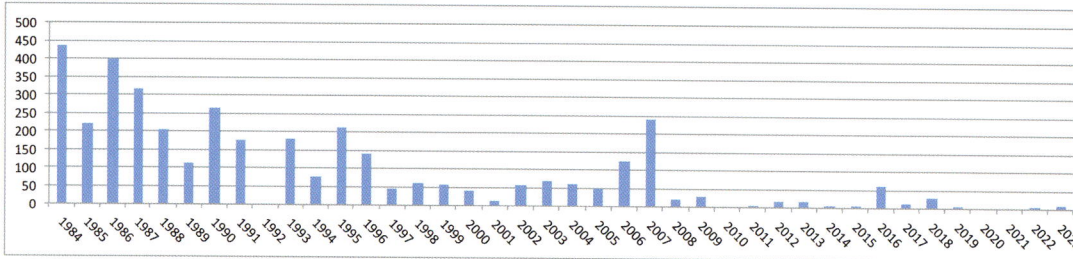

Fig. 18.19: Scaup numbers on Lough Swilly from 1984/5 to 2023/4

highest modern count was 435 on 23 February 1985. Over the next decade, numbers declined to under 100. After a brief surge to 240 in 2007/8, the decline continued to only about a dozen turning up in recent years – below the threshold for National Importance (Figure 18.19). Durnesh Lough also had a good flock in the not too distant past. The highest count there was 162 in the mid-1980s[38], but now only one or two Scaup can be expected, and that only on a casual basis.

The Internationally Important figure (1% of the flyway population) is 3,100.

The Nationally Important figure (1% of the national population) is 25.

Qualifying Sites: None

Scaup is Red-listed in BirdWatch Ireland's 'Birds of Conservation Concern in Ireland 4: 2020–2026' with a population decline of 58 per cent over the last twenty-five years.[39]

Lesser Scaup

Aythya affinis | Mionlacha iascán

RARE VAGRANT

Lesser Scaup is a North American species, and one of the more recent to set up a fairly regular pattern of trans-Atlantic vagrancy. There are records of six in Donegal. The first to be seen was at Inch Lough on 5 December 2003 (A. Ó Dónaill et al.). Since then there have been five more – a male at Lough Fern on 21 October 2007 (D. Breen) which was probably the bird that appeared at Inch Lough six days later and remained until December, a female at Inch Lough on 8 October 2008 (G. Campbell), a female which was constantly in the company of a pair of Ring-necked Ducks at Dunfanaghy

New Lake on 18 February 2014 (A. McMillan and M. Boyle), and a male off Farland Bank, Inch on 9 to 11 May 2018 (C. Ingram). Dunfanaghy New Lake attracted the final bird, from 29 October to at least 5 December 2022 (D. Brennan). The last three have yet to be verified by the IRBC.

King Eider

Somateria spectabilis | Éadar taibhseach

RARE VAGRANT

This is a high-arctic species with a circumpolar range. It doesn't breed closer than arctic Russia, or winter closer than the northern coast of Norway. Yet there are records of seven King Eider in Donegal. The first was at Rossbeg on 21 April 1974 (H. and R. Northridge). What was presumed to be the same bird returned each winter (apart from 1974/5) until December 1982, and was also seen on the north coast of Northern Ireland in spring 1971, spring 1972 and summer 1982.

Six other birds have been recorded. These were a first-winter male at Roaninish on 19 May 2012, a female at Portnoo on 17 March 2014, an eclipse male off Murvagh beach 24 July to 22 September 2016, returning on 13 October 2017, a male off Sheskinmore on 24 October 2016, and two females at St John's Point on 28 November 2020 which were last seen on 13 April 2021.

It is curious how many of these birds have been long-staying, or seen in late spring and summer. Could it be that having strayed so far from their normal range, they are a bit hesitant about making the return journey?

Eiders, north Fanad.

Eider

Somateria mollissima mollissima | Éadar

RESIDENT AND WINTER VISITOR
RED-LISTED BOCCI-4

The earliest record of Eider from Donegal is of two birds at Inishtrahull, in February 1890.[40] The species first bred in Ireland in 1912, on a small island off west Donegal, thought to be Roaninish.[41] It has slowly consolidated its toehold, and now breeds commonly on the small uninhabited islands and more locally along the rockier shores of the mainland. Its breeding range includes Northern Ireland but it is still very scarce further south.

Our local breeders are very sedentary, but they are joined in winter by immigrants from the north. These have not been monitored well, so some sites that should probably be rated as Nationally Important are simply lacking enough data. Malin Head, Trawbreaga Bay and north Fanad coast have all had counts of over 400, but half that would probably be more typical. Inishtrahull, Loughros More Bay and Killybegs have all had 150 or more.

There doesn't appear to have been any significant change in either breeding or wintering numbers over the last few decades, but this is a very subjective judgement.

The Internationally Important figure (1% of the flyway population) is 9,800.

The Nationally Important figure (1% of the national population) is 55.

Qualifying Sites: north Fanad coast

The species is Red-listed in BirdWatch Ireland's 'Birds of Conservation Concern in Ireland 4: 2020–2026', due to its declining global numbers.[42]

Subspecies

Northern Eider

Somateria mollissima borealis | Éadar tuaisceartach

SCARCE VAGRANT

There have been at least twelve records of this race. Northern Eider breeds in north-east Canada and west Greenland, and was first recorded at Fanad Head on 14 March 2004 (R. Millington). But there are undocumented reports from at least 2002, and most recently at Tory in May/June 2024. It is thought that some birds may be resident.

Subspecies

Dresser's Eider

Somateria mollissima dresseri | Éadar Dresser

RARE VAGRANT

Dresser's Eider is from further south in arctic Canada than Northern Eider. The first record for the Western Palaearctic was of a bird at Glashagh Bay, north Fanad from 2 January to 21 February 2010 (D. Charles, W. Farrelly et al.). What is presumed to be the same bird returned to Pollan Bay on the Inishowen north coast from 8 to 19 June 2011 and to Malin Head from 3 to 16 June 2012 (R. McLaughlin et al.). There has only been one other confirmed Western Palaearctic record since then, in County Dublin.[43]

Dresser's Eider (foreground), Doagh Isle, Inishowen.

AIDAN KELLY

Surf Scoter

Melanitta perspicillata | Scótar toinne

SCARCE WINTER VAGRANT
DECREASING

The Surf Scoter is one of our more frequent trans-Atlantic vagrants. Forty birds have been recorded. The first was an adult male at Bundoran on

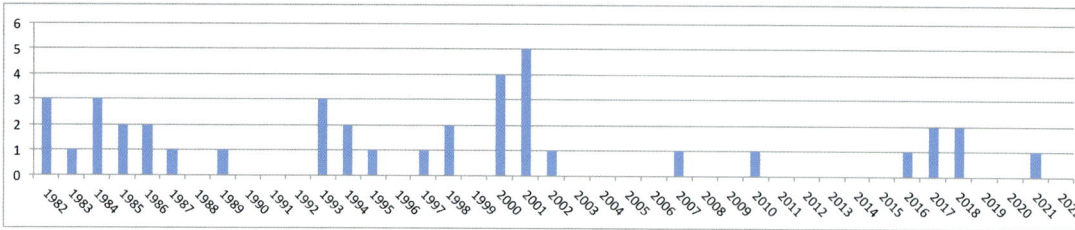

Fig. 18.20: Surf Scoter numbers per year from 1982/3 to 2022/3

13 March 1977 (C. Moore and K. Mullarney). More than one is not unusual. They almost all turn up in the large Common Scoter flocks found at several locations within greater Donegal Bay – nineteen times at Murvagh and nine at Bundoran. The only sites beyond Donegal Bay to have had any records are Lough Swilly with three and north Fanad with two. Figure 18.20 shows that the records are declining in frequency. This is in contrast to some other vagrant American ducks, of which Ring-necked Duck (Figure 18.17) is the most striking. It is of course quite possible that some of the birds prior to the millennium returned, perhaps more than once, which would reduce the sharpness of the decline, if not entirely eliminate it (see also Smew).

Velvet Scoter

Melanitta fusca | Sceadach

SCARCE WINTER VISITOR

INCREASING

RED-LISTED BOCCI-4

Donegal is beyond the western limit of the Velvet Scoter's normal wintering range on the Irish east coast, but close enough for 110 birds to have ex-plored this far west, and for one or more birds to be recorded almost every year. The first record was on Lough Swilly, in January 1890.

Like the Surf Scoter, Velvets are usually associated with Common Scoter flocks, where they can at times be just a little stand-offish, and with a tendency to feed closer to shore. Donegal Bay is the prime area, with sixty-six records – three to six birds in a good year, which can be dispersed among the sub-sites, or all together. The highest count is of eight birds at Rossnowlagh on 28 March 1992. Lough Swilly had twenty-nine records up to 2022, but since then there has been a regular and increasing presence, with a peak of

A female Velvet Scoter and two female King Eiders, St John's Point.

KIM PERIERA

six birds on 7 February 2024. They are always to be found in the area of deep water off the west of Inch Island and Ballymoney and usually in the company of other sea ducks, divers and grebes.

It would be expected for a winter visitor, as opposed to a vagrant, that some of these birds could return in subsequent winters. A bird in Killybegs Harbour in March 2007 and throughout the following January is evidence of this, as is the presence over the past decade of a few birds most winters in Lough Swilly. Regardless of that possibility, the annual trend is slightly upwards.

The species is Red-listed in BirdWatch Ireland's 'Birds of Conservation Concern in Ireland 4: 2020–2026', due to its declining global numbers.[44]

Common Scoter

Melanitta nigra | Scótar

WINTER VISITOR

The main haunt of Common Scoter is Donegal Bay, with sub-flocks in good years off the sandy beaches at Bundoran, Rossnowlagh and Murvagh. Following severe winter storms the bay can be almost deserted for a few years, presumably until decimated stocks of shellfish can re-build their numbers. So the Common Scotor's numbers are erratic, and estimates are also unreliable except in very calm conditions. Bearing those caveats in mind,

Bundoran occasionally has more than 500, Rossnowlagh more than 1,000 and Murvagh more than 1,500. The highest estimate was of 3,000 birds at Murvagh in January 2010.

When there was a significant breeding population at Lower Lough Erne in County Fermanagh, a post-breeding moulting flock of about 300 birds could be found at Murvagh in late July. The breeding population peaked at about 150 pairs in the late 1960s, but from then it faded to final extinction in 1993.[45]

Lough Swilly has typically held about 20 birds, with high counts of 100 in 1984/5 and 78 in 2012/13. A breakthrough came in 2023/4 when numbers rose to a peak of 250 birds on 7 February (see also Velvet Scoter). Other haunts with small, less regular flocks include Loughros More Bay (50) and Sheep Haven Bay (50).

The Internationally Important figure (1% of the flyway population) is 7,500.

The Nationally Important figure (1% of the national population) is 110.

Qualifying Sites: Donegal Bay

Long-tailed Duck
Clangula hyemalis | Lacha earrfhada
UNCOMMON WINTER VISITOR
DECREASING
RED-LISTED BOCCI-4

Long-tailed Duck is an abundant summer visitor to the arctic, wintering mostly in marine habitats to the south. Our birds are most likely from the Icelandic population.[46] In Ireland this shellfish-eating diving duck tends to visit similar haunts to those used by Eider and the scoters, but it has its own favourites. Most of the Irish population is in Donegal, but the sites have not been counted frequently enough, with the result that Lewis, Burke et al. concluded that no site regularly supports in excess of twenty birds.[47] This is certainly not the case. While numbers at each site ebb and flow, we usually have two or three sites that peak each year at comfortably more than twenty. There is a hint that this would have been normal historically, from a count of forty to fifty birds at Bundoran on 11 April 1938 (F. Egginton).

Long-tailed Ducks turn up in ones and twos widely around the coast, and very occasionally inland at Loughs Eske and Fern, and on Lough Swilly at Blanket Nook and Inch. But the vast majority are faithful to a few sites.

Inishfree Bay was the best site from the late 1970s to the 1990s, with frequent counts of 50–100 birds, and a peak count of 119 on 1 March 1993. The only counts at or above fifty since 2000 have been fifty-eight on 3 December 2005 and fifty on 28 December 2017. Peaks in the thirties are now normal.

North Fanad has always rivalled Inishfree Bay, and peaked at eighty-six on 10 December 1991. Flocks of more than fifty could be expected up to about 2010. The highest count since then was thirty-eight on 9 March 2018, with twenty to thirty being normal.

Portnoo had up to seventy birds from the 1960s to '90s, but they then collapsed to single figures. However, that minimal presence can still be expected each winter. Donegal Bay also attracted a lot of attention in earlier years, when fifty or sixty birds could be expected. In the 1970s, the Erne estuary had fifteen to twenty-five most winters, peaking at thirty-five in 1979. Bundoran was favoured in the 1980s with counts in excess of thirty being frequent, with the peak being sixty-plus on 18 December 1983. Since then it is Mountcharles that is the go-to location, with annual peaks of about twenty to twenty-five birds.

Minor sites, with up to about a dozen birds, have been Tullagh Bay in Inishowen from the 1960s to the '80s, Magheraroarty in the north-west from about 2000 to 2015. Since 2015 their presence on Lough Swilly has become regular with the peak of twelve birds in the most recent season of 2023/4.

It is clear that both Inishfree Bay and the north Fanad coast should prove to be of National Importance, if counted regularly – and maybe also Donegal Bay.

On grounds of an 82 per cent short-term decline in its wintering population, Long-tailed Duck is Red-listed in BirdWatch Ireland's 'Birds of Conservation Concern in Ireland 4: 2020–2026'.[48]

Goldeneye

Bucephala clangula clangula | Órshúileach

WINTER VISITOR
DECREASING
RED-LISTED BOCCI-4

Goldeneye is one of the later winter visitors to arrive. It is found on large lakes, both coastal and inland, but usually in small numbers. The exception is Inch Lough, where the highest count was 217 in the winter of 1996/7. Blanket Nook can also hold up to fifty birds. At Lough Swilly, the numbers rose and fell in a regular cycle until 2011, since when they have been erratic (Figure 18.21). Overall, the long-term trend is slightly downward. The national trend is more strongly downward.

The Internationally Important figure (1% of the flyway population) is 11,400.

The Nationally Important figure (1% of the national population) is 40.

Qualifying Sites: Lough Swilly

Goldeneye is Red-listed in the 'Birds of Conservation Concern in Ireland 4: 2020–2026' assessment by BirdWatch Ireland, with a population decline of 68 per cent over the last twenty-five years.[49]

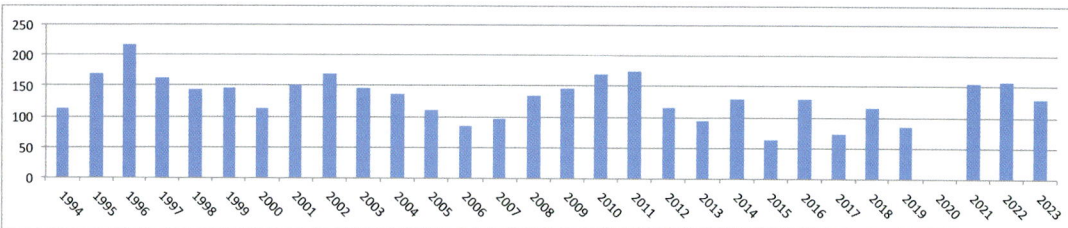

Fig. 18.21: Goldeneye numbers on Lough Swilly from 1994/5 to 2023/4

Smew

Mergus albellus | Síolta gheal

VAGRANT AND SCARCE WINTER VISITOR
DECREASING

Smew breeds in the taiga and forest-tundra zone from north Scandinavia eastwards across Eurasia. Its wintering range is mainly around the Baltic

Smew, Blanket Nook.

RICHARD SMITH

and North Seas.[50] The first Donegal record was of a bird shot on Lough Swilly in January 1891.[51]

There have been over 400 recorded sightings of Smew in Donegal. By eliminating what appear to be repeat sightings within each season, the total comes down to only seventy-three birds. Repeat visits in separate seasons were not taken into consideration for that calculation, so the actual number of birds involved is smaller again. What seems to be happening is that a vagrant bird arrives on Lough Swilly, likes it, and returns with one or two friends the following winter. If they also like it, a small group returns each winter until the last one dies – about a decade later – and Smew returns to its default status of vagrant. This has now happened on two occasions, making Lough Swilly (on average) the best place in Ireland to see this beautiful duck. They were absent in only two years between 1993 and 2018.

Blanket Nook was the place to look in the 1990s. It was graced with five Smews in 1994/5, then in the following years to 1998/9 there were seven, five, nine and four. On Inch Lough there was a male present throughout every winter from 2008/9 to 2017/18, with the occasional presence of up to two extra birds on the first five of those winters (and with only a single sighting of one bird in 2012/13). This works out as thirty bird-seasons in the Blanket Nook period, and eighteen at Inch.

As well as Lough Swilly, there have been two birds at Sessiagh Lough, Dunfanaghy, one at Malin Head and one on Lough Foyle west bank.

Hooded Merganser

Lophodytes cucullatus | Síolta chochaill

RARE VAGRANT AND ESCAPE

There is one acceptable record of this North American species, which is of a pair recorded on Tory from 19 to 23 May 2015 (A. Meenan et al.). They were accepted as genuine vagrants from North America – the fifth and sixth records for Ireland. Until recently, any Hooded Mergansers seen in Ireland were regarded as likely to be escapes from captivity – a bird at Kinny Lough in north Fanad on 1 September 2002 (W. Farrelly et al.) was one such.

Goosander (Common Merganser)

Mergus merganser merganser | Síolta mhór

SCARCE WINTER VISITOR AND RARE BREEDER
STABLE

The first instance of Goosanders breeding in Ireland was at Glenveagh in 1969.[52] This pair, or their descendants, bred almost continuously (not seen in 1972 or 1973) until 1977, and was also present in 1978. Another site at about 15 km distance was located in 1978, but no proof of breeding was obtained. Two sites were occupied in 1981. A pair bred at an undisclosed location in 1989 (T. Cooney), and birds were present at potential breeding locations from 2003 to 2005 and in 2019. It is difficult to be sure how many sites have been used or prospected, but it is at least five.

Given that early start, it may seem odd that Goosanders have not become established in Donegal, like the population in Wicklow which first bred much more recently. The limiting factor is almost certainly nest sites. Mature riparian woodland, with trees large enough to provide nest holes, is very rare, and the ubiquitous conifer plantations are usually felled long before the trees have a chance to develop potential nesting holes or crevices. It will probably need a nest box scheme to secure a future in Donegal for this impressive bird – 'duck' doesn't quite do them justice.

As winter visitors, Hart regarded them as rare in the nineteenth century.[53] The first site where small numbers could be found in some years was the River Foyle from St Johnstown downstream into Northern Ireland. The highest count there was twenty-three, on 14 March 1993, but the site hasn't been used in recent years. In the south-east, various lakes east of Ballintra and Ballyshannon are occasionally used, with the highest count being ten at Rath Lough on 14 January 2001. Lough Fern has always attracted occasional birds, and since four birds arrived in the winter of 2018/19 it has become the regular haunt of several. Subsequent annual peak counts were five in 2019/20 (at the same time as there were five on Lough Roshin, in the Ballintra area), seven in 2020/1, six in 2021/2, and in 2023/4 there was a peak of four. Inch Lough and Blanket Nook frequently attract one or two birds.

Red-breasted Merganser

Mergus serrator | Síolta rua

RESIDENT

INCREASING

Small numbers of Red-breasted Mergansers breed in the county, mainly near the north coast and around Donegal Bay. The west coast was also used in the 1960s.[54]

Late summer moulting flocks are sometimes recorded – notably sixty on the Erne estuary on 25 July 1991 and 120 at Murvagh on 19 August 1994. The largest winter flock seen had 151 birds. It was on 8 December 2012 at the Leannan estuary on Lough Swilly.

In winter, Lough Swilly usually has several small flocks totalling over 100. Numbers are currently growing slightly, having been fairly stable over the longer term (Figure 18.22). The situation on Donegal Bay is similar, with a peak of 156. Small parties of up to a dozen are generally widespread around the coast, mainly in the bays and estuaries.

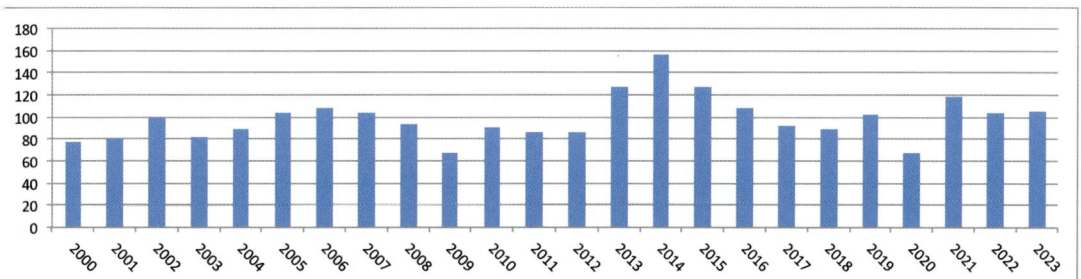

Fig. 18.22: Red-breasted Merganser peaks on Lough Swilly, from 2000/1 to 2023/4

The Internationally Important figure (1% of the flyway population) is 860.

The Nationally Important figure (1% of the national population) is 25.

Qualifying Sites: Lough Swilly, Donegal Bay

Ruddy Duck

Oxyura jamaicensis jamaicensis | Lacha rua

RARE VAGRANT

The Ruddy Duck, native to North America, had become naturalised around Lough Neagh, in Northern Ireland, since first recorded breeding there in 1973. It is presumably from that stock that one male moved to Inch Lough on 27 August 1986, and a second two days later (T. Campbell and R. Sheppard). This is the only record from Donegal. As a result of culling in the UK, the species is now very rare in Ireland and Britain, so further records emanating from this side of the Atlantic are unlikely.

Capercaillie

Tetrao urogallus urogallus | Capall coille

EXTINCT

Until its bones were finally identified from excavations in 'Viking' Dublin, the very existence of Capercaillie in Ireland was doubted. The growing evidence from various counties now makes it clear that it would also have lived in Donegal, but confirmation is so far rather shaky. This giant grouse, which is normally resident in coniferous forests, is known to accept deciduous forests in Spain, and had clearly adapted in Ireland to the gradual demise of our native Scots pine habitats. The Ordnance Survey Memoires from 1833 to '35 mention Capercaillie (under the old name of Cock-of-the-Wood) as 'extirpated'.[55] This is thought by Gordon D'Arcy as credible evidence that the author must have 'been aware of it being formerly found (perhaps in living memory)'.[56]

Red Grouse (Willow Ptarmigan)

Lagopus lagopus scotica | Cearc fhraoigh

RESIDENT

STABLE

RED-LISTED BOCCI-4

Red Grouse are scattered thinly throughout the mountainous areas of the county. They have decreased over the past half century as a result of habitat changes, most notably the decline in heather. Overgrazing by sheep and uncontrolled wildfires are probably the most important factors involved in reducing heather cover. A national survey from 2006 to 2008 found an estimate of 2,038 males in the north-west (mainly Donegal, but including small areas in Leitrim, Cavan and Monaghan).[57] This was a decline from sixty-seven occupied 10 x 10 km squares in 1968–72 to forty-five, but was still 48 per cent of the national total.[58]

The Red Grouse is Red-listed in BirdWatch Ireland's 'Birds of Conservation Concern in Ireland 4: 2020–2026' on grounds of a 50 per cent long-term decline in its population.[59]

Red Grouse, Blue Stack Mountains.

ROBERT VAUGHAN

Grey Partridge

Perdix perdix perdix | Patraisc

EXTINCT RESIDENT

RED-LISTED BOCCI-4

The Grey Partridge is a native Irish species which almost died out in Ireland in recent times. But it had already been judged by Thompson to have declined prior to the 1840s, possibly as a result of the practice of using poisons on seed wheat.[60] Ussher and Warren confirmed continuing decline, and thought it 'scarce in the west of Connacht and Donegal, where moors and mountains prevail'. They describe the preferred habitat as 'small farms where there is always some untidy tillage, for instance potato-fields in which Persicaria and similar weeds are plentiful'. Such conditions cannot have been in short supply in Donegal at that time. Kennedy et al. reported that the national decline continued to 1920, but was reversed by re-introductions and restrictions on shooting.

In Donegal, they faded away during the middle years of the twentieth century, until the last birds were seen in the 1970s. The intensification of agriculture was undoubtedly to blame. Wildflowers which had been abundant in potato and other arable crops were eliminated, and with them the Partridge. In 2013 the Inch Island Gun Club, with the support of many organisations and agencies, launched the Inch Island Partridge Project, to re-introduce the species. Habitat was created, broods were reared and released, but after eight years it had become clear that the birds were not surviving in the wild and the project was abandoned.[61]

On grounds of historical decline, and a 95 per cent long-term decline in breeding range, Grey Partridge is Red-listed in BirdWatch Ireland's 'Birds of Conservation Concern in Ireland 4: 2020–2026'.[62]

Pheasant (Common Pheasant)

Phasianus colchicus ssp. | Piasún

RESIDENT

DECREASING

This species was introduced to Ireland from central Asia as early as the late sixteenth century.[63] It remained largely in managed estates, but by the end of the nineteenth century there was a wild population in every county.[64] It was always assumed that without continuous reinforcement, this wild population would soon die out. While this is not certain, the decline in organised shooting in recent years, and the hand-rearing that went with it, probably accounts for a general thinning out of its presence in the wild. Pheasants appear to be absent from only the far south-west of the county.

Quail (Common Quail)

Coturnix coturnix coturnix | Gearg

RARE SUMMER VISITOR
DECREASING
RED-LISTED BOCCI-4

Quail migrate from Africa to Europe south of Scandinavia. Huge numbers are slaughtered for food and sport in countries bordering the Mediterranean, which may have impacted on breeding numbers here. Agricultural changes can also account for reducing the species from its status in earlier days, when Ussher and Warren said it was common until 1850. During the twentieth century, records were few and erratic. In 1989, ten birds were heard in the arable east of the county, between Lifford and Inch Island – a relative invasion. Two birds were recorded in 1995, and there were singles in 1970, 1993 and 2000. Finally, a bird was heard on Inch Levels on 23 June 2022. The irruptive pattern of their migrations leaves us with some expectation that they will continue to be heard at rare intervals, but the restoration of their status as a regular summer visitor seems to be now beyond hope.

On grounds of historical decline, Quail is Red-listed in BirdWatch Ireland's 'Birds of Conservation Concern in Ireland 4: 2020–2026'.[65]

Nightjar (European Nightjar)

Caprimulgus europaeus europaeus | Tuirne lín

RARE SUMMER VISITOR
PROBABLY EXTINCT
RED-LISTED BOCCI-4

Donegal has never had a good population of Nightjars. At the end of the nineteenth century Hart considered it rare, but was aware of its presence at Greencastle, Fahan and Lough Eske. In the first half of the twentieth century the species was reported to be plentiful in the area between Mulroy Bay and Kilmacrenan, and Ruttledge still noted it as breeding in the north of the county in the 1960s.[66] There were a few records from the south during the survey for the first breeding atlas,[67] and also claims for the Marblehill area in north Donegal in the 1980s. But since then, the few birds that have been reported have not been confirmed. These could well have been displaying Common Snipe making their drumming sound, which is similar to the song of a Nightjar.

The Nightjar is Red-listed in BirdWatch Ireland's 'Birds of Conservation Concern in Ireland 4: 2020–2026' on grounds of a 95 per cent long-term decline in its breeding distribution.[68]

Alpine Swift

Tachymarptis melba | Gabhlán Alpach

VAGRANT

The Alpine Swifts which breed in southern Europe are migrants that have wintered in Africa. In March 2023 there was an exceptional influx to Ireland of overshooting birds. Most were seen around the coast from Dublin to Cork. One bird was seen in Kilcar on 18 March (J. O'Boyle).

Swift (Common Swift)

Apus apus apus | Gabhlán Gaoithe

SUMMER VISITOR

DECREASING

RED-LISTED BOCCI-4

Swifts are one of the last of our common summer visitors to arrive, and among the first to leave. They breed only on tall buildings in towns and villages, and their numbers have declined in recent years. The reasons undoubtedly have to do with the collapse of insect populations throughout Europe, and possibly also to conditions in Africa. One other contributing

Swift over Ardnageer (626 m), Blue Stack Mountains.

JOHN CROMIE

factor that is being successfully addressed is the lack of suitable nest sites in newer buildings. Nest box schemes have started in Buncrana, Letterkenny and Dunfanaghy and have already had some success.

During the breeding season, Swifts are known to navigate whole weather systems in search of food, which will at times bring them over mountain tops, as shown in the photo. They will also congregate in early May before the start of nesting, and prior to departing in August, at places like Inch Lough where swarms of aerial insects are available.

The Swift is Red-listed in BirdWatch Ireland's 'Birds of Conservation Concern in Ireland 4: 2020–2026' on grounds of a 56 per cent short-term decline in its population.[69]

Yellow-billed Cuckoo

Coccyzus americanus | Cuach ghob-bhuí

RARE VAGRANT

There are a relatively large number of Western Palaearctic records for this North American species, but only one for Donegal. A bird spent three days on Tory from 6 October 1989 (N. O'Neill). It was the seventh record for Ireland.

Cuckoo (Common Cuckoo)

Cuculus canorus canorus | Cuach

SUMMER VISITOR

STABLE

For many people, the Cuckoo still heralds the arrival of summer, but summer is still with us when the adults depart once again for their African winter quarters – their chicks remain, in the care of their foster parents, to find their own way later on. In Donegal it is invariably in the nests of Meadow Pipits that Cuckoos lay their eggs, and therefore it is in their habitats that Cuckoos will be heard, but they do also seem to have a preference for marginal habitats with some trees.

These conditions are readily found around the edges of conifer plantations, especially where there are young or scattered trees. So the atlas projects and the Countryside Bird Survey have detected no major change in the status of the Cuckoo in Donegal.[70] However, while the Cuckoo had been common in living memory throughout lowland Donegal, most of those would have been parasitising Dunnocks, and that relationship has long since ended. Hence the Cuckoo's presence in the lowland habitats of east Donegal is now only a fading memory.

Cuckoo (female), Tory.

ANTON MEENAN

Pallas's Sandgrouse

Syrrhaptes paradoxus | Gaineamhchearc Pallas

RARE VAGRANT

These exotic dove-like birds erupt on rare occasions from their Central Asian home. There have been two such events that delivered birds to Ireland, in 1863 and 1888. None reached Donegal on the second invasion, but a flock of at least thirteen was recorded in June 1863. Initially, two birds were shot from a flock at Naran. Then what was presumed to be the same flock, with thirteen or fourteen birds, was seen in unnamed mountains and later on the sands at Drumbeg (Inver Bay). From this flock three birds were shot. One of them was only wounded, and was presented by William Sinclair to the Regent's Park Gardens, now the London Zoo.[71]

Rock Dove / Feral Pigeon

Columba livia livia | Colm aille

RESIDENT
DECREASING

The semi-domesticated Feral Pigeon is common throughout the county in towns and farmyards, although the decline of widespread mixed farming has reduced the rural population. The true Rock Dove is found mainly in small colonies on the wilder cliff-bound parts of the coast and islands. Unfortunately, these colonies all have some birds exhibiting the mixed plumages of feral birds, and even those with virtually pure Rock Dove plumage mostly prove, on closer examination, to be of feral origin. But pure, uncontaminated Rock Doves can still be found at the more remote colonies on Tory, Horn Head, etc. Natural selection will be working to eliminate the artificial features of the feral birds and restore the populations to their natural state, but this will only come to fruition if the supply of feral birds dries up.

Rock Doves were members of the bird community supported by small mixed farms with weedy organic gardens growing potatoes and other staple vegetables – typical of coastal districts up into the 1960s and '70s, and widespread in the decades up to that. Rock Doves would commute from their nearby breeding cliffs, and can still be found feeding in the tightly grazed grassland near settlements. In the past, those settlements would also have been attractive to Corncrakes, Chough, Tree Sparrows, Twite, Corn Buntings and Yellowhammers – a community of birds (most of them now rare or completely gone) that helped to define a distinctive part of both our natural and our cultural heritage.

Stock Dove

Columba oenas oenas | Colm gorm

RESIDENT
DECREASING
RED-LISTED BOCCI-4

Stock Doves first appeared in Ireland as recently as 1875.[72] They never managed to build up a strong base in Donegal before the decline in tillage made that unlikely to happen now. They are currently restricted to the lowland east of the county between Lough Swilly and the River Foyle. This small cross-border pocket is isolated from the core of the national population in the south-east of Ireland. Tree Sparrow and Yellowhammer have similar distributions, and all three species are threatened here in the north-west. Stock Doves nationally are experiencing a severe decrease – the second largest decline of any species in the Countryside Bird Survey.[73]

Favoured haunts in Donegal are in the farmland near the south bank of Lough Swilly. Scattered birds can still be found in winter flocks elsewhere, for example near Convoy, but they are easily overlooked if Woodpigeons and Feral Pigeons are also present. As the Stock Dove is sedentary, it is always worth checking winter sites in summer for evidence of breeding.

Stock Dove is Red-listed in the 'Birds of Conservation Concern in Ireland 4: 2020–2026' assessment by BirdWatch Ireland, with a twenty-five-year short-term population decline of 54 per cent.[74]

Woodpigeon (Common Woodpigeon)

Columba palumbus palumbus | Colm coille

ABUNDANT RESIDENT
STABLE

Woodpigeon is one of our most abundant and widespread species – and given its bulk, it probably constitutes the greatest bio-mass of any wild bird species in the county. Its numbers are greatest in the arable east. The shift from spring- to autumn-sown cereal crops, which has had such a detrimental impact on seed-eating birds, has been a boon to the leaf-eating Woodpigeon. And the shift from arable to grass has not been a problem, as the clover content in grazing or silage mixtures suits them well. Not surprisingly, Woodpigeon can often do serious damage to crops, with shooting them being the natural reaction. But it was shown in 1965 that shooting only killed 'birds that would die anyway through other causes … it would be necessary to kill more birds in a year than could be replaced by reproduction in one breeding season'. The way to control numbers was to organise crop rotation so that there was a break in the year when Woodpigeons would have nothing to eat.[75] But crop rotation is currently out of favour, so it is fortunate that the Woodpigeon is not currently regarded as a serious pest in Donegal, despite its large population.

The Countryside Bird Survey has recorded a steady increase in their numbers.[76]

Turtle Dove (European Turtle Dove)

Streptopelia turtur turtur | Fearán

SCARCE VAGRANT

STABLE

RED-LISTED BOCCI-4

There are forty records of Turtle Dove for Donegal. The catastrophic decline in numbers throughout Europe since the 1970s does not appear to have significantly altered its status here. It was a fairly frequent vagrant at places like Tory in the 1950s, with one or two birds turning up most years. It was then not seen at all between 1963 and 1997, but that was a period of reduced observation. Increased vigilance in recent years has shown that it does still visit us, with a bird seen in most years since 2013, and three birds in 2017.

Tory, with twenty-five, has had most birds. Malin Head and Inishtrahull have had seven between them. There have been four at the south-west sites around Glencolumbkille, and the remaining four records are from

Inishbofin, Gweedore, Arranmore, and most recently at Fanad Head from 24 to 25 September 2022.

Most birds have been in September, with a total of twenty. The peak spring month, May, has had seven and there were four in June. Three have turned up in both July and August. October has had two and April one.

The Turtle Dove is Red-listed in BirdWatch Ireland's 'Birds of Conservation Concern in Ireland 4: 2020–2026' on grounds of a 90 per cent long-term and 94 per cent short-term decline in its breeding range.[77]

Collared Dove

Streptopelia decaocto decaocto | Fearán baicdhubh

RESIDENT

STABLE

Collared Doves were not found west of the Balkans until the early 1930s. They advanced rapidly westwards, reaching Britain in 1955 and Ireland in 1959. The first record for Donegal was at Inishtrahull, on 18 July 1961 (D. O'Sullivan). Breeding was first suspected in 1965 near Malin and Bundoran. By 1968 to 1972, the first breeding atlas found them to be widespread in the north and east of the county.[78] Lowland gaps in the south-west have since been filled in and their population and distribution now seems to be stable. They are found mainly in towns and villages, and in farmyards. They seem not to have caused any disruption to the existing avifauna, and are a welcome addition to our biodiversity. This is an exception to the usual rule that invasives detract from local biodiversity – probably because this invasion is genuinely natural, and not determined by human interference.

Water Rail

Rallus aquaticus aquaticus | Rálóg uisce

RESIDENT AND PASSAGE MIGRANT

This is a very elusive bird, more often heard than seen. Those familiar with the sound of a squealing pig should be alert to the nocturnal calls of the Water Rail. It is likely that they are present wherever there is fairly extensive swamp vegetation or reedbeds, but it has turned up at a wide range of sites, some of them quite small. Most records are for well-watched places like Inch Lough and Blanket Nook. Winter records are very few, but the resident population is nonetheless thought to increase somewhat in winter. Arranmore records have been in late October or winter.

Subspecies

Icelandic Water Rail

Rallus aquaticus hibernans | Rálóg uisce Íoslannach

PASSAGE MIGRANT AND POSSIBLY WINTER VISITOR

Migrant Water Rails at Tory and elsewhere are presumed to be Icelandic birds,[79] and the subspecies is believed to winter in Ireland,[80] but the case has not yet been proven. They still occur regularly on Tory in both autumn and spring.

Corncrake (Corn Crake)

Crex crex | Traonach

SUMMER VISITOR

STABLE

RED-LISTED BOCCI-4

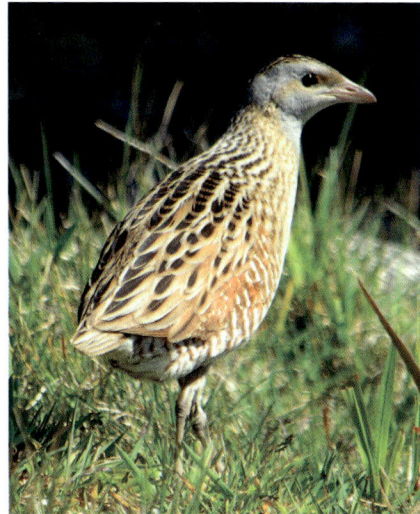

Corncrake, Tory.

In the memory of our current older generation, Corncrakes were ubiquitous in farmland. On one night in the late 1960s, I was able to count ten birds calling within earshot from the house where I now live, in inland east Donegal. In even earlier decades, that would have been the general experience, when chicks were all safely fledged by the time the hayfields were cut in August or September. But haymaking was abandoned in the 1960s and '70s, and farmers now save their grass only as silage. A first cutting is able to advance to earlier in the summer, leaving time for a second cut. Add to that the speed of modern machinery, and it was inevitable that nests and chicks would seldom survive. The

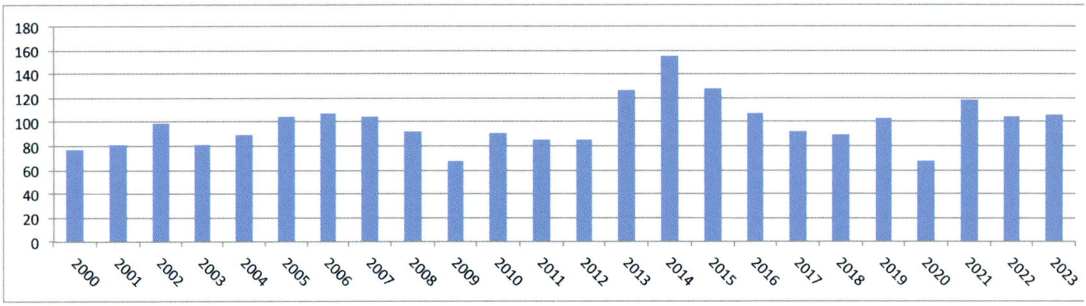

Fig. 18.23: Number of calling male Corncrakes each year from 2000 to 2023

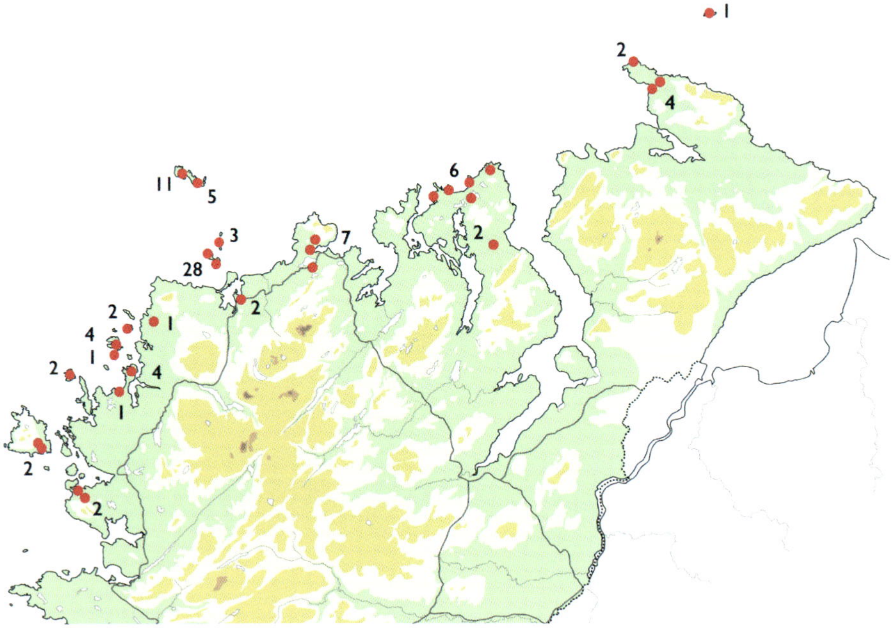

Map 18.1: Calling Corncrakes in 2018

species was very rapidly confined to marginal areas of wet land, or small rocky fields where making silage is not practical. The continuous rasping call of the Corncrake may have prevented sleep, but it is sadly missed as an evocative sound of summer.

The first breeding atlas, of 1968–72, showed a retreat from the Irish south coast counties.[81] That continued, and a national survey found only 174 calling males in Ireland, almost all in the Shannon Callows, the Moy Valley in County Mayo, and north Donegal.[82] A succession of support schemes for farmers halted the decline in Donegal, where most of the Irish birds have been located since 2000. However, these schemes have had little success in recovering lost ground, and the population has hovered around

100 calling birds ever since (Figure 18.23). Map 18.1 shows what is, more or less, the current distribution of Corncrakes in Donegal. Since 2018 the average total number of territories (rather than calling birds, although the difference should be slight) has remained at 100, with 28 per cent being on Inishbofin and 19 per cent on Tory (NPWS). The numbers on other islands are small, and on the mainland they are scattered in pockets around the north and north-west coast from Malin Head to Maghery.[83]

The recent establishment of dedicated Corncrake SPAs should make it easier to provide the necessary support for farming communities to make space for breeding birds. Unfortunately, in some of the target areas where birds had recently bred in good numbers, such as Meenlaragh and Falcarragh on either side of Ballyness Bay, the housing footprint has expanded to almost suburban density, and farming has been largely abandoned. It is difficult to implement Corncrake support measures in such areas.

The Corncrake is Red-listed in BirdWatch Ireland's 'Birds of Conservation Concern in Ireland 4: 2020–2026' due to an 83 per cent long-term (since 1980) decline in its population.[84]

Spotted Crake

Porzana porzana | Gearr breac

VAGRANT SUMMER VISITOR

There are two old records of Spotted Crakes from the nineteenth century. The first was a bird obtained near Bogay House (probably Port Lough), at Newtowncunningham, in 1828.[85] There is one modern record of a bird calling at Inch Lough from late June to 27 July 1985, and a tape recording was made (D. McLaughlin et al.). Given such an extended stay, the possibility that there was a breeding pair cannot be ruled out.

Moorhen (Common Moorhen)

Gallinula chloropus chloropus | Cearc uisce

RESIDENT

DECREASING

Moorhens are widespread throughout the county, in lakes, rivers and marshes. They tend to be dispersed, so numbers in any density can only be found at especially good sites, like large coastal lagoons with marshy margins. There is good evidence that they have suffered from the predations of escaped mink, particularly along rivers. There is also now the suspicion that an accommodation has been reached – perhaps Moorhens are choosing more secure nest sites, or mink are becoming more specialised in their prey.

Coot (Eurasian Coot)

Fulica atra atra | Cearc cheannann

RESIDENT AND WINTER VISITOR

DECREASING

Coot are confined to a small number of shallow eutrophic lakes. They are relatively sedentary, but numbers are boosted in early winter by immigrants.

On Lough Swilly they breed at both Inch Lough and Blanket Nook, but they soon depart from Blanket Nook as winter arrives. At Inch, numbers reach their maximum in autumn and early winter and have been stable for the last twenty years, at 500–700 (Figure 18.24). Those high peaks soon fade and Coot can disappear entirely in mid-winter. These movements do not seem to be responding exclusively to hard weather, so it is more likely that there is a regular partial migration. The highest count at Inch Lough was 1,130 on 25 October 2002.

Lough Fern's Coot are more erratic, and are often absent. Their peak count was 800 on 28 October 1984 – a date on which Pochard and Tufted Duck were also present in their highest numbers. Although early-winter peaks are still impressive, there appears to have been a general contraction in the winter population of the county. This is certainly indicated by the most recent atlas survey.[86] Coot also appear to be declining at previously good sites, or have already disappeared.

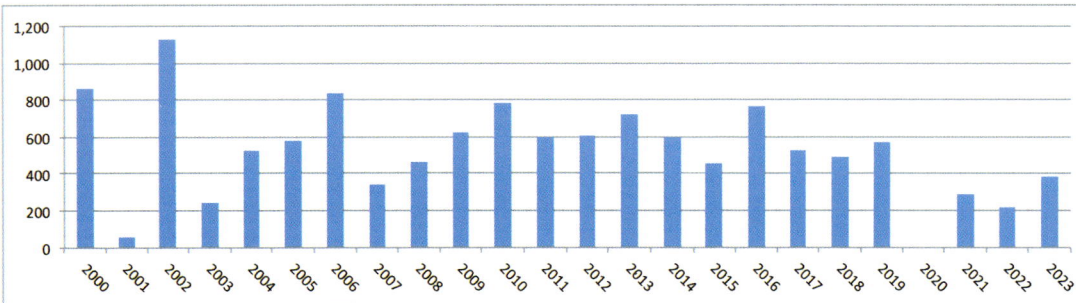

Fig. 18.24: Coot peak counts on Lough Swilly from 2000/1 to 2023/4

Coot flock, Inch Lough.

Dunfanaghy New Lake had a mean of 283 in the 1980s but only 102 in the late 1990s. At Durnesh Lough they have almost gone. This prime site held a mean peak of 210 birds in the winters from 1984/5 to 1986/7, but a peak of only 50 in the period from 1994/5 to 2000/1. There was a further drop to a mean of only three and a peak of seventeen from 2007/8 to 2017/18. Lough Fern had 800 Coot in late 1984, at a time when there were even larger numbers of Pochard and Tufted Duck present, but there were only small numbers in the next two seasons. The mean for that period at Lough Fern was 283. By the late 1990s the peak was only three, and Coot have not been recorded in recent seasons.[87]

Coot generally breed at sites with wintering flocks, but the numbers would not match those present in winter. Minor sites with a few breeding pairs include Rosepenna Lough, Lough Akibbon, Lough Alaan and Clooney Lough.

The Internationally Important figure (1% of the flyway population) is 15,500.

The Nationally Important figure (1% of the national population) is 190.

Qualifying Sites: Lough Swilly

Crane (Common Crane)

Grus grus | Grús

RARE VAGRANT

There are four records of Crane in Donegal. All involved birds first seen in summer. The first was shot on Inch Levels on 23 June 1896. Two records were of long-staying birds at Blanket Nook, albeit more than a decade apart. The first of these birds was present from mid-June to mid-September 1961 (J. Breslin and W. Bigger). The record was published as near Newtowncunningham.[88] Although correct, the actual site was Blanket Nook.[89] The other one was present from 24 May to at least mid-November 1975 (B. McAuley et al.). The most recent bird was at Kiltooris Lough from 19 to 20 May 2012 (C. Ingram).

Little Grebe

Tachybaptus ruficollis ruficollis | Spágaire tonn

RESIDENT
STABLE

Little Grebe is often the only waterfowl species to be seen at small upland or coastal lakes. They can even be seen some distance offshore. Although widespread, their numbers are usually small – one or two is not unusual, or a single brood. Post-breeding totals at Lough Swilly, usually in September, can be much higher, with a peak count of 159 on 7 September 2008 (Figure 18.25), and since about 2005/6 the peaks appear to be fairly stable. These high totals are halved over the winter months. Most of the Little Grebe on Lough Swilly will always be found at Inch Lough and Blanket Nook.

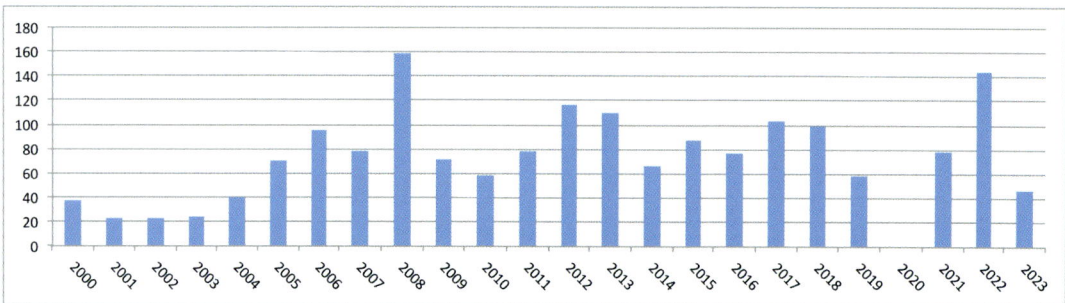

Fig. 18.25: Little Grebe peak counts on Lough Swilly from 2000/1 to 2023/4

Donegal Bay had a peak of seventy between 2007 and 2018, with a mean peak of twenty-nine.[90]

The Internationally Important figure (1% of the flyway population) is 4,700.

The Nationally Important figure (1% of the national population) is 20.

Qualifying Sites: Lough Swilly, Donegal Bay

Pied-billed Grebe

Podilymbus podiceps | Foitheach gob-alabhreac

RARE VAGRANT

The only sighting of this North American relative of Little Grebe was at Lough Anarget, near Glencolumbkille on 23 April 1988 (J. O'Boyle). It was the second Irish record.

Red-necked Grebe

Podiceps grisigena grisigena | Foitheach píbrua

RARE WINTER VAGRANT

There are eleven records of Red-necked Grebe. The first was a bird shot on Lough Swilly in 1875. The skin was acquired by the Derry Museum, but following closure of the museum it is likely to have ended up in the Ulster Museum. The second bird was shot at Inver Bay and was sent to the National Museum on 23 November 1887. It was thought to have been killed some years previously.

The first twentieth-century bird was at Sheep Haven Bay on 29 December 1977 (R. Sheppard). There have been five on Lough Swilly. One was present at Blanket Nook from 26 March to 15 April 1989. There was one at Inch on 10 December 2009, one at Shellfield on 15 January 2012, a long-staying bird at Inch from 9 February to 26 March 2014, and one off the west bank on 14 February 2015. Donegal Bay has had two – the first from 20 March to 10 April 1994 at Ball Hill, and the second from 15 to 17 March 2020. There has been a single record from Arranmore, which was on 5 December 1999, and a record of two at Loughros More Bay on 3 January 2021.

With birds emanating from eastern Europe, it is not surprising that those reaching Donegal do so from December onwards.

Great Crested Grebe

Podiceps cristatus cristatus | Foitheach mór

RESIDENT

STABLE

In the nineteenth century, Hart saw a specimen shot at Ray, on Lough Swilly, in April 1879, and thought the species probably visits some of the lakes.[91] But his contemporary, Professor Leebody, was unaware of it on Lough Swilly.[92] Early protection when it was facing extinction in the UK has allowed Great Crested Grebes to multiply since then, and they now breed at many of the lowland lakes. These include Lough Eske, Lough Alaan, Inch Lough and Blanket Nook, Loughs Gartan and Akibbon, Lough Fern, Dunfanaghy New Lake, Durnesh Lough, and some of the many lakes between Durnesh and Pettigo.

In winter, the Great Crested Grebe is largely found on salt water, and is widespread around the coast in sheltered waters. It may be that they are cyclical in their use of Lough Swilly, but the long-term trend is clearly down, although the recent trend since the turn of the century seems to be stable (Figure 18.26). The highest count was 329, on 27 November 1997. After Dublin Bay, Lough Swilly is the most important wintering site in Ireland.[93] In Donegal Bay they are just short of the national threshold.

The Internationally Important figure (1% of the flyway population) is 6,300.

The Nationally Important figure (1% of the national population) is 30.

Qualifying Sites: Lough Swilly

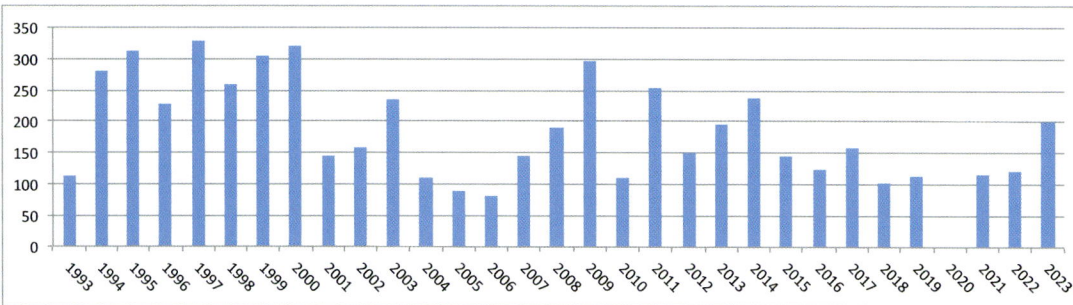

Fig. 18.26: Great Crested Grebe peaks on Lough Swilly, from 1993/4 to 2023/4

Slavonian Grebe (Horned Grebe)

Podiceps auritus auritus | Foitheach cluasach

SCARCE WINTER VISITOR

DECREASING

RED-LISTED BOCCI-4

Slavonian Grebes are summer visitors to the taiga zone from Scandinavia eastwards, with a tiny western outpost in Scotland. They winter around the coasts of western Eurasia. In 1892, Leebody noted that one or two spent each winter in the bay on the south side of Inch.[94] They are now mainly seen well offshore in the deeper waters of Lough Swilly, which may have been inaccessible to Leebody. Detection is greatly increased when the water is calm, so the higher figures are usually the most accurate. The highest count was forty-two, on 21 January 2006, but a count of thirty-four on 12 January 2008 was surprising, in that it was recorded during the only year (apart from the pandemic year of 2020/1) when none were recorded on I-WeBS counts (Figure 18.27). Since 2000, their numbers have varied quite widely, but there has been a clear reduction in those calm-water peaks. One or two can appear elsewhere on rare occasions – nine in Donegal Bay during the winter of 2010/11 was exceptional.

The Slavonian Grebe is Red-listed in BirdWatch Ireland's 'Birds of Conservation Concern in Ireland 4: 2020–2026'on the grounds of being globally vulnerable.[95]

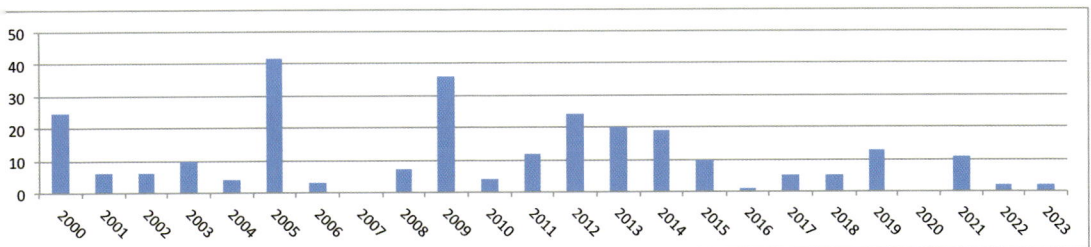

Fig. 18.27: Slavonian Grebe annual peaks on Lough Swilly from 2000/1 to 2023/4

Black-necked Grebe

Podiceps nigricollis nigricollis | Foitheach píbdhubh

RARE WINTER VAGRANT

Black-necked Grebe, Lough Swilly.

DEREK BRENNAN

Black-necked Grebe winters mainly in southern and coastal Europe, breeding in central Europe in the zone to the south of Slavonian Grebe. Eight have been recorded in Donegal. The first was a bird shot on 6 February 1893, at Killybegs. It is preserved in Barrington's collection.[96]

The first modern sighting was of a bird at Greencastle from 2 to 9 January 1982 (C. Brewster). One was seen at Bundoran on 18 March 1984. Two birds were seen on Lough Swilly in 1988 – one was present from 24 January to 1 April, and another at Blanket Nook on 16 and 17 October. At Dunfanaghy New Lake there was one on 22 November 2010. Lough Swilly had one at Inch Lough from 23 to 24 February 2013 and one at Mill Bay, off Inch Island, on 17 March 2021.

Chilean Flamingo

Phoenicopterus chilensis | Lasairéan Sileach

ESCAPE

An undoubted escape – one bird was seen at Inch on 14 June 1980 (C. Brewster).

Stone Curlew (Eurasian Stone-curlew)

Burhinus oedicnemus oedicnemus | Crotach cloch

VAGRANT

There is a single record of this striking large wader, which winters in North Africa and Iberia, and breeds mainly in southern Europe, with a small northerly outpost in England. It was a bird shot near Gweedore on 12 October 1903.[97]

Oystercatcher

Haematopus ostralegus ostralegus | Roilleach

RESIDENT AND WINTER VISITOR

STABLE

RED-LISTED BOCCI-4

Oystercatchers breed in small numbers all around the coast, but mainly on the offshore islands.[98] They mostly use rocky or shingle shores, with only a very few of them in machair. In addition, non-breeding birds from the winter migrant population can be found in estuaries and bays during summer months. This is because this long-lived specialist feeder needs several years to master the techniques needed to put on sufficient fat reserves to migrate with the adults to the breeding grounds.[99]

In winter, large numbers arrive from Scotland and Iceland. The population on Lough Swilly has shown a significant drop in recent years, but has been fairly stable in the longer term (Figure 18.28). The peak count was 2,823 birds on 15 November 2014. They are always found on the open shores, but large numbers also come to roost at Blanket Nook. Donegal

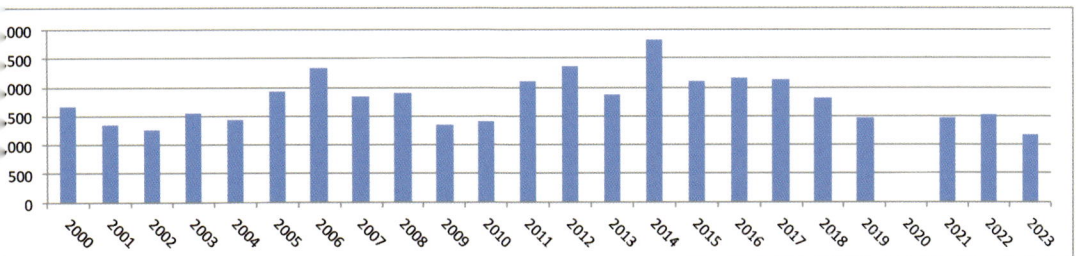

Fig. 18.28: Oystercatcher annual peak numbers on Lough Swilly, from 2000/1 to 2023/4

Bay normally has 800 to 1,200 birds, with the highest count being 1,266 in January 2017.

Lesser sites, with peaks during the 1990s, are Trawbreaga Bay (250), Ballyness Bay (160), Killybegs Harbour (123), Fanad north coast (83), Trawenagh Bay (77), Dunfanaghy Harbour (69) and Gweebarra Bay (54).[100]

The Internationally Important figure (1% of the flyway population) is 8,200.

The Nationally Important figure (1% of the national population) is 610.

Qualifying Sites: Lough Swilly, Donegal Bay

The Oystercatcher is Red-listed in BirdWatch Ireland's 'Birds of Conservation Concern in Ireland 4: 2020–2026' on the grounds that its global conservation status is of concern.[101]

Black-winged Stilt

Himantopus himantopus | Scodalach dubheiteach

RARE VAGRANT

Blank-winged Stilt is a summer migrant from Africa to southern Europe, but one that is spreading slowly northwards. There is a single record of a bird shot on Tory by James Dixon in April 1916 at a small pool near the lighthouse. It was presented to the National Museum by Mrs R.M. Barrington.

Avocet (Pied Avocet)

Recurvirostra avosetta | Abhóiséad

RARE VAGRANT

There are two records of Avocet. The first was a bird shot at Buncrana in mid-October 1911. The second was a record of two birds at Sheep Haven Bay on 22 July 1956 (J. and W. Gentleman). Its continuing spread as a resident species in Great Britain should bring more stray birds to Donegal – but obviously, not yet.

Lapwing (Northern Lapwing)

Vanellus vanellus | Pilibín

RESIDENT AND WINTER VISITOR

DECREASING AS RESIDENT, STABLE IN WINTER

RED-LISTED BOCCI-4

It is not so many years ago that Lapwing bred widely throughout Ireland on marshy ground, permanent pasture with bare patches, and ploughed land. The intensification of agriculture that led to the extermination of the Grey Partridge in Donegal has now reduced Lapwing numbers to a perilous state. The main loss occurred between the period of the first and second breeding bird atlases (1968–72 and 1988–91) when they retreated from the better farmland in the east of the county. That they still survive at all is due to their ability to nest in marshy ground that has not yielded to modern farming techniques, mainly in the northern coastal zone of the county. Recent intensification of farming on Inch Levels, which was already intensive by normal standards, exactly coincides with the decline in Lapwing breeding there (Table 18.2).

Year	Lapwing Territories	Productivity (chicks per nest)
2017	43	1.68
2018	23	0
2019	7	0

Table 18.2: Recent Lapwing breeding success on Inch Levels

Tory remains a good site, with fifteen or more pairs breeding each year. On machair sites, the number of territories rose from 112 in 1985[102] to 150 in 1996,[103] and then dropped in more recent years to 79 in 2017,[104] 90 in 2018,[105] 86 in 2019[106] and 79 in 2020[107] (see Chapter 10). The most recent three totals include fifty, forty-nine and fifty at only two sites (Rinmore Point on the Fanad peninsula, and Magheragallon on the Derrybeg estuary) where Birdwatch Ireland has erected predator-proof fences. So the general decline at the unprotected sites is much greater than the machair survey totals reveal, but it does at least show that predator-proof fences can make a huge difference.

In winter, immigrants from continental Europe and Great Britain come in large numbers, and are widespread around the coast. The recent trend

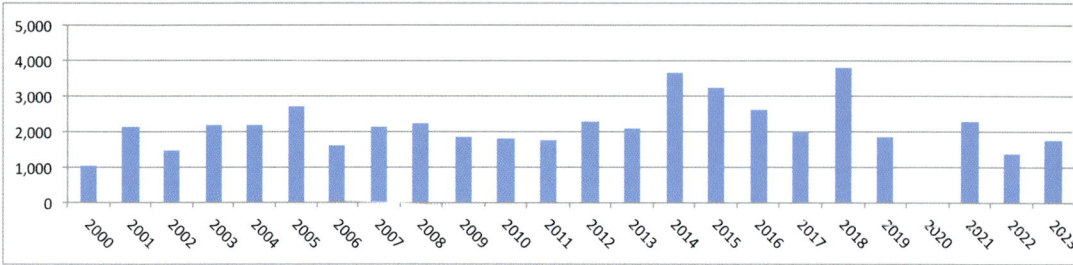

Fig. 18.29: Lapwing annual peak numbers on Lough Swilly, from 2000/1 to 2023/4

(from 2014) in their numbers on Lough Swilly is a bit erratic, but it has been fairly steady over the long term (from 2000), with a peak count of 3,829 on 15 December 2018 (Figure 18.29). Unlike Oystercatchers, which are in similar numbers, about a third of the Lapwings on Lough Swilly are usually found at Inch.

Other good sites are River Foyle (570), Donegal Bay (242), Trawbreaga Bay (460), Ballyness Bay (110) and Fanad North Coast (100).[108]

The Internationally Important figure (1% of the flyway population) is 73,300.

The Nationally Important figure (1% of the national population) is 850.

Qualifying Sites: Lough Swilly

The Lapwing is Red-listed in BirdWatch Ireland's 'Birds of Conservation Concern in Ireland 4: 2020–2026' on the grounds of a 74 per cent long-term and 95 per cent short-term decline in its breeding population, and a 67 per cent long-term and 58 per cent short-term decline in its wintering population.[109]

Golden Plover (European Golden Plover)

Pluvialis apricaria | Feadóg bhuí

RESIDENT AND WINTER VISITOR

DECREASING

RED-LISTED BOCCI-4

There is some evidence that Golden Plover were locally common in the nineteenth century. Writing in the 1890s, Hart said, 'about Killybegs Mr Brooke finds them far less common than they were twenty years ago, though still abundant, winter and summer'.[110] This contrasts with several other contemporary observations of a few birds here and there. Recent surveys estimate the Irish population at only 150 pairs, with all birds now restricted to Donegal, west Mayo and Connemara.[111]

Golden Plovers breed in the hills, usually on large expanses of grassy or heathery moors with short and sparse vegetation. Both the first breeding atlas, from 1968 to 1972,[112] and the Upland Bird Survey of 2002 were episodes of detailed surveying, but the 2002 survey was much more focused than the atlas work, so it is not surprising that it found increased occupancy

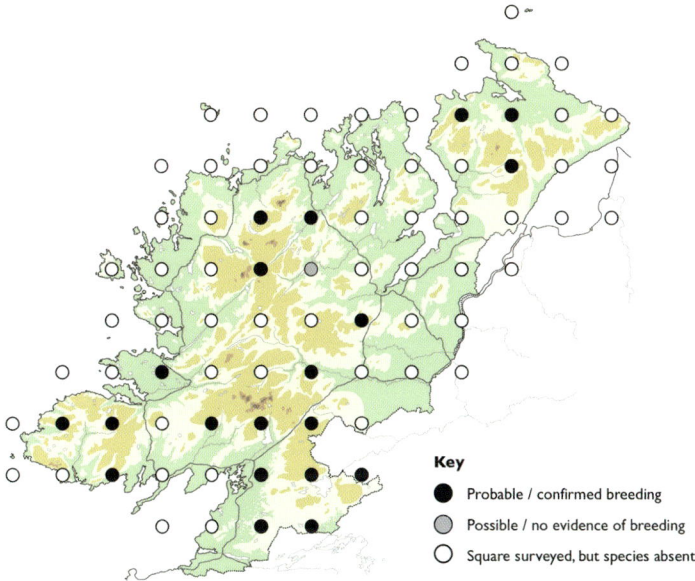

Map 18.2: 10 km² distribution of Golden Plover in Donegal in 1968–72 (top), and in 2002 (bottom)

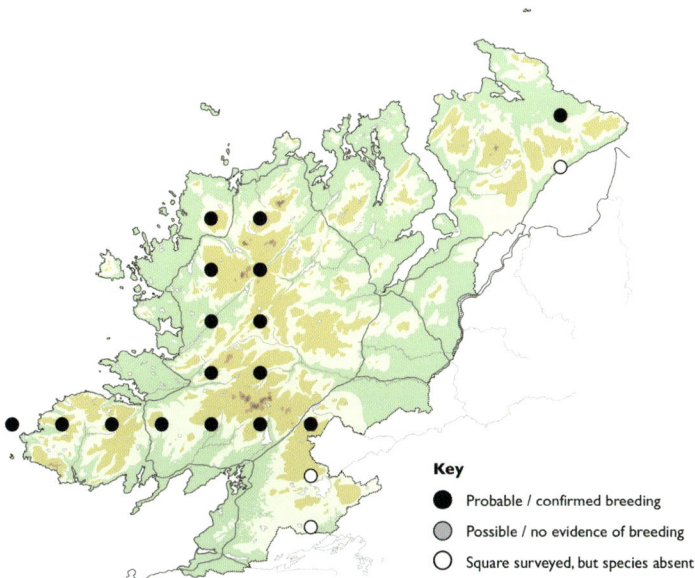

Key

● Probable / confirmed breeding

● Possible / no evidence of breeding

○ Square surveyed, but species absent

Key

● Probable / confirmed breeding

● Possible / no evidence of breeding

○ Square surveyed, but species absent

in the western areas that it covered (Map 18.2). Of more consequence is that it gathered enough information pointing to the absence of birds in the eastern half of the county, apart from a very few in Inishowen, to justify not surveying there. That also included the eastern half of the main north and central uplands. The 2002 final estimate of seventy to ninety pairs in the county is probably as good as was possible.[113]

One of the best areas was the Lough Nillan Bog SAC, which held seventeen pairs. This area was re-surveyed in 2020 over a more limited season, but only two pairs were found.[114] So the decline seems to be continuing.

The decline in density and range is almost certainly attributable to a combination of the following factors:

- Afforestation in the hills has reduced the area of available habitat, and has fragmented much of what is left into areas that are too small for Golden Plover.

- Forest blocks and their roadways provide both cover and access for nest predators, like foxes.

- Wind farms are a recent addition to the landscape. Along with their access roads and power lines, they add to the fragmentation of habitat.

- Fire. Whether ignited spontaneously or by misguided landowners and arsonists, climate change is likely to increase the frequency and severity of hill fires during the breeding seasons. Despite the apparent quick recovery of plant growth, each time burning occurs it is a less diverse vegetation that returns. And along with it, the invertebrate fauna is also reduced.

- Golden Plover in Ireland are at the southern limit of their global range. It is likely that they are naturally retreating towards their northern heartlands.

Large flocks of Golden Plover arrive from the north to winter on our estuaries and sea loughs. Small flocks that used to appear on inland farms seem to have disappeared over the last few decades. The main site is Lough Swilly. Annual totals fluctuate, but five-year means which had been stable at about 2,000 since 2010 have slipped below 1,000 over the last five years for the first time since 2000. Although erratic from year to year, numbers

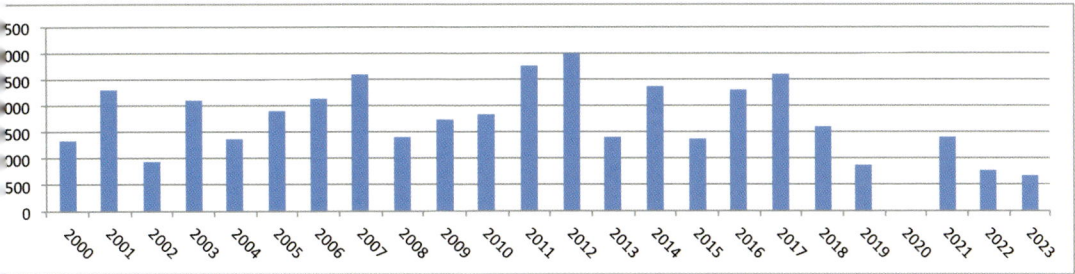

Fig. 18.30: Annual winter peaks of Golden Plover on Lough Swilly, from 2000/1 to 2023/4

had been stable until about 2017 (Figure 18.30). The River Foyle, with similar habitat, is also home to a good wintering flock of about 1,000 birds. Peak counts at the lesser sites are Ballyness Bay (530), Donegal Bay (211) and Trawbreaga Bay (120).[115]

American and European Golden Plovers, Derrybeg.

The Internationally Important figure (1% of the flyway population) is 9,300.

The Nationally Important figure (1% of the national population) is 920.

Qualifying Sites: Lough Swilly, River Foyle

Golden Plover is Red-listed in BirdWatch Ireland's 'Birds of Conservation Concern in Ireland 4: 2020–2026' as a result of the long-term decline in its breeding population of 84 per cent since 1980.[116]

American Golden Plover

Pluvialis dominica | Feadóg bhuí Mheiriceánach

SCARCE VAGRANT

There have been eighteen records of this relatively frequent trans-Atlantic vagrant. The first was at Inch Lough on 8 October 1988 (R. Sheppard). They have been evenly spread around a few preferred sites, with seven birds on Lough Swilly, five on Tory, five at the Derrybeg estuary in Gweedore or thereabouts, and one at Keadew in the Rosses.

One additional bird was accepted as either American or Pacific Golden Plover. It was seen at Rocky Point on 9 March 1997 (J. O'Boyle).

Grey Plover (Black-bellied Plover)
Pluvialis squatarola squatarola | Feadóg ghlas
SCARCE WINTER VISITOR
STABLE
RED-LISTED BOCCI-4

Grey Plover is a common winter visitor to east coast sites and to Great Britain, from the arctic tundra. Individuals turn up unpredictably all around the Donegal coast, typically at small sandy beaches with a good supply of washed-up seaweed, where they would mix with Sanderling, Ringed Plover, Oystercatcher and Rock Pipits. Currently the only reliable site apart from Lough Swilly is Inishfad, on Donegal Bay, where they have a choice of sand or rock, and where up to a dozen are to be found each winter.

Grey Plover can turn up at various places within Lough Swilly, favouring the Leannan estuary in the 1980s, and now the Swilly estuary. Numbers have always been small, and the maxima of fifty-one in 1986/7 and fifty in 1996/7 were exceptional. Such small numbers can easily be overlooked among the huge flocks of other waders, but they have been recorded every year since 2010, so the overall stability during that period is probably genuine (Figure 18.31).

The Grey Plover is Red-listed in BirdWatch Ireland's 'Birds of Conservation Concern in Ireland 4: 2020–2026' on the grounds of a 54 per cent long-term decline in its wintering population.[117]

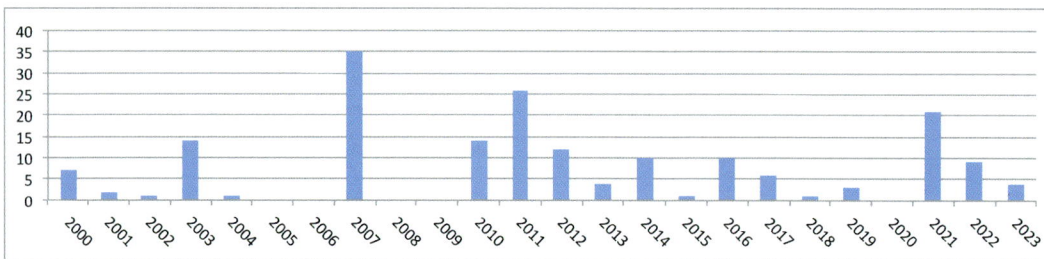

Fig. 18.31: Grey Plover peaks on Lough Swilly, from 2000/1 to 2023/4

Ringed Plover (Common Ringed Plover)

Charadrius hiaticula hiaticula | Feadóg chladaigh

RESIDENT, PASSAGE MIGRANT AND WINTER VISITOR

STABLE

Ringed Plover are common breeding birds around all our sandy and shingly beaches. On machair they are widespread and stable, but the numbers recorded have never been more than five pairs at any one site, with the exception of Rinmore Point, where thirteen pairs were present in 1996.[118]

The winter flocks are thought to be largely from western Europe including southern Scandinavia, and their numbers nationally have been declining since 2008/9. Lough Swilly had a recorded peak of 386 on 15 November 2014, and a five-year mean from 2011/12 to 2015/16 of 228. Donegal Bay had a peak of 237 in November 2015 and a five-year mean of 161.[119] Other sites that have counts exceeding 120 are Trawbreaga Bay with 350 on 11 January 2003, Ballyness Bay with 340 in September 1970, Rossnowlagh with 254 on 16 November 1998, and Fanad North Coast with 138 for the 1994/5 to 2000/1 period.[120] Nationally important sites are limited to data for their winter populations, excluding flocks on migration. It is nonetheless still likely that some of these other sites should also qualify. More frequent counting is probably all that is needed.

The Internationally Important figure (1% of the flyway population) is 540.

The Nationally Important figure (1% of the national population) is 120.

Qualifying Sites: Lough Swilly, Donegal Bay

Subspecies

Ringed Plover (Greenland)

Charadrius hiaticula psammodromus | Feadóg chladaigh (Graonlainne)

PASSAGE MIGRANT

Numbers of Ringed Plover are boosted on migration by smaller, dark-mantled birds, of a separate race *Charadrius h. psammodromus*, intermediate between the other two races. They breed in Iceland, Greenland and the north-eastern Canadian islands, and stop off at Donegal en route to winter quarters along the Atlantic coast of Africa.

Subspecies

Ringed Plover (Northern)

Charadrius hiaticula tundrae | Feadóg chladaigh (nuaisceartach)

RARE VAGRANT

This third race of the Ringed Plover to occur in Donegal breeds eastwards from arctic Scandinavia. It is smaller than the other two, and like the Greenland race is also dark-mantled. Three trapped birds, on 24 April, 31 August and 10 September 1960, were identified on Tory from their wing measurements (TBO). At the time, *Charadrius h. psammodromus* was not recognised as separate from the nominate form (and still not by some authorities). But as *C. h. tundrae* is thought to largely bypass Great Britain on its southward migration, it is not likely to be anything other than a rare vagrant to Ireland. So it is possible that these three Tory birds would now be identified as *psammodromus*.

Semipalmated Plover

Charadrius semipalmatus | Feadóg mionbhosach

RARE VAGRANT

The first Irish record of this North American counterpart of our Ringed Plover was seen at Leabgarrow, Arranmore on 10 October 2003 (R. Mundy). It was a juvenile bird.

Little Ringed Plover

Charadrius dubius curonicus | Feadóigín chladaigh

RARE VAGRANT

This summer visitor from northern Africa to most of Europe has only visited Donegal twice. The first time was at Inch Lough on 3 and 4 July 2021 (D. Brennan), and the second was at the same place on 21 May 2022 (D. Brennan et al.). However, the species is spreading northwards in Great Britain and appears to be in the process of colonising the south of Ireland, so more records in Donegal might be anticipated.

Killdeer

Charadrius vociferus vociferus | Feadóg ghlórach

RARE VAGRANT

Kildeer is another North American plover which rarely reaches Ireland. The only one seen in Donegal was at Killybegs on 29 March 2013 (G. Thomas).

Dotterel (Eurasian Dotterel)

Charadrius morinellus | Amadán móinteach

RARE PASSAGE MIGRANT

The Dotterel has a small breeding presence on Scottish mountain summits, and migrates to winter in North Africa. However, birds moving through Ireland are as likely to belong to the much more abundant Scandinavian population. The first record in Donegal is of a bird 'shot on top of one of the highest mountains' prior to being exhibited on 4 March 1854.[121] This record is in keeping with a known pattern of spring migration via mountain summits in Great Britain and occasionally in Ireland. Another was shot on Donegal Bay, on 29 November 1905.

Since then, a further twenty-five birds have been seen in Donegal – including a remarkable flock of eighteen at Rocky Point from 2 to 3 May 1992 which made the day for George Gordon and John O'Boyle. The remaining seven sightings cover a wide scattering of dates, spanning April, July, September, October and late November. They are all singles, the first being near Horn Head on 27 July 1956, then Ballyness Bay on 23 November 1980, Arranmore on 3 October 1994, Rocky Point on 28 September 2002, Sheskinmore on 29 April 2006, Carrickfinn airport on 11 September 2011 and Malin Head on 4 October 2011. The Sheskinmore bird was recorded as a male.

Whimbrel (Eurasian Whimbrel)

Numenius phaeopus islandicus | Crotach eanaigh

PASSAGE MIGRANT
STABLE

The May Bird, as it is often called, passes through Donegal en route between its taiga zone breeding grounds and the coasts of Africa where it spends

KIM PERIERA

Whimbrel, St John's Point.

the winter. They are most reliably present here in April/May and August/ September. Whimbrel are mostly seen in small parties, with the largest on record being 237 birds at Rossan Point on 29 August 2015. Coastal headlands, like Rossan/Rocky Point, are typical, and flocks have been seen regularly at Bloody Foreland and Malin Head. Tory has a steady trickle of birds passing through. Lough Swilly also has its share – flocks at Blanket Nook can be up to 150. There are fewer than ten winter records, mostly singles, but including twenty birds in Derrybeg from 29 to 31 December 2008.

Curlew (Eurasian Curlew)

Numenius arquata arquata | Crotach

RARE RESIDENT AND COMMON WINTER VISITOR
DECREASING AS A RESIDENT, STABLE IN WINTER
RED-LISTED BOCCI-4

In living memory the bubbling song of the Curlew (surely one of nature's most beautiful sounds) was heard on bogs and hills throughout Ireland. But in the forty years between the first and third breeding atlas periods (1968 to 1972 and 2007 to 2011) it decreased by 78 per cent. By 2019 it was down to about 150 pairs.[122] In Donegal, it was in almost every 10 km² in the earlier survey, apart from the far south-west and north-west corners.[123] By 2019 it was down to five pairs in four locations, from which only two chicks were raised. This was despite the best efforts of BirdWatch Ireland in managing the habitat for those last pairs, and protecting the sites against predators.[124]

Curlews require wide-open spaces of wet grassland and bog. These are almost all now fragmented into smaller units by tracks, peat cuttings, new forests and developments of all sorts. So predators have routes, vantages and safe refuges from which they can penetrate, although in most cases now the Curlews won't even try to nest (see also under Golden Plover).

The plight of Curlews on the breeding grounds is not to be confused with the winter situation. Although the large breeding population across northern Europe is also declining, there are still sufficient numbers to populate our estuaries and coasts in winter. Inland flocks have largely disappeared. As with Golden Plover, this is most likely in response to land drainage and other agricultural changes. The national winter population is also declining. But remarkably, the winter numbers in Lough Swilly have remained stable, albeit with a very recent dip that may well prove to be lasting (Figure 18.32). The annual totals of 1,700 to 2,000 birds reached their highest point of 2,571 on 21 January 2006. Curlew are to be found all around Lough Swilly and on adjacent fields, but curiously, they mostly shun both Inch and Blanket Nook, even for roosting.

Donegal Bay's winter total is equally stable, at 300 to 450, with a peak of 531 in January 2018. The cross-border Lough Foyle as a whole is listed as of national importance, and its Donegal west bank could qualify on its own. The cross-border River Foyle should also qualify if it could be monitored more frequently, though it is doubtful if the Donegal side would qualify on its own.

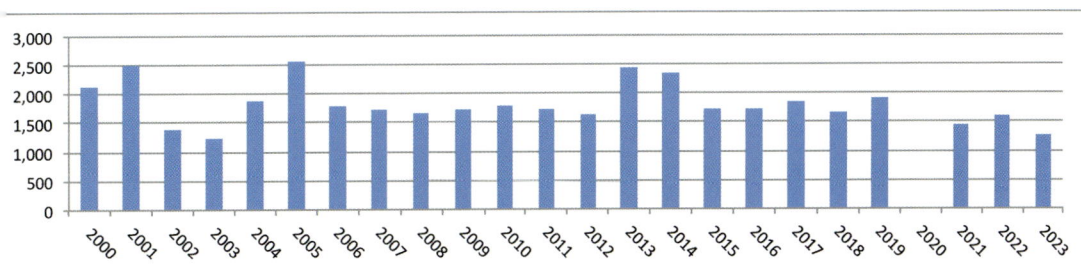

Fig. 18.32: Curlew peaks on Lough Swilly, from 2000/1 to 2023/4

Other good sites, with peaks for the 1994/5 to 2000/1 period, were Trawbreaga Bay (273) and Ballyness Bay (201).[125]

The Internationally Important figure (1% of the flyway population) is 7,600.

The Nationally Important figure (1% of the national population) is 350.

Qualifying Sites: Lough Swilly, Donegal Bay, Lough Foyle

Curlew's breeding population nationally has declined by 86 per cent over the short term, and by 98 per cent over the long term. Its breeding range has declined by 73 per cent short term and 78 per cent long term. Its winter population has declined by 65 per cent over the short term. On all of these counts the Curlew justifies its Red-listing in BirdWatch Ireland's 'Birds of Conservation Concern in Ireland 4: 2020–2026'.[126]

Bar-tailed Godwit

Limosa lapponica lapponica | Guilbneach stríocearrach

WINTER VISITOR

INCREASING

RED-LISTED BOCCI-4

Bar-tailed Godwits breed in the tundra zone from Europe east to Alaska. They have recently become famous for one record-breaking bird which has been logged (twice!) making a non-stop flight from Alaska to New Zealand (eight days to cover the 13,000 km). Fortunately for us, some of the population that breeds in northern Fennoscadia are content to fly no further than Donegal. The wintering flocks are mainly found in sheltered estuaries and bays, but a few can be seen almost anywhere around the coast on muddy and sandy shores.

As a whole, the cross-border Lough Foyle is of International Importance for Bar-tailed Godwit, but most of those birds are on the Northern Ireland side. However, although it is less often surveyed, the County Donegal side could probably qualify as Nationally Important on its own. For example, there were 250 birds at Whitecastle on 6 October 1993, at a time when Lough Swilly held only about 50 to 150 birds. That has now increased to a much higher level (Figure 18.33). In Donegal Bay, they are often found at Mountcharles, and more specifically at Jack's Quay, at the head of the bay where you might have expected Black-tailed Godwits. The mean total for

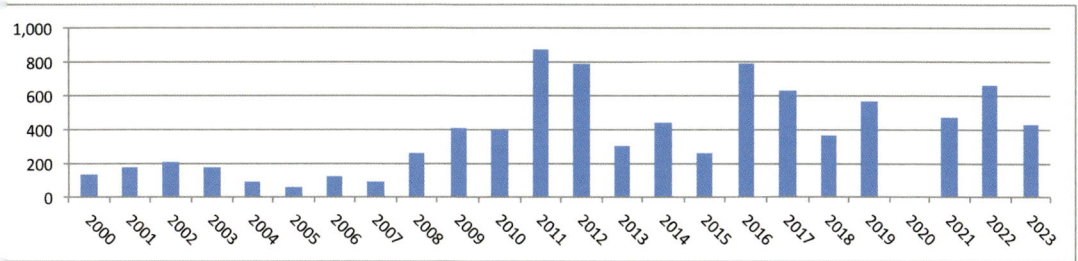

Fig. 18.33: Bar-tailed Godwit peaks on Lough Swilly, from 2000/1 to 2023/4

the whole of Donegal Bay is 69, with the maximum being 259 in January 2011.

Other good sites are Trawbreaga Bay (66), Dunfanaghy Harbour (59) and Ballyness Bay (29).[127]

The Internationally Important figure (1% of the flyway population) is 1,500.

The Nationally Important figure (1% of the national population) is 150.

Qualifying Sites: Lough Swilly

The Bar-tailed Godwit is Red-listed in BirdWatch Ireland's 'Birds of Conservation Concern in Ireland 4: 2020–2026'on the grounds that its global conservation status is of concern.[128]

Black-tailed Godwit

Limosa limosa islandica | Guilbneach earrdhubh

WINTER VISITOR AND PASSAGE MIGRANT
INCREASING
RED-LISTED BOCCI-4

The Black-tailed Godwits wintering mainly in Ireland are from the isolated race that breeds in Iceland. There is no reference to the species having occurred in Donegal by either Ussher and Warren (1900) or Kennedy et al. (1954). Likewise, Hart (1881–92) had nothing to say, but Leebody (1882) did note that it was an occasional visitor to Lough Swilly. The Winter Wetland Survey of 1984/5 to 1986/7 did occasionally record very small numbers there in autumn and winter, to a maximum of thirty, but usually none at all.[129] The period of January-only counts that followed for the seven years to 1993/4 showed that mid-winter stays had not yet become normal.

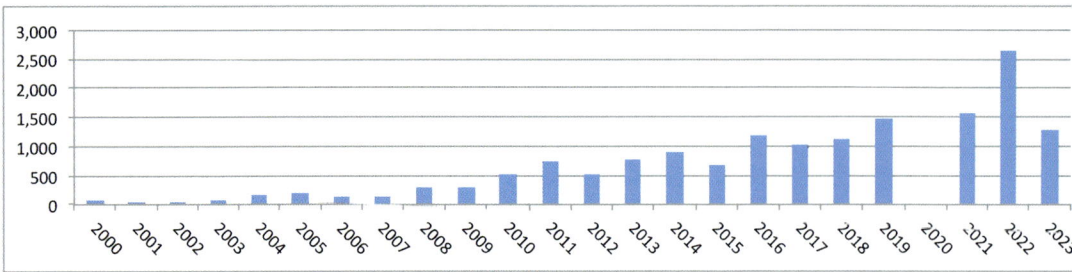

Fig. 18.34: Black-tailed Godwit winter peaks on Lough Swilly, from 2000/1 to 2023/4

They remained erratic until about 2003/4, when numbers approaching 100 birds became normal, and following 2008/9 they expanded steadily (Figure 18.34).

There is an obvious spring migration of birds moving north to Iceland from their main haunts on the south coast of Ireland. Although that starts on the south coast in mid-winter,[130] it does not reach Donegal until April/May, well after winter numbers here have faded away. A single flock of 1,500 birds at Inch Lough as far back as 13 April 2013 contrasts with the peak in that winter season of only 508 birds in the whole of Lough Swilly. Spring numbers comparable to those in 2013 are now normal, and non-breeding flocks also hang on into June and July at Blanket Nook and Inch. The exceptional winter peak in the 2022/3 season (Figure 18.34) was on 19 March 2023, and would be an early spring movement.

The species is rarely seen elsewhere in the county in numbers of any significance, apart from occasional small parties in Donegal Bay and Sheep Haven Bay.

The international and national thresholds for Black-tailed Godwit have been rising with each revision, so that despite the steady growth in numbers on Lough Swilly, they continually hover around, or just below, the current international threshold.

The Internationally Important figure (1% of the flyway population) is 1,100.

The Nationally Important figure (1% of the national population) is 200.

Qualifying Sites: Lough Swilly

The Black-tailed Godwit is Red-listed in BirdWatch Ireland's 'Birds of Conservation Concern in Ireland 4: 2020–2026'on the grounds that its global conservation status is of concern.[131]

Turnstone, St John's Point.

KIM PERIERA

Turnstone (Ruddy Turnstone)

Arenaria interpres interpres | Piardálaí trá

WINTER VISITOR

STABLE

Although Turnstone is officially a winter visitor, a few birds can be seen at any time of the year. But don't be fooled by seeing them in June or July in their full breeding regalia – Donegal is too far from where ringing returns show most Irish Birds breed (in Greenland or arctic Canada) for these birds to make any breeding attempts here.[132]

Turnstones are found all around the coast, on rocky shores and on sandy beaches with a good line of rotting wrack to forage in. For Donegal Bay, the mean count for the 2007 to 2018 period was seventy-four, with a peak count of 155. For Gweebarra Bay the peak count for 1994/5 to 2000/1 was fifty-eight.[133] Small flocks come to Lough Swilly to feed on the mussel beds, where the highest count is 126, on 10 March 2007. The five-year mean peaks are currently at about eighty, having declined from a high point of over 100 a decade ago. There is no estimate of the total size of the much larger coastal population, but it appears to be stable in Donegal and the neighbouring north-west counties.[134]

The Internationally Important figure (1% of the flyway population) is 1,400.

The Nationally Important figure (1% of the national population) is 95.

Qualifying Sites: Donegal Bay

Knot (Red Knot)

Calidris canutus islandica | Cnota

WINTER VISITOR

INCREASING

RED-LISTED BOCCI-4

The subspecies of the circumpolar Knot which winters in Britain and Ireland has its breeding grounds in arctic Canada and Greenland. Ringing returns suggest that they first arrive on the southern North Sea coasts and then move westwards.[135] However, the long series of counts on Lough Swilly, from the Winter Wetland Survey and I-WeBS, doesn't hint at a late arrival.

Numbers were under 500 until 2007/8, when a larger flock seem to have discovered the site. Two years later a high count was recorded – 1,982 birds, on 24 January 2010. Numbers then settled back to between 500 and 1,000 each winter, at which level they were stable – until an extraordinary peak of 2,970 was recorded on 10 December 2023 (Figure 18.35). It remains to be seen if that was a one-off, or as before, an event that heralds a general increase. They are usually to be found on the Swilly estuary as far north as Castlewray, or on the extensive mud around the north of Big Isle.

Lough Swilly is the only site in the county with a flock of any significance. Very small numbers of early autumn migrants can be seen widely around the coasts and estuaries, including a few birds in their red summer plumage (probably failed breeders migrating early).

The Internationally Important figure (1% of the flyway population) is 5,300.

The Nationally Important figure (1% of the national population) is 160.

Qualifying Sites: Lough Swilly

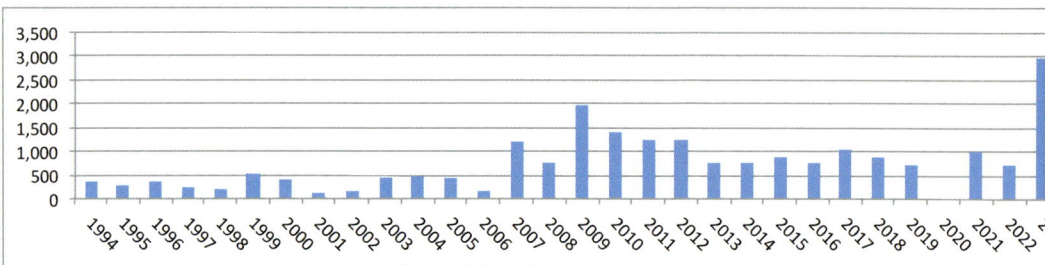

Fig. 18.35: Knot annual peaks on Lough Swilly, from 1994/5 to 2023/4

The Knot is Red-listed in BirdWatch Ireland's 'Birds of Conservation Concern in Ireland 4: 2020–2026' on the grounds that its global conservation status is of concern.[136]

Ruff

Calidris pugnax | Rufachán

SCARCE PASSAGE MIGRANT

INCREASING

Ruff is one of the scarce autumnal migrants from further north in Europe – more reliably encountered than most. Numbers are usually less than a dozen, with Inch Lough and Blanket Nook being the regular haunts – often one or the other, depending mainly on water levels. The combined total for the two sites over the last decade has increased slightly, and the normal peak is now between ten and twenty birds. However, the largest number seen was before that increase, on 31 August 1996, when there were thirty-two birds at an ephemeral grassland pool on the Inch Levels. On that occasion the pool had already persisted for more than a year, creating a new and highly productive bird habitat, which was unfortunately then drained.

The first winter record was of three birds at Inch on 1 December 1999, and five on the 19th at Blanket Nook. In most years now a few (less than

Ruff, Blanket Nook.

DEREK BRENNAN

ten) hang on at Inch through the winter months. Since 2009 a very few started appearing on spring passage, and since 2014 about three or four can be expected each year, including an occasional individual male dressed in his flamboyant summer finery.

Away from Blanket Nook and Inch there have been eight sightings of one or two birds at Thorn, and one on the Leannan estuary. But away from Lough Swilly, Ruff are rarely encountered.

There was a good flow of records from Tory between 1958 and 1964, but few since then, with the most recent being one on 18 August 2012, joined by a second the following day. Elsewhere around the coast there is a record of a male at Sheskinmore on 4 May 1982. There were two on the Erne estuary on 16 August 1996, and in the same year one on north Fanad on 21 September. In 1997 there was one at Rocky Point on 5 September and another at Bundoran golf course on the same day.

Stilt Sandpiper

Calidris himantopus | Gobadán scodalach

RARE VAGRANT

A single Stilt Sandpiper from North America has been recorded in the county. It stayed at Inch Lough from 27 September to 14 October 2018 (T. Campbell et al.).

Curlew Sandpiper

Calidris ferruginea | Gobadán crotaigh

UNCOMMON PASSAGE MIGRANT

STABLE

RED-LISTED BOCCI-4

This is a species that can turn up widely in autumn on open sandy shores and muddy estuaries, en route from the Siberian tundra to sub-Saharan Africa. Appearances are erratic, and they can miss some years completely (see also Little Stint). The total number of individuals recorded is 365 (Figure 18.36), although it could be higher, as the peak number is what is usually recorded when several birds are present over an extended period – if

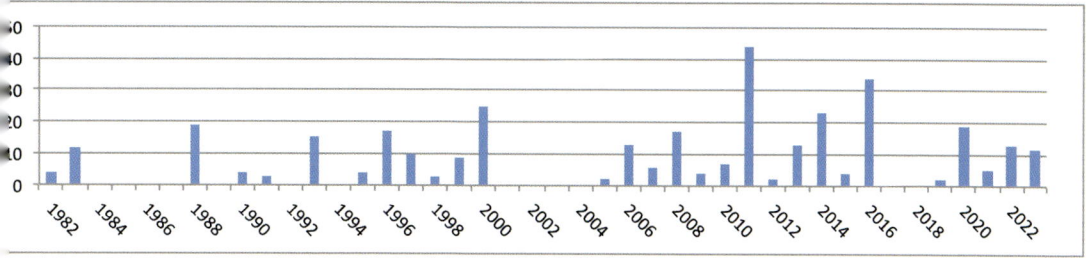

Fig. 18.36: Annual totals of Curlew Sandpiper, from 1982 to 2023

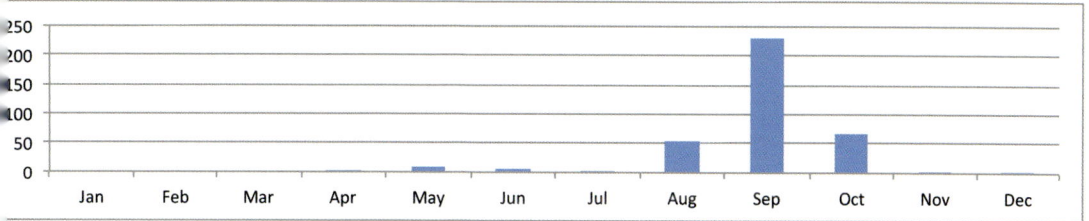

Fig. 18.37: Monthly totals of Curlew Sandpiper

some of these are birds that pass quickly through, the real total would be higher. There are seventeen spring records (April to June) and one each in November and December (Figure 18.37). The most favoured sites are all on Lough Swilly – Blanket Nook with 130, Inch with 79 and Thorn with 67. Thorn could well be the most important site, but access is a problem, so it doesn't get the attention it deserves. Various sites in the north-west, from Dungloe to Bloody Foreland, have had thirty-one birds between them. See also Little Stint.

Curlew Sandpiper is Red-listed in BirdWatch Ireland's 'Birds of Conservation Concern in Ireland 4: 2020–2026' on the grounds that its global conservation status is of concern.[137]

Temminck's Stint

Calidris temminckii | Gobadáinín Temminck

RARE VAGRANT

Although breeding as close as southern Norway (and occasionally Scotland), only one Temminck's Stint has strayed to Donegal on its autumn migration to sub-Saharan Africa. This was at Blanket Nook on 5 October 2007 (D. Charles et al.).

Sanderling (juvenile), Magheraroarty

MOLLY BELL

Sanderling

Calidris alba alba | Luathrán

WINTER VISITOR

STABLE

Sanderlings breed in the very high arctic, and winter on coasts almost globally. Our birds come from Greenland, Svalbard and western Siberia.

A good flock of 100 to 200 birds can usually be found on most of the larger beaches, and the smaller ones might have twenty to fifty. They are here all winter, and the early arrivals in August sometimes include a summer-plumaged adult. The distinctive juveniles soon follow. The highest count on Lough Swilly was 226 birds on 15 November 2014. Recent mean peaks (2011/12 to 2015/16) are 98 on Lough Swilly and 140 on Donegal Bay.[138] But neither of these sites would be among the best. As the county has more sandy beaches than any other, an increase in monitoring would result in many more sites being designated as nationally important (or cause the national threshold to be increased).

The Internationally Important figure (1% of the flyway population) is 2,000.

The Nationally Important figure (1% of the national population) is 85.

Qualifying Sites: Donegal Bay, Lough Swilly

Dunlin

Calidris alpina alpina | Breacóg

WINTER VISITOR AND PASSAGE MIGRANT

STABLE

RED-LISTED BOCCI-4

This is the race comprising most of the wintering Dunlin population. It breeds from eastern Greenland to the tundra of European Russia. These birds are widespread among small flocks of other waders around the coast, and numerous in the estuaries and larger bays. Lough Swilly is a prime site. Almost all the Dunlin are found in huge flocks on the vast expanse of mud around Big Isle and the opposite shore along the Swilly estuary. Their aerial manoeuvres, in the right light, are dazzling, with the whole flock flicking in unison from the dark upperside view to reveal their gleaming white undersides. Numbers peak in mid-winter and fall off fairly quickly. They used to total around 7,000 birds in the 1990s, with the highest count being 9,151 on 14 January 1995, but they dropped in the next decade to an average of 4,000 to 5,000 each winter. However, they stabilised at that level and in the past few years they are back again to 6,000 (Figure 18.38). Smaller flocks make use of all the other parts of the lough, with Leannan estuary, Inch Lough, Blanket Nook and Lisfannan each occasionally holding up to 500 birds.

Other peak counts are Trawbreaga Bay (550), Donegal Bay (515), Ballyness Bay (320), Trawenagh Bay (200).[139]

The Internationally Important figure (1% of the flyway population) is 13,300.

The Nationally Important figure (1% of the national population) is 480.

Qualifying Sites: Lough Swilly, Trawbreaga Bay, Donegal Bay

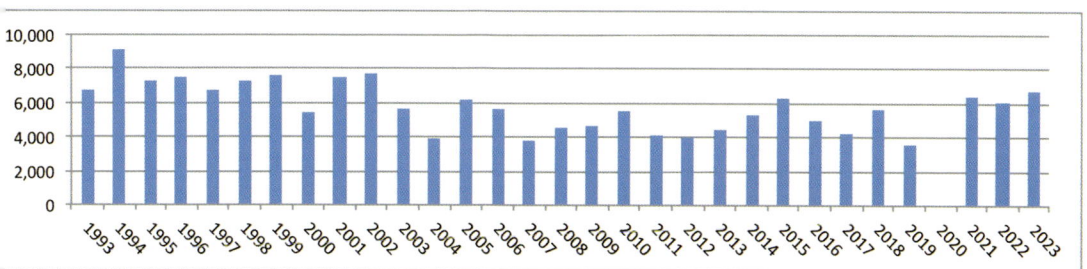

Fig. 18.38: Dunlin totals on Lough Swilly, from 1993/4 to 2023/4

With a short-term decline in its wintering population of 62 per cent, and a long-term decline of 54 per cent, Dunlin is Red-listed in BirdWatch Ireland's 'Birds of Conservation Concern in Ireland 4: 2020–2026'.[140]

Subspecies

Dunlin (Arctic)

Calidris alpina arctica | Breacóg (Artach)

PASSAGE MIGRANT

The high-arctic race of the Dunlin passes through in fairly large numbers – 500 is not unusual in Lough Swilly. They have usually gone by the time the winter visitors of the previous race have arrived. They are fewer on spring passage, but more easily distinguished from other Dunlin in the small parties that move along the coast in breeding plumage.

Subspecies

Dunlin (British)

Calidris alpina schinzii | Breacóg (Bhriotanach)

RARE RESIDENT

DECREASING

RED-LISTED BOCCI-4

This third race is the one that is found here as a breeding species, but it is rare and declining. It uses two quite different habitats – short wet grassland, particularly machair, and areas of wet blanket bog with dystrophic pools.

A recent survey of the upland SPAs designated, in part, to protect breeding waders found only two pairs of Dunlin, but in 2002 there were twenty birds on two sites.[141]

On the coastal machair at Sheskinmore in 1969 there were at least six pairs, and possibly up to twelve. The series of breeding wader surveys on machair sites found seventeen pairs in 1985, twelve in 1996, but none in 2009. It is clear that machair should still support good populations of breeding waders, but unfortunately, both the area and the quality of most sites are steadily declining. Even where good habitat still exists, the management targets may no longer prioritise Dunlin – perhaps partly because botanical goals are more easily achieved.

Dunlin, Tory.

GRACE MEENAN

Dunlin have long been noted in the breeding season on Tory, but with large gaps between observations – displaying in 1954, at least one pair breeding in 1961, and three or more pairs in 1993. Then one chick was raised from two pairs in 2014. Tory is now known as probably the most stable breeding site in the county, with a maximum of seven pairs in 2022, and typical numbers at four to five pairs.[142]

This resident breeding population of Dunlin has declined nationally by 87 per cent over the short term (twenty-five years), and by 93 per cent long term (forty years), so it is Red-listed in BirdWatch Ireland's 'Birds of Conservation Concern in Ireland 4: 2020–2026'.[143]

Purple Sandpiper

Calidris maritima | Gobadán cosbhuí

WINTER VISITOR

STABLE

RED-LISTED BOCCI-4

Purple Sandpipers are visitors from the arctic zones on both sides of the Atlantic, but there is some evidence that most of the Irish birds are from eastern Greenland and some of the Canadian arctic islands.[144] The numbers in Donegal are small, and scattered around rocky shores. The north Fanad coast, and Rinmore Point in particular, normally attracts around twenty to thirty birds but in a good year there can be up to fifty. The highest count was at Ballyhiernan Bay where there were sixty-four on 4 March 2023. Other reliable haunts include Inishfree Bay in the west where seventy were recorded on 11 February 2023. Muckross Head and Fintragh Bay in the south have smaller numbers. Donegal Bay has several locations used by Purple Sandpipers – Kildoney Point, Bundoran, Rossnowlagh and Doorin Point. More often than not, none are recorded, but they may be hidden from view beneath some stretches of cliff. When seen, twenty-five is about what is expected, but the true total for all the Donegal Bay sites is likely to be somewhat greater. The population, although very small, appears to be stable in Donegal and the neighbouring north-west counties, and is a significant proportion of the total national estimate of 640 birds.[145]

Purple Sandpipers have declined by 61 per cent over the short term, and by 56 per cent over the long term, so the species is Red-listed in BirdWatch Ireland's 'Birds of Conservation Concern in Ireland 4: 2020–2026'.[146]

Baird's Sandpiper

Calidris bairdii | Gobadán Baird

RARE VAGRANT

There are six records of this small North American wader that have so far been accepted. The first was seen on Tory on 20 October 1992 (M. O'Donnell

Baird's Sandpiper, Carrigart.

DEREK CHARLES

et al.).The others are a juvenile at Inishfad from 20 to 21 September 2005, one from 14 to 17 September 2011 at Downings/Carrigart, one on 26 October 2013 at Fanad, one from 15 to 16 October 2018 at Inch, and one on 9 September 2020 at Trabeg on the north Fanad coast.

In addition, there are four reports of five birds which have not been submitted to the IRBC, or for which verdicts have not yet been published – one from 17 July 2010, two from 22 to 24 September 2013, one from 7 to 11 September 2016, and one on 20 September 2022 – all at Blanket Nook.

Little Stint

Calidris minuta | Gobadáinín beag

SCARCE PASSAGE MIGRANT
STABLE

Like the Curlew Sandpiper, the Little Stint is an erratic passage migrant, missing some years altogether and coming in very small numbers in others. Although breeding in arctic Europe, the birds coming here are from the main breeding concentrations in western Siberia. These are juveniles, arriving later than European birds, and only in years with good breeding success. The total number recorded is 147, with the average of those years with records being five. The best years were mostly towards the end of the last century – fifteen in 1983, eleven in 1993, twenty in 1995 and thirteen in 2022.

Inch has had fifty-six birds and Blanket Nook forty-eight. All but one of Tory's thirteen birds have been singles, while Trawbreaga Bay's eight birds were all on 19 September 1993. The best year was 1995, when all of the twenty birds were at Inch and Blanket Nook on and around 16 September.

As the western Siberian tundra is also the breeding ground of the Curlew Sandpiper, it may explain why the two species tend to appear here in the same years, when breeding conditions would have been good for both.[147] Although there is an obvious similarity in the pattern of annual occurrence of the two species, it is worth pointing out that prior to 2000, the numbers of both species were broadly similar, but since 2000, Little Stints totals are definitely smaller (Figures 18.36 and 18.39). The seasonal distributions of the two species are almost identical (Figures 18.37 and 18.40).

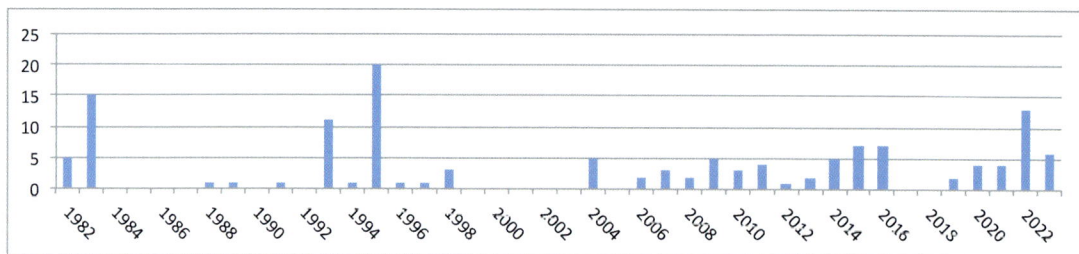

Fig. 18.39: Annual totals of Little Stint, from 1982 to 2023

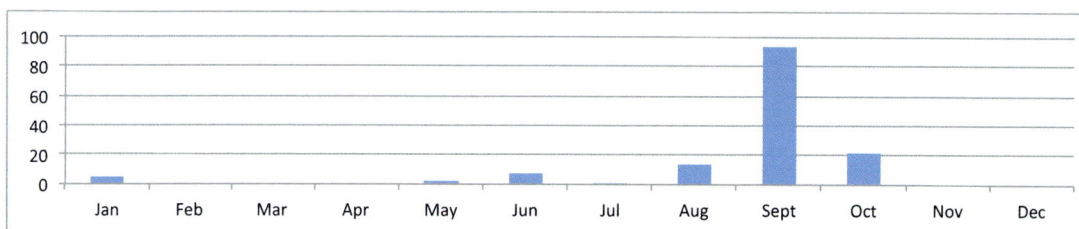

Fig. 18.40: Monthly totals of Little Stint

White-rumped Sandpiper

Calidris fuscicollis | Gobadán bánphrompach

RARE VAGRANT

There are reports of ten White-rumped Sandpipers having reached Donegal from North America, of which only four have so far been verified by IRBC.

The first verified record was of a bird at Malin Head on 27 August 2008 (D. O'Hara). That was followed on 28 September 2011 by one at Inch (C. Ingram). The third and fourth were also at Inch – one on 21 October 2017 (C. Ingram) and the other on 15 October 2018 (R. Vaughan).

Records not yet assessed include the first two reported for Donegal, comprising three sightings between 17 September and 8 October 2000 (A. McGeehan et al.). The circumstances of these sightings have been published in an article in *British Birds*.[148] Other un-assessed records are of a juvenile bird at Blanket Nook on 21 September 2014 (C. Ingram) and two days later at Inch, another at Blanket Nook on 6 September 2020 (G. Mitchell), a bird at Inch on 15 and 16 October 2022 (D. Brennan et al.) and another from 2 to 5 November (D. Brennan et al.).

Buff-breasted Sandpipers, Tory.

PETER PHILLIPS

Buff-breasted Sandpiper

Calidris subruficollis | Gobadán broinn-donnbhuí

SCARCE VAGRANT

This is one of the more distinctive of the autumnal North American waders to have reached our shores, not only in its appearance but also in its preference for short wet grass and maritime heath, rather than tidal mud or lagoons. Twenty-six birds have been seen, the first two of them being on Tory on 8 September 1963, followed by a third on the 13th (TBO). All but two at Inch Lough have been around the north and west coasts. Listing from the north-east, they were at Ballyliffin (1), Glashagh Bay (1), Tory (10), Bloody Foreland (4), Bunbeg (1), Kincaslough (1), Keadew (1), Arranmore (1), Naran (1) and Rocky Point (3). Apart from two birds in August and one in October, all have been seen in September.

Pectoral Sandpiper

Calidris melanotos | Gobadán uchtach

SCARCE VAGRANT

Pectoral Sandpiper is one of the most numerous of the North American sandpipers to reach Donegal. There have been 31 birds in total. The first was shot at Ardara on 21 October 1930, and sent to the National Museum. The next was on Tory on 14 September 1960 (TBO). Most of them have been on Tory (12), followed by Blanket Nook (8), Inch Lough (7), Derrybeg (2) and one each at Ardara and Rossbeg. The earliest autumn arrival was a non-breeding plumage adult at Inch from 11 to 19 July 1984. The most recent bird was at Blanket Nook from 12 to 16 June 2023. There is a fairly wide spread of dates normally spread out from the peak in September. The number of birds in each month have been May (2), June (2), July (4), August (5), September (11), October (5) and November (2).

Pectoral Sandpiper, Tory.

VICTOR CASCHERA

Semipalmated Sandpiper

Calidris pusilla | Gobadáinín mionbhosach

RARE VAGRANT

There have been seven Semipalmated Sandpipers (one of the North American stints) recorded in Donegal, and the first of them was at Malin Head on 25 August 1989. This bird re-appeared at Inch and was seen there from 2 to 4 September (C.A. Brewster et al.). Others were at Glashagh Bay on 29 September 1991, Tory from 14 to 15 September 1996, Tory from 14 to 15 September 2002, Inch Lough from 3 September to 4 October 2011, Derrybeg from 28 September to 5 October 2013, and Inch Lough from 14 to 22 October 2016.

Long-billed Dowitcher

Limnodromus scolopaceus | Guilbnín gobfhada

RARE VAGRANT

There has only been one Long-billed Dowitcher seen in Donegal. That was on Tory, on 5 May 1962 (TBO). This was the first spring record in Britain or Ireland for what is now known to be a relatively frequent vagrant from North America.

Woodcock (Eurasian Woodcock)

Scolopax rusticola | Creabhar

RESIDENT AND WINTER VISITOR

DECREASING

RED-LISTED BOCCI-4

Unique among our waders, Woodcock are birds, as their name reveals, of woodlands rather than wetlands, although they have to feed in soft, usually very damp soil.

Woodcocks' cryptic plumage and secretive habits ensure that they will very rarely be spotted on the ground, so flushing them in winter, and hearing or seeing their breeding display flights, are the best ways to record them.

The first breeding atlas (1968 to 1972) found Woodcock scarce in Inishowen, absent from the far south-west and north-west, but otherwise widespread, mainly in deciduous woodland.[149] It was thought then to be increasing, but by 1988 to 1991 there were only eight 10 km squares occupied, with breeding proven in seven of them.[150] At the time of the third atlas (2007 to 2011) Woodcock was found in seven squares, with no proof of breeding in any of them.[151] This apparent crash is reflected nationally so is likely to be real, but targeted surveying for this secretive species is needed to quantify its true extent. One such survey is ongoing but has yet to report.

In winter, Woodcock are much more numerous, although still far from abundant, and found throughout the county in all sorts of woodland. Incoming migrants from northern Europe are regularly seen on Tory.

The breeding range of Woodcock in Ireland has shrunk by 73 per cent over the long term, since 1980, so it is Red-listed in BirdWatch Ireland's 'Birds of Conservation Concern in Ireland 4: 2020–2026'.[152]

Jack Snipe

Lymnocryptes minimus | Naoscach bhídeach

WINTER VISITOR

Even less likely to break cover than the Woodcock, the Jack Snipe is so rarely seen that its true winter population can only be guessed at. Evidence from Great Britain suggests that it is local, and probably absent from uplands.[153] The shooting fraternity claims more evidence of its presence than birdwatchers, who record only occasional sightings in Donegal. So we can only state that it is probably widespread in winter in very small numbers. Ringing returns show that the population wintering in Britain and Ireland breeds mostly in Fennoscandia. On migration two or three are recorded most years on Tory Island, in September, October, April or May.

Common Snipe

Gallinago gallinago gallinago | Naoscach

RESIDENT AND WINTER VISITOR
STABLE
RED-LISTED BOCCI-4

The Common Snipe breeds widely in Donegal wherever there is marshy ground. Intensification of agriculture, and land drainage in particular, has reduced their numbers in the lowland east of the county.

The number on selected wet machair sites was eighteen pairs in 1985,[154] five in 1996,[155] eight in 2009,[156] twelve in 2017,[157] twenty-three in 2018,[158] twenty in 2019[159] and fourteen in 2020.[160] Fewer sites were occupied by the time of the later surveys, but they included three sites with predator-proof fences which have boosted the number of successful pairs. At Inch Lough, where conditions have not changed significantly, there were eight pairs in 2017, five in 2018, six in 2019 and four in 2020.[161] Our breeding birds are thought to be largely sedentary.

Snipe, Lough Swilly.

RICHARD SMITH

In winter, the numbers are greatly boosted by immigrants from Europe north of the Alps. Flushing of individual birds or small wisps gives little indication of the true numbers present. They can sometimes be seen in the open at favoured sites, like Blanket Nook, but there are no credible estimates of the totals (see also Faroe Snipe, below).

The breeding population of Common Snipe has decreased by 50 per cent over twenty-five years, and by 87 per cent since 1980, so it is Red-listed in BirdWatch Ireland's 'Birds of Conservation Concern in Ireland 4: 2020–2026'.[162]

Subspecies

Faroe Snipe

Gallinago gallinago faeroeensis | Naoscach Fharó

WINTER VISITOR

This race of the Common Snipe breeds mainly in Iceland, but also in the Faroe, Shetland and Orkney Islands. It is thought that most of the population of about half a million birds winters in Ireland.[163] They are not distinguishable in the field, but there are shooting and ringing returns which verify their presence in Donegal.[164] On 14 October 1995 several birds arrived at Glencolumbkille in circumstances that pointed to their identify as Faroe Snipe.[165]

Wilson's Phalarope

Phalaropus tricolor | Falaróp Wilson

RARE VAGRANT

There are two records of this North American wader. The first was at Inch on 1 and 2 December 2009 (D. Allen et al.). The second was an adult female at the Falcarragh machair from 16 to 19 April 2024 (A. Moroney).

There were nineteen Irish records at the last national assessment,[166] so clearly this is a vagrant that manages to target a southern landfall. That might also explain why our winter and spring arrivals don't fit in with the pattern of records nationally. In 1989, Hutchinson charted forty-nine records which were all in the autumn months of August to October,[167] and

Wilson's Phalarope, Falcarragh.

DEREK BRENNAN

of the eighty-four birds listed by the Irish Birding website since 2007, all of them are autumn arrivals – apart from these two Donegal birds.

Red-necked Phalarope

Phalaropus lobatus | Falaróp gobchaol

RARE SUMMER VISITOR AND PASSAGE MIGRANT
RED-LISTED BOCCI-4

In 1900, Ussher and Warren were only aware of a single Irish record of this species.[168] But they did note a report of one bird out of a party of three being shot by Major Sinclair at Drumbeg (Inver Bay) in Donegal in August 1869. C.V. Stoney found a pair on Roaninish in 1916, followed by a breeding pair at the same site in 1924.[169] Breeding took place again in 1927, and probably also in 1929. Stoney found a small marsh with a single pair in 1930, 'but as the marsh had dried up, the birds have not nested there since 1934'.[170] Frank Egginton found two or three pairs nesting every year prior to 1932 in the machair swamp at Dunfanaghy.[171] Although the dates are similar, this is almost certainly a different site from the one found by Stoney.

Ten single birds have been seen since then in the modern era. Two were off Inishtrahull in October 1965, and one off Rossan Point in July 2015. There was an autumn sighting on Tory, on 21 September 2013, and a bird seen from the ferry in Tory Sound on 1 October 2015. The remainder have all been in June, on Tory in 2012, 2016 and 2021, and at Blanket Nook in 2007, 2008 and 2020. Although these are most likely to be spring migrants on their way to Scotland and beyond, and Blanket Nook would not be suitable for breeding anyway, it does remind us that re-colonisation of Donegal by breeding birds is still a possibility, given suitable management and security at potential sites.

Until very recently it was assumed that Icelandic-, Scottish- and Irish-bred Red-necked Phalaropes migrated overland across Europe to the Indian Ocean, where they joined with birds from northern Scandinavia that had crossed Russia. Tagging with geo-locators has now revealed the truly astonishing fact that Scottish birds migrate across the Atlantic, the Caribbean and Central America to winter in the Pacific Ocean off South America.[172] We can safely assume that our tiny Irish population of breeding birds do likewise – more justification, not that it is needed, for making every effort to ensure their survival.

The Red-necked Phalarope is Red-listed in BirdWatch Ireland's 'Birds of Conservation Concern in Ireland 4: 2020–2026' on the grounds of its historical decline.[173]

Grey Phalarope

Phalaropus fulicarius | Falaróp gobmhór

UNCOMMON PASSAGE MIGRANT
STABLE

Grey Phalaropes breed at even higher latitudes than their red-necked relatives, and winter in the south-east Atlantic. Ussher and Warren were aware of eight occurrences in Donegal, including one

Grey Phalarope, Tory.

or more specimens obtained in 1891. That was a year in which small flocks were met with in Donegal and Derry after south-westerly gales – which implies rather more than eight individuals.

In the next half-century, between 1900 and 1952, there was a minimum of twenty-eight birds recorded, and a maximum of eighty-one (depending on how 'birds seen daily' over a period of several days is interpreted). These totals include a bird shot on Tory on 15 November 1911 and one found dead at Dunkineely on 8 November 1952.[174]

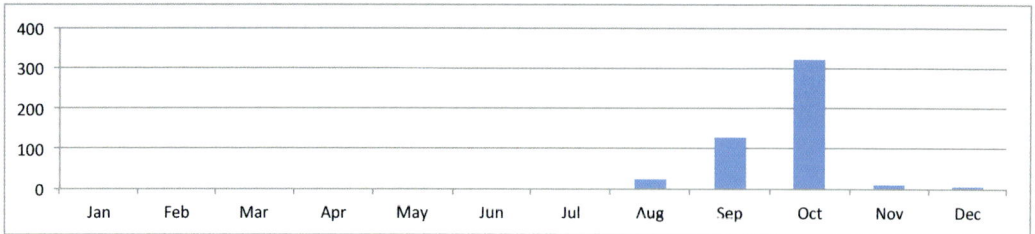

Fig. 18.41: Grey Phalarope monthly totals

There are now records of 494 birds seen since 1952, of which 116 are from Tory and 49 from Malin Head during the time of observatory manning in the early 1960s. That was a period when seawatching was a priority, but even so, it is a very high percentage of the total. The dominance of the more oceanic sites has remained, with Tory, Arranmore and Malin/ Inishtrahull having 331 birds between them, whereas the more inshore but much more heavily watched sites of Fanad, Melmore and Bloody Foreland have only 85. Most birds have been seen in October (Figure 18.41). The only spring record is of a bird at Tory on 14 May 2017. Winter records (ten in November and eight in December) tend to be of birds driven ashore by storms. Five birds first seen in the harbour on Tory on 5 December 1953 stayed for a few days.

There have been eighty-two sighted on the timed seawatches that re-started in the 1990s. They have mostly been in winds between north and south-south-west – at an average of one every ten hours. Many more have been recorded incidentally, and on untimed seawatches. The average number of birds from sixty-four records between 1953 and 1998 was five, whereas it was only two for the fifty-seven records from 2000 to 2020 – perhaps evidence of a decline, but more likely a change in observer bias towards the less-favoured sites (see above). Two birds have been seen on Inch Lough – on 10 October 1988 and 20 October 2017. One was seen about 40 km west of Arranmore Island on 6 August 1988.

Common Sandpiper

Actitis hypoleucos | Gobadán coiteann

SUMMER VISITOR

STABLE

Common Sandpipers are to be found around lake shores and upland rivers. They are absent from the lowland east and scarce in Inishowen. Otherwise,

Common Sandpipers, Mourne Beg River.

LEE DARK

the three breeding atlas projects show that they are widespread. A survey of other species of upland birds found fourteen pairs of Common Sandpiper at twenty-two Ring Ouzel survey sites, and eleven pairs at forty Golden Plover survey sites.[175] As there is no suggestion that Common Sandpipers have any affinity with the habitats of these other species, it is a fair indication of their relative ubiquity in the uplands. A more recent survey of breeding waders at five key upland sites found eleven territories – more than for any other wader species.[176] However, ten of the territories were on one site, so it is possible that the species is no longer as widespread as had been thought.

Common Sandpipers breeding here spend the winter in sub-Saharan Africa. Small numbers turning up on migration on the coast and at wetland sites like Lough Swilly probably include birds dispersing from breeding sites in the local hills.

Spotted Sandpiper

Actitis macularius | Gobadán breac

RARE VAGRANT

There is only one record of this North American relative of the Common Sandpiper. It was seen on Tory, on 5 to 6 October 2013 (T. Murphy et al.).

Green Sandpiper

Tringa ochropus | Gobadán glas

SCARCE PASSAGE MIGRANT

In theory, Donegal is within the wintering zone of Green Sandpipers that breed in northern Europe – in practice, they have never taken up the option. It seems that we are just too far west for regular visits. But migrants en route to or from southern Europe and beyond do call in occasionally. There are three nineteenth-century records. Of forty-four birds since then, four were in the spring months of April and May, and three in June were probably also spring migrants. The rest were autumn migrants in the months of July to October. They tend to turn up more widely than other migrant waders of comparable frequency. Twenty-two were at Blanket Nook, with only nine at Inch and six on Tory. Three were at the coastal sites of Tullagh Point, Arranmore and Durnesh Lough. One was along the tidal Isle Burn about 3 km inland from Big Isle on Lough Swilly. And there were other inland birds. One on 20 July 1988 was well upstream on the River Deele at Ballindrait, one on 21 August 1994 was in the Rosses at Lough Nafullanrany, and one of the spring birds, on 2 April 2002, was at Lough Namafin, closer to the Blue Stack Mountains than the coast.

Lesser Yellowlegs

Tringa flavipes | Mionladhrán buí

RARE VAGRANT

This North American relative of the Redshank has been seen here on only five occasions, but three of those have been in 2023. The first record was on Tory on 8 and 9 September 2007 (J.F. Dowdall et al.), the second at Blanket Nook from 6 to 19 September 2014 (B. Robson et al.). The three most recent records have all been at Inch – from 16 to 17 May (D. Brennan), 16 to 21 September (D. Brennan) and 15 to 17 October 2023 (J. Larkin).

Redshank (Common Redshank)

Tringa totanus totanus | Cosdeargán

VERY SCARCE RESIDENT, COMMON WINTER VISITOR AND PASSAGE MIGRANT
STABLE
RED-LISTED BOCCI-4

The population of Redshanks breeding in Britain and Ireland (which some authorities treat as a separate race) is largely sedentary, wintering on nearby coasts. So our few Donegal breeding birds will be joined by some of the birds nesting in Scotland and northern England. The much larger numbers breeding across northern Europe largely bypass Donegal.

As a breeding species in Donegal the Redshank is hovering on the brink of extinction, but there is no evidence that there has been any decline in recent years. It can be found in wet grasslands, especially where salt marsh is also available for feeding. At the best site, Tory Island, there are about fifteen to twenty pairs.[177] Inch Lough had an average of six pairs breeding between 2017 and 2020. On machair there were twelve pairs in 1985,[178] eight in 1996,[179] one in 2009,[180] three in 2017,[181] three in 2018,[182] five in 2019[183] and five in 2020.[184] The slight upturn in numbers since 2018 is due to the erection of predator-proof fences at Rinmore Point and Magheragallon. There is also the possibility that one or two pairs still breed in the lake-studded zone in south-east Donegal, between Lough Eske, Bundoran and Pettigo. However, this area has not been specifically surveyed, other than

Redshank in breeding plumage, Tory.

PETER PHILLIPS

for the breeding atlases, which recorded breeding on the first two (1968–72 and 1988–91) but not on the final one (2007–11).

The breeding population of Redshank in Ireland has decreased by 50 per cent over twenty-five years, and by 94 per cent since 1980, so it is Red-listed in BirdWatch Ireland's 'Birds of Conservation Concern in Ireland 4: 2020–2026'.[185]

Subspecies

Redshank (Icelandic)

Tringa totanus robusta | Cosdeargán (Íoslannach)

WINTER VISITOR AND PASSAGE MIGRANT
STABLE

The Redshanks we are mostly familiar with are the wintering birds of the discrete race *robusta* which come here in large numbers from Iceland and the Faroe Islands, to mix with unknown numbers of the nominate race from Scotland. It is thought that the majority of the Icelandic birds winter in Ireland.[186]

Redshank is one of the most widespread waders on Lough Swilly, with the outer sections of Rathmullan, Ray, Lisfannan and Ballymoney the most likely to have few birds or none at all. The totals are erratic, which is likely to be because at least one important roost at Thorn is often uncountable. However, the trend, which had been fairly stable and hovered around the Internationally Important threshold, is no longer within reach of that (Figure 18.42). The maximum number recorded was 2,819 on 7 September 2008.

Although being almost ubiquitous around the coast, even the better sites have numbers well below the nationally important threshold. They are River

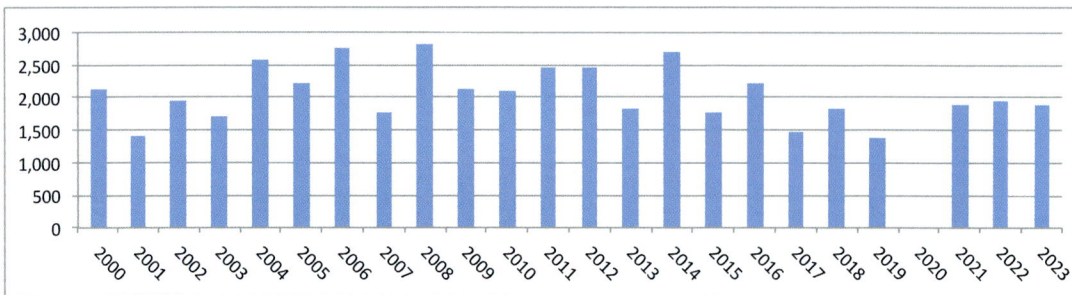

Fig. 18.42: Redshank numbers on Lough Swilly from 2000/1 to 2023/4

Foyle (165), Donegal Bay (159), Trawbreaga Bay (77), Killybegs Harbour (74) and Ballyness Bay (48).[187]

Numbers on Lough Swilly often peak in September. These peaks would include migrants passing south, probably travelling to the south and east coasts of Ireland, with the Bay of Biscay as their southern limit.

The Internationally Important figure (1% of the flyway population) is 2,400.

The Nationally Important figure (1% of the national population) is 240.

Qualifying Sites: Lough Swilly

Wood Sandpiper

Tringa glareola | Gobadán coille

SCARCE PASSAGE MIGRANT

There are nineteen records of Wood Sandpiper in Donegal. The birds that are seen here could breed as close as Scotland, but the bigger population in Fennoscandia and the Baltic states is the more likely source. They winter in sub-Saharan West Africa. The first bird recorded was shot at Brenagh, Upper Lough Swilly on 17 August 1910. From the second record in 1982 sightings remained very infrequent until 2014 when the eighth bird was seen. Since then they have been almost annual, but that has a lot to do with the recent intensity of observation at Inch Lough and Blanket Nook. All birds have been on Lough Swilly, of which eleven were at Inch, six at Blanket Nook and one elsewhere.

Wood Sandpipers are largely early autumn migrants – six birds in July, five in August, three in September and only one in October. One bird first seen on 16 August 1997 remained at Blanket Nook until 12 December. This is a rare occurrence for the species. There have been four spring records – on 20 June 1982, 8 May 2020, and 28 May and 6 to 7 June 2023, all at Inch. The only record with more than one bird was on 27 July 1989, when there were two together at Inch. There were two separate birds in 2021, and also in 2023.

Spotted Redshank

Tringa erythropus | Cosdeargán breac

SCARCE PASSAGE MIGRANT
INCREASING

The population of Spotted Redshank that concerns us breeds in Fennoscandia and north-western Russia, and migrates to sub-Saharan Africa, largely bypassing Ireland apart from the south-east. We have had fifty birds so far in Donegal, so given that the European population has been stable throughout,[188] you would have thought the earliest record would have been a bird shot in the nineteenth century. In fact the first bird was seen on Trawbreaga Bay as recently as 17 August 1959 (W. Finlay), quickly followed by another at the same site on 13 September, and two days later by two at Inch. This was the year which Ruttledge reported as the start of a sharp increase in the size of flocks being seen in south-east Ireland, and only five years before I was recording a passage at Donnybrewer in County Derry, near the shore of Lough Foyle, which over three seasons averaged four birds per day and peaked at eleven. That purple patch lasted only three seasons, until Donnybrewer Lough was drained. Records in north-west Ireland were then few and far between until 2004, since when they have been almost annual, and slowly increasing (Figure 18.43). Six of the fifty Donegal birds have been in winter, six in spring and the rest in autumn – twenty-five were at Blanket Nook, nine at Inch, five on Tory, three at Trawbreaga Bay, and the remainder mainly elsewhere on Lough Swilly.

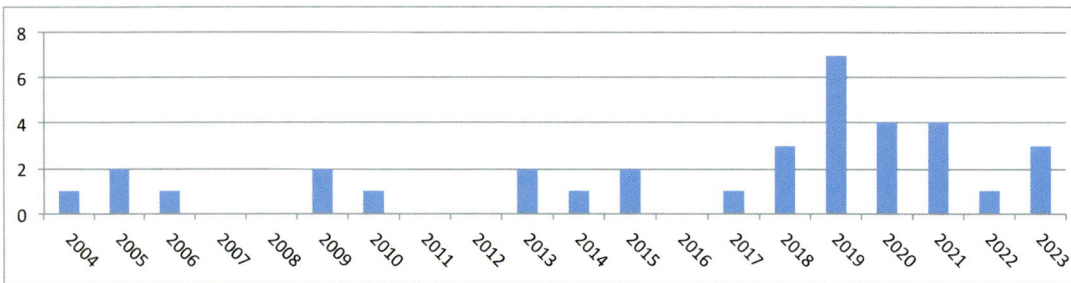

Fig. 18.43: Spotted Redshank totals, from 2004 to 2023

Greenshank

Tringa nebularia | Laidhrín glas

WINTER VISITOR AND PASSAGE MIGRANT

INCREASING

A few Greenshanks among a flock of Redshank will brighten any day, and is a common enough sight around the main estuaries and bays. Migrants arrive early, and the annual peak is usually in August or September. Numbers in Lough Swilly have been increasing, reaching their all-time high of 140 birds on 20 September 2019, and with most years now having a peak of over 100 – up from under 50 at the turn of the century (Figure 18.44). Greenshanks occur all around the lough, but Blanket Nook is undoubtedly the favoured spot. It had an exceptional total of forty-four birds on 17 September 2020. The peak on Donegal Bay is fifty and the mean is twenty-four. No other site is nationally important, although Ballyness Bay comes close, with a peak of nineteen birds.[189]

The Internationally Important figure (1% of the flyway population) is 3,300.

The Nationally Important figure (1% of the national population) is 20.

Qualifying Sites: Lough Swilly, Donegal Bay

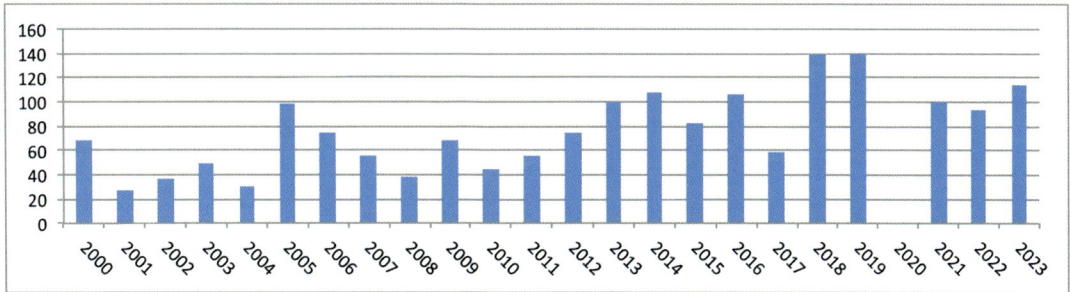

Fig. 18.44: Greenshank totals on Lough Swilly, from 2000/1 to 2023/4

Greater Yellowlegs

Tringa melanoleuca | Ladhrán buí

RARE VAGRANT

There is a single record of this North American counterpart of our Greenshank – the sixth record for Ireland. It was seen on Tory on 5 October 1964 (TBO).

Collared Pratincole

Glareola pratincola pratincola | Pratancól muinceach

RARE VAGRANT

So far, Ireland has had only six of these beautiful tern-like waders, and Donegal is fortunate to have been the destination for the fourth. It was identified as being second calendar-year or older, and was present at Blanket Nook from 22 to 23 July 2019 (R. Vaughan et al.). Collared Pratincoles of this race breed in southern Europe and central Asia, and winter in the Sahel zone of North Africa.

Black-winged Pratincole

Glareola nordmanni | Pratancól dubheiteach

RARE VAGRANT

The Black-winged Pratincole breeds in the Eurasia steppes, from Ukraine eastwards to Kazakhstan, and winters in southern Africa. Its global population has plummeted in recent years, and only two individuals have previously reached Ireland. The third was at Blanket Nook (which is garnering a reputation for pratincoles) and stayed from 14 July to 2 September 2023 – time enough to be seen by a large number of visiting birdwatchers.

Kittiwake (Black-legged Kittiwake)

Rissa tridactyla tridactyla | Saidhbhéar

RESIDENT AND PASSAGE MIGRANT

DECREASING

RED-LISTED BOCCI-4

This common species is not usually to be seen with the everyday gulls, being strictly maritime. It breeds colonially on vertical sea cliffs at places like Horn Head, where in the nineteenth century Ussher and Warren wrote that the cliff was occupied for miles for 200 feet above sea level.[190] Operation Seafarer in 1969–70 found 10,860 pairs of Kittiwakes breeding in the county.[191] This was more than twice that of any other county in Ireland, even when only the identifiable colonies are reported, as in the summary table (Table 18.3). The follow-up surveys also detailed in the table below show a major decline, which is mirrored by the large quantified national declines in Ireland and Great Britain over the past couple of decades, to the point where the species is now Red-listed internationally by IUCN.

Counts at Horn Head need to be treated with caution. Although Kitti-wake is one of the easier seabirds to census, it is still a daunting task where so much of the occupied cliffs can only be counted from the sea. In 1969, the estimate of 7,500 had very wide margins of confidence, at ± 2,500. Watson and Radford did a whole colony count in June 1980 when there were 4,544 nests, with the north-facing sections (two west of the Horn and one east) holding most birds, but they considered that most of the nests would be only countable from the sea.[192] Their monitoring plots were a useful attempt to standardise future counting. A land and sea count in 1987 had 4,246 nests, with the nests counted from land considered minimal.

The next best site after Horn Head in Ussher and Warren's time, and until recently, was Tormore, including adjacent mainland cliffs between Port and Glenlough along with the sea stack. The decline in the fortunes across

Sites (from NE to South)	1954	1969	1980	1987	1993	2000	2005	2015
Kinnagoe		55						
Stookaruddan (area)		35		540	130			
Inishtrahull						58		7
Garvan Isles				251				
Malin Head								170
Horn Head		7,500	4,544	4,256		3,854		1,820
Tory	508	361		591		408		340
Arranmore (two sites)		36				121		161
Tormore (area)		1,090		600		394		213
Slieve League				280				
Tawny		470				530		143
Muckross Head (area)		57		322			50	64
Coolmore		304				372		
Totals for Full Counts		9,908		6,840		5,737		2,918

Table 18.3: Kittiwake colonies (pairs or nests)

all species at this once-teeming site has been most striking. Some of the lesser sites have fared better, but Inishtrahull, which for a while had taken some of the fallout from the collapse of the Stookaruddan and Garvan Isles colonies, had only seventeen nests in 2021.

In winter, Kittiwakes are usually out at sea, but a few will come in with the fishing fleets to harbours like Killybegs. On autumn migration they are a constant feature at all the headlands around the coast, but significant numbers are usually seen late in the season, in October and November. In good conditions they can then pass at rates of 150 to 200 birds per hour, peaking at 302 per hour on 1 November 2021 at Bloody Foreland. An exceptional event was witnessed at inner Donegal Bay, when about 1,000 Kittiwakes were present from 15 to 20 September 1993, feasting on a huge shoal of sprat. They were joined by many local people, who only needed a bucket to join in the harvest.

The Kittiwake is Red-listed in BirdWatch Ireland's 'Birds of Conservation Concern in Ireland 4: 2020–2026' on the grounds that its global conservation status is of concern.[193]

Ivory Gull

Pagophila eburnean | Faoileán eabhartha

RARE VAGRANT

There is a single record of this high-arctic species. It was an immature female shot at Teelin pier on 13 March 1913.[194] If our shores are ever graced by a second bird it will as likely be seen feeding on the carcass of a beached cetacean, as on the Killybegs sea-food smorgasbord.

Sabine's Gull

Xema sabini | Sléibhín Sabine

UNCOMMON PASSAGE MIGRANT
STABLE

Sabine's Gull is a circumpolar species, but any Irish birds are trans-Atlantic migrants from the Canadian arctic archipelago, passing Ireland on their way to winter off South Africa. The first on record was a first-year bird shot at Donegal Bay on 19 September 1878.[195] A second bird was not recorded until 29 September 1963, at Malin Head. It was not until the interest in seawatching revived in the 1980s that sightings of this species started to pick up, and since the 1990s they are now annual.

A total of 106 have been logged on the timed seawatches, which is small compared with the 249 birds recorded in total. Mostly it is just single birds that are seen. Counts above ten include twenty-five at Rocky Point on 11 September 1994, twenty-one at Bloody Foreland on 7 October 2011 and twelve at Melmore Head on 25 September 2007. The annual totals in Figure 18.45 show little evidence of any decline. A graph of national numbers between 1999 and 2008 shows peaks in 2001 and 2007, the second of which is mirrored in Donegal.[196]

Sightings are almost all from August to October (Figure 18.46), with singles in March, April, May and June, two in July and three in November. The only birds not seen on passage were at Blanket Nook, where one stayed from 3 to 6 July 2009 and another appeared on 18 June 2017.

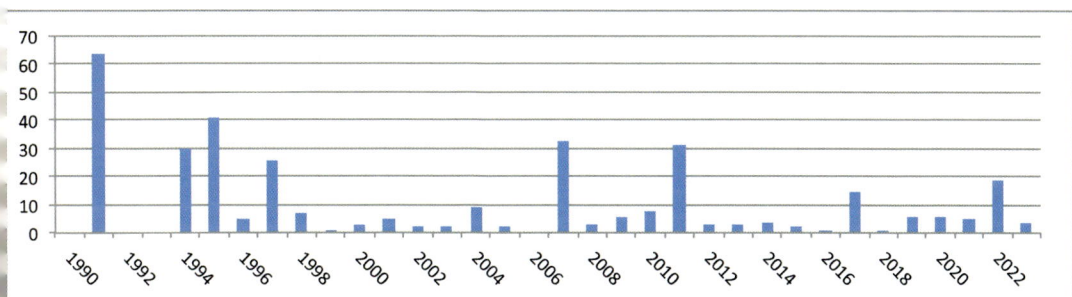

Fig. 18.45: Sabine's Gull annual totals for the last thirty years

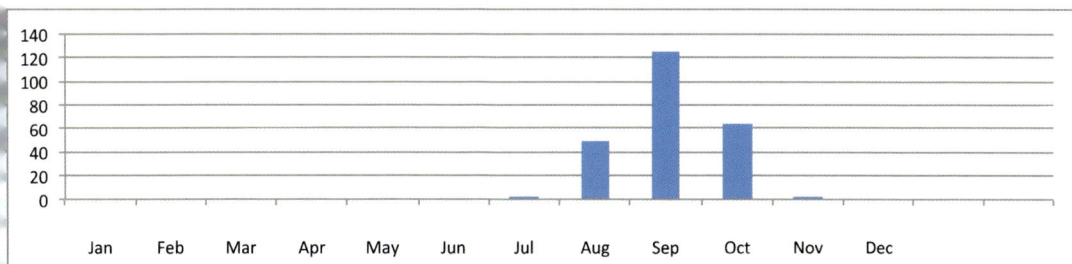

Fig. 18.46: Sabine's Gull monthly totals

Bonaparte's Gull

Chroicocephalus philadelphia | Sléibhín Bonaparte

RARE VAGRANT

This North American equivalent of our Black-headed Gull was first seen at Killybegs from 12 February to 9 March 2018 (R. Bonser et al.).

There was a first-winter bird at Killybegs from 15 to 30 March 2019 (G. Mitchell et al.). The third bird was first calendar-year, again at Killybegs, on 8 December 2019 (D. Charles).

Black-headed Gull

Chroicocephalus ridibundus | Sléibhín

RESIDENT AND WINTER VISITOR

INCREASING

Black-headed Gulls nest colonially, mainly on inland lakes. These have been comprehensively surveyed twice, in 1978 and 1992.[197] The total in 1978 was 1,090 individuals, with significant colonies at Lough Ultan (100), Lough Sessiagh (150) and Lough Kindrum (105). In 1992, the total was 718. On this occasion only three sites were occupied – Lough Golagh (512 individuals, or 256 nests), Sessiagh Lough (206 individuals) and the islets around Roshin Point in Gweebarra Bay (50 pairs). Greer's Island in Mulroy Bay held between 200 and 400 pairs in the same period, although often with no output to speak of due to predation by mink. Carlan Isles in Mulroy Bay appear (from the mainland) to have been re-colonised in 2022, but no visit has yet taken place to establish how many pairs are nesting. A colony on Tory had a peak of thirty-two individuals in 2016.

The best site currently is the small island in Inch Lough, which has had between 300 and 900 occupied nests on most years between 1993 and 2003, rising to 1,450 in 2016. This colony and Lady's Island Lake in County Wexford together now hold half of all the breeding pairs in Ireland.[198]

Our breeding birds are thought to remain within Ireland through the winter, when they are joined by large numbers from northern Europe. The uplands are avoided, but otherwise birds are found feeding throughout the area – around the coast, and in wet grassland or ploughed fields. Not as many are seen following the plough as thirty or forty years ago. The west

Black-headed Gulls (and Whooper Swans) at Inch Levels in flood.

bank of Lough Foyle has the highest peak figure – 3,739 on 15 November 1990. The highest daytime counts on Lough Swilly would normally be 1,000 to 2,000, but flooded fields always bring in dense flocks, with 3,000 at Inch Levels on 17 March 2019 contributing to the record Lough Swilly total of 3,267. River Foyle supports around 500 birds, and there are smaller numbers at other regular coastal or estuarine sites.

Wherever they are in the county, Black-headed Gulls will always travel at dusk to roost communally on the estuaries.

Little Gull

Hydrocoloeus minutes | Sléibhín beag

SCARCE PASSAGE MIGRANT

STABLE

A small population of Little Gulls now winters in the Irish Sea, so the prospect for visitors to Donegal should be improving. However, there is little evidence yet of any change (Figure 18.47). The first known bird recorded

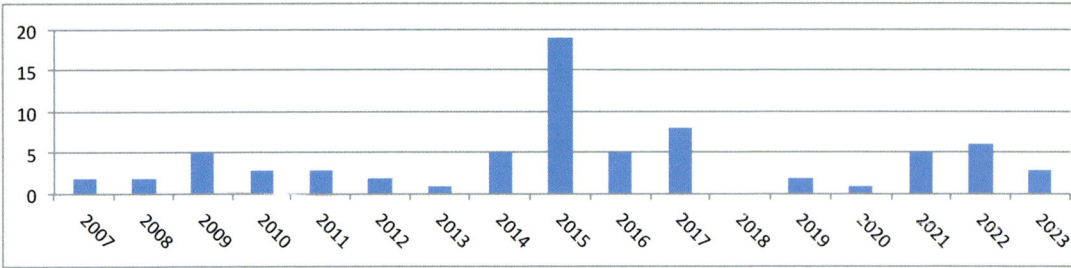

Fig. 18.47: Annual totals of Little Gull, from 2007 to 2023

here was one that was shot on Tory in 1954. There have been eighty-five birds recorded so far. They can be seen at almost any time of year, with only December having no recent records. Of the seventy-three birds seen from 2004 to 2023, autumn passage accounts for thirty-one, but that total includes a single flock of thirteen seen passing Bloody Foreland on 9 August 2015. Spring and summer (April to July) are slightly more popular, with a total of thirty-four birds so far. Only eight have been seen in winter – despite that being the season when most Irish birds are seen.

Many Little Gulls have had an extended stay, and many have been sub-adults. But as they don't breed in significant numbers any closer than Finland and Russia, apart from the occasional pair in Great Britain, there is no suspicion of potential breeding here in the near future.

Ross's Gull

Rhodostethia rosea | Faoileán Ross

RARE VAGRANT

Ross's Gull is a high-arctic species wintering along the edge of the pack-ice. A single bird seen at Killybegs on 3 April 1983 (A. McGeehan, K. Mullarney) was the sixth record for Ireland. With that number having increased to twenty-two Irish records (in 2019), Donegal is certainly due some more sightings.

Laughing Gull

Leucophaeus atricilla | Sléibhín an gháire

RARE VAGRANT

Two or three Laughing Gulls have been seen in Donegal. The first of this North American species was at Bunbeg on 28 May 2006 (J. Larkin). It was a first summer bird, and could have been the same individual that was seen at Dungloe on 7 and 8 June (D. Charles), although these are taken as two separate birds by IRBC. Whether one or two birds, these two sightings belong with a spate of eleven Laughing Gulls which arrived in Ireland with Hurricane Wilma in the autumn of 2005. Another one was seen at Moville on 23 December 2018 (T. Campbell et al.).

Mediterranean Gull

Ichthyaetus melanocephalus | Sléibhín Meánmhuirí

SCARCE VISITOR AND RARE BREEDER

INCREASING

Home base for the vast majority of the confusingly named Mediterranean Gull – is the Black Sea. Over the past fifty years it has slowly spread across Europe, leading in time to the establishment of scattered breeding colonies, including a couple in Ireland.

The first Mediterranean Gull to be seen in Donegal was a second-winter bird at Killybegs on 2 January 1984 (C. Brewster and W. Laird).

From a total of 116 birds, there have been 62 first recorded in winter, 32 in autumn, and 22 spring migrants. Killybegs has had twenty-one birds, all of them singles apart from two sightings of two birds each, and one of three. The numbers were stable for a long time but are now increasing (Figure 18.48).

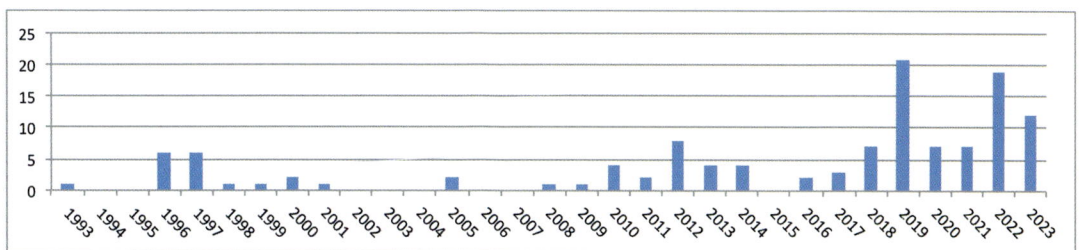

Fig. 18.48: Annual totals of Mediterranean Gull, from 1993 to 2023

Rather than Killybegs, as might be expected, it is Kerrykeel in Mulroy Bay that is shaping up as the venue of choice for a regular flock. The first two were seen there as long ago as 1 September 2000, and since then one or two were seen irregularly until nine appeared on 8 January 2019. This was followed by five on 13 January 2021, and nineteen on 21 February 2022. The peak in the following season was thirty, on 27 September 2022, and on 11 July 2023 there were nine.

The hazards of attempting to arrive at the total number of a scarce species like this are well illustrated by a clearly identified individual – in this case a bird which got its colour-rings as a nestling in The Netherlands on 22 May 1995. It was first seen in Donegal at Bundoran on 9 August 1997, and last seen at Killybegs on 4 April 1998. In between, it was at Bundoran in September, and at Killybegs in all subsequent months except January. Without the rings, it would have been recorded as two birds, at the very least.

In addition to the totals above, two adults first appeared in 2017 at the Black-headed Gull breeding colony on Inch Lough. One adult and at least two other birds were present in 2018. Two adults were paired in 2019, but didn't breed. In 2020, a pair bred, and fledged a single chick. Birds have been present each year since then, but with no further proof of successful breeding.

Common Gull (Mew Gull)

Larus canus canus | Faoileán bán

RESIDENT AND WINTER VISITOR
STABLE

Most Common Gulls are expected to breed inland. In 1977–8 Tony Whilde found eight colonies with a total of 230 birds (equivalent to 115 pairs) in attendance.[199] These were mostly at fairly large lowland lakes. Continuing to divide total birds by two, the number of pairs at other sites were: Gartan Lough (25), Lough Derg (19), Glen Lough (12), Lough Greenan (12), Kindrum Lough (10), Mullaghderg Lough (7). There was a further twenty-eight pairs at smaller sites.

A repeat survey in 1992 found seven colonies with only thirty-three pairs.[200] Lough Fad was the only significant colony, with fifteen pairs, and

Common Gulls, Tory.

PETER PHILLIPS

it is an upland site. Glen Lough had four pairs, Loughs Eske, Fern, Sessiagh and Hanane had three each and Mullaghderg Lough had two.

On coastal sites Common Gulls are less obvious among the hundreds and thousands of other seabirds, but the cumulative total is greater than at the inland colonies. The recent seabird survey spanning 2015 to 2018 has found 537 pairs, with Tory accounting for 406. Of the remainder, the northeast Inishowen sites from Glengad Head to Malin Head, and including the offshore islands, had thirty-two pairs, as had Dunaff Head on its own. Inishdooey had sixteen pairs, Illancrone twelve and Inishbarnog eleven.

Large numbers of birds arrive from northern Europe in autumn, spending the winter around the coast, where they characteristically can be found feeding on the breaking waves along shores churned up by storms. They also feed in fields, often in the company of Black-headed Gulls. The peak count on Lough Swilly is 3,899 on 23 October 2011, but the annual peak is usually under 2,000, and overall shows no significant trend since gulls were first counted there in 1996/7. The Donegal Bay peak count from 2007 to 2017 was 614, with the mean for all counts being 311.

Ring-billed Gull

Larus delawarensis | Faoileán bandghobach

SCARCE VAGRANT

STABLE TREND

Ring-billed Gull is one of the more frequent vagrant species from North America, with forty birds seen so far. A close relative of the Common Gull,

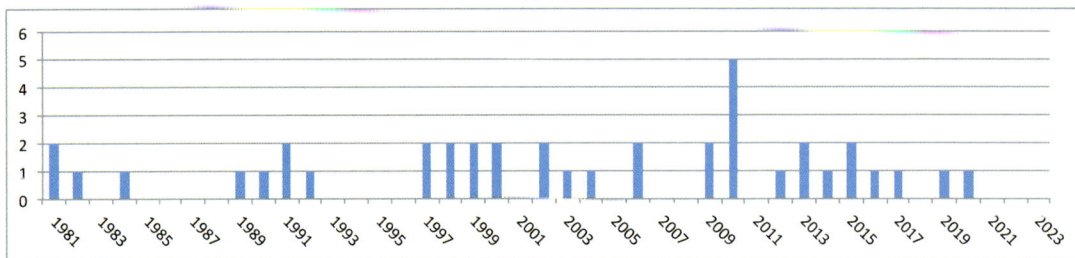

Fig. 18.49: Annual totals of Ring-billed Gull, from 1981 to 2023

it is usually seen among flocks of Common and Black-headed Gulls. The first was an adult in Donegal Town on 14 November 1981 (R. Sheppard), followed quickly by one found dead at Gweebarra Bay on 28 December. That bird had been ringed in the nest the previous summer, on 21 June 1980, at Lake Champlain, New York, USA.

The forty birds have been widely spread around, with nine each seen at Killybegs and Donegal Bay, eight each on Lough Swilly and the west coast, five on the north coast and one on Tory. Six have first appeared in spring/summer, five in autumn, and the remaining twenty-nine birds in winter. Although the trend is stable overall, there is just a hint that the frequency of records may be starting to decline, with no birds recorded since the single bird in January 2020 – which would be in line with the wider national trend (Figure 18.49).

Great Black-backed Gull

Larus marinus | Droimneach mór

RESIDENT AND RARE PASSAGE MIGRANT

STABLE

One or two pairs of Great Black-backed Gulls, probably predatory birds, can be found nesting at most colonies of other seabird species. Where there are multiple pairs, the numbers are usually small. The exceptions are three islands in the south-west of the county where they have fairly large colonies of their own – Roaninish, Rathlin O'Birne and Inishduff. Roaninish had 250 pairs in 1986 but only 29 in 2000 and 58 in 2015–18.[201] Rathlin O'Birne had 52 in 1987, increasing to 145 in 2015–18. Inishduff had 30 pairs in 1961,[202] 60 in 1969–70,[203] and 135 in 1985. In the 2000 survey Inishtrahull had forty pairs.

Overall, the main census periods of 1969–70, 1985–7, 1998–2002 and 2015–18 fail to reveal a consistent trend, although currently the species seems to be on the rise (Table 18.4). The national understanding was that the species had increased greatly in the first half of the last century. At that time, it did not breed on Tory. The decline on Gola is probably connected to

Sites (from NE to South)	1961	1969–70	1981	1984–7	1996	1998–02	2015–18
Inishowen Head		6					
Kinnagoe (area)		6					
Stookaruddan (area)		4					
Inishtrahull		8				40	14
Garvan Isles				6			3
Malin Head		5					7
Dunaff Head		14				2	8
Tormore (Melmore)		3				26	
Horn Head				4		5	3
Inishdooey						8	18
Tory				36		2	8
Inishsirrer						4	
Gola		15				10	
Arranmore (Illanaran)		79				10	1
Roaninish		5		250	150	29	58
Inishbarnog		11					12
Tormore (Port)		5		3		4	3
Rathlin O'Birne				52	75		145
Slieve League		2		10		3	
Tawny				10			
Muckross Head				4			
Inishduff	30	60	150	135			45
Other Sites		100		20			24
Totals for Full Counts		323		530		143	349

Table 18.4: Great Black-backed Gull colonies (pairs)

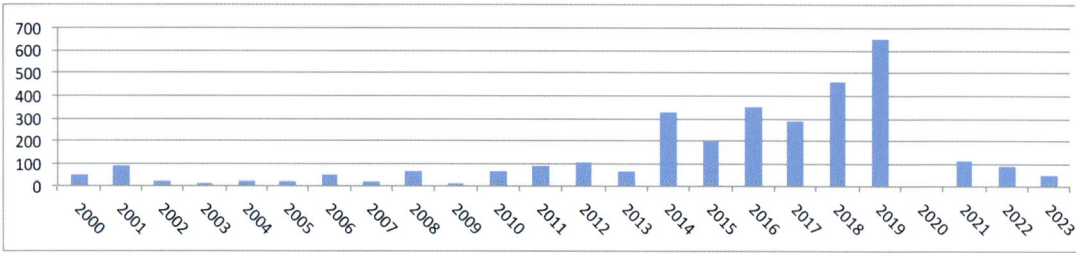

Fig. 18.50: Winter peaks of Great Black-backed Gull on Lough Swilly, from 2000/1 to 2023/4

the re-settlement of the island by humans. The declines at Arranmore and Inishduff are harder to explain.

Winter numbers on Lough Swilly have generally remained below 100. The peak count of 653 in October 2019 may have been related to a glut of fish, coinciding with record numbers of both common and grey seals. However, it came at the end of a surge in numbers between 2014/15 and 2019/20 which was more likely a response to increased feeding opportunities from the dredging of Pacific oysters (Figure 18.50). The west bank of Lough Foyle has had 300 to 500 birds on a number of occasions in the 1990s, in association with the Culmore landfill site (now closed) just across the border. At Killybegs the Great Black-backed Gulls tend not to join the Herring Gulls at the mass assaults on loading sea-food lorries. They are rarely counted there, but 200 on 13 February 1988 and 415 on 10 February 2002 give some idea of the numbers that can be expected. River Foyle has had a peak of 138 and Trawbreaga Bay 81 in 1993/4 to 2000/1.

Migration has been noted at Malin Head, but there is little evidence of significant immigration in winter.

Birds occasionally gather at unexpected places. There were 150 at Tullagh Bay on 6 January 1995, 100 on Glashedy Island on 16 November 2002, and Tory had 50 daily in September 1999.

Glaucous Gull

Larus hyperboreus hyperboreus | Faoileán glas

UNCOMMON WINTER VISITOR

DECREASING

Most Glaucous Gulls don't stray in winter too far south from their arctic breeding grounds, but small numbers just about reach our latitude in

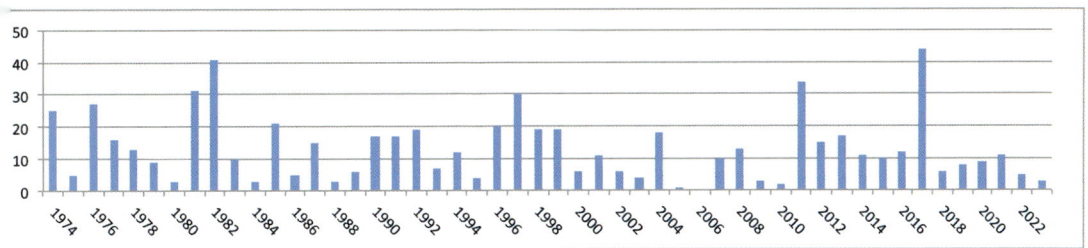

Fig. 18.51: Winter totals of Glaucous Gull, from 1974/5 to 2023/4

Glaucous Gull (second-winter) with Herring Gulls, Killybegs.

search of easy pickings. They are mainly juvenile birds from the Western Palaearctic population of the nominate race. Ussher and Warren had records of seven nineteenth-century Donegal records, out of the national total of sixty-five.[204] In 1954, Kennedy et al. said that the Glaucous Gull appeared to be of regular winter occurrence on the Donegal coast, but gave no further details. There were only eight birds in the twenty years following that publication, until the logjam was broken on 9 February 1975 with the discovery of fourteen birds at Killybegs. The winter totals (Figure 18.51) are from August to July, with most birds occurring between December and early April. The seasonal peaks at Killybegs, which are most frequently in February, are taken as winter totals – although a throughput of short-staying birds, as well as the obvious build-up of long-staying birds, means that these peak counts are under-estimates of the true total. Figure 18.51 shows that the overall trend is a slow decline – probably due in part to changes in the suitability of Killybegs, and also to climate change.

The total estimated for the county is 670 – almost all since 1974. Of the 119 individual birds recorded in the last ten years, Killybegs has had 65, Tory 9, Magheraroarty 6 and Lough Swilly 5. A long list of coastal sites have had one or two birds each – especially harbours like Rathmullan and Burtonport. Quite a few birds have lingered into May. Beyond that, a summer bird was present at Magheraroarty on 9 June 2012, and in the same year there was another at Trawbreaga on 24 June. Inishbofin had one on 6 July 2018, and one was at Inch from 4 to 5 June 2021.

Iceland Gull

Larus glaucoides glaucoides | Faoileán Íoslannach

UNCOMMON WINTER VISITOR

DECREASING

Having its breeding grounds in Greenland, the Iceland Gull is almost as badly named as the Mediterranean Gull. But it does at least winter commonly in Iceland, with smaller numbers of immature birds straying further south, like the Glaucous Gull. The early history in Donegal of both species is very similar. Ussher and Warren recorded eleven Iceland Gulls in Donegal out of the nineteenth-century national total of sixty-eight.[205] The next publication, in 1954, doesn't enumerate the records for Donegal, confining itself to a comment about Donegal having had its quota.[206]

From then until the early 1990s Iceland Gull was scarcer here than Glaucous Gull, but after a decade of parity, it has been the more numerous. Since the turn of the millennium it has also had several exceptional influxes, in 2001/2, 2011/12 and 2017/18 (Figure 18.52), the last two of those being in the same years as Glaucous Gull influxes. While the long-term trend has been upwards, there has been a very slight downturn in recent

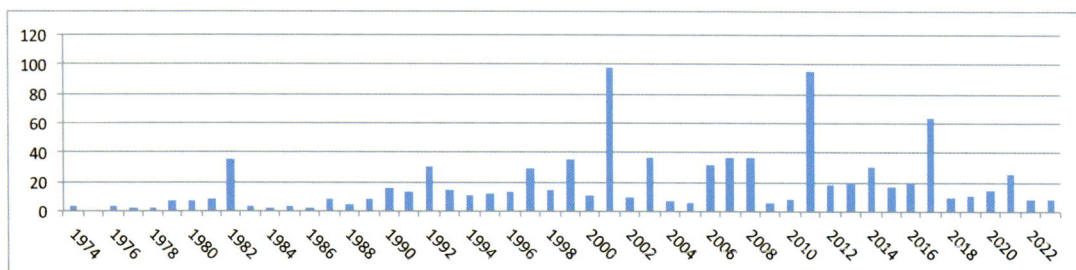

Fig. 18.52: Winter totals of Iceland Gull, from 1974/5 to 2023/4

Iceland Gull, Donegal Town.

years, probably due to changes in the management of Killybegs Harbour, and to climate change – as with the Glaucous Gull.

The total number of birds on record is approximately 873. As with the Glaucous Gull, tracking records of individual birds at Killybegs would be impossible, so the seasonal peak is used instead, which must understate the true number to some degree. Of the 221 individual birds seen in the last ten years, 142 were at Killybegs, 14 each at Tory and Malin Head, and 10 at Greencastle/Moville. Many other sites can claim one or more birds. Only four birds have stayed as long as June – one at Ramelton on 30 June 2000, one on Tory/Magheraroarty from 1 to 24 June 2013 and another on 12 June 2016, and a strikingly white bird stayed at Magheraroarty from 17 April 2022 to at least 30 August. There are only three inland records. The first of these was a bird following the plough on my own farm near Convoy on 17 April 1979, the second was at Lough Shivnagh on 6 January 2012, and in the same year there was another bird at Carndonagh on 10 March.

Subspecies

Kumlien's Gull

Larus glaucoides kumlieni | Faoileán Kumlien

SCARCE WINTER VISITOR

This race of the Iceland Gull breeds to the west of Greenland, in southern Baffin Island. The first to be identified in Donegal (the fourth in Ireland) was at Killybegs from 30 January to 12 February 1983 – a year when there was a major influx of both Iceland Gulls of the nominate race, and Glaucous Gulls. The second for Donegal was from 16 to 23 February 1991, followed by two in February 1995. From 2003/4 they have been of annual occurrence (Figure 18.53). Only one bird was seen in Ireland in 2010, and

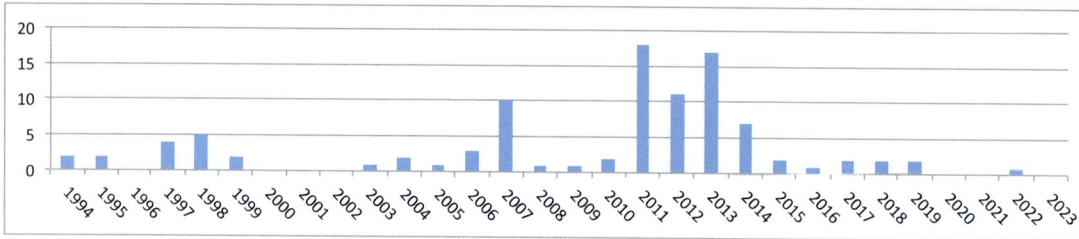

Fig. 18.53: Kumlien's Gull annual totals, from 1993/4 to 2023/4

Kumlien's Gull,
Killybegs.

DEREK CHARLES

that was at Killybegs in February. There was a boom in the years from 2011/12 to 2014/15, coinciding with a series of good years for both Iceland Gulls of the nominate race and Glaucous Gulls, after which they have again become very scarce. The total number of birds recorded is 101. Killybegs, as would be expected, has the lion's share, with up to fourteen birds in a single season. Most of the remainder have been cumulative totals of two or three birds each in harbours like Greencastle, Moville and Burtonport.

Subspecies
Thayer's Gull

Larus glaucoides thayeri | Faoileán Thayer

RARE VAGRANT

Thayer's Gull is a controversial taxon. Some regard it as a full species, and others as a race of American Herring Gull. Most authorities, including the IOC, treat it as a race of the Iceland Gull. It breeds in the high-arctic

Thayer's Gull, Killybegs.

DEREK CHARLES

Canadian archipelago to the north and west of Kumlien's Gull, and seems to be linked to the nominate Iceland Gulls in Greenland via the Kumlien's Gulls in Baffin Island.

The first to be seen in Donegal (the third in Ireland) was at Killybegs from 22 February to 15 March 1998 (A. McGeehan and R. Millington).[207] There have been two others since then, also at Killybegs – a first-winter bird from 2 February to 11 March 2003 (D. Hunter and R. Steele), and an adult on 16 December 2013 (D. Charles).

Herring Gull

Larus argentatus argenteus | Faoileán scadán

COMMON RESIDENT AND WINTER VISITOR
DECREASING AS RESIDENT, INCREASING IN WINTER

Our familiar 'seagull' is a race of the widespread Herring Gull resident in Britain and Ireland. It breeds on headlands and offshore islands. The estimates of breeding pairs show a serious collapse in numbers, from 5,380 in 1969–70 (Operation Seafarer)[208] to 2,068 in 1986/7 (Seabird Colony Register),[209] and finally to 778 in 1999/2000 (Seabird 2000).[210] A fuller survey by NPWS which came up with a total of 1,413 pairs for the 2015 to 2018 period gives little comfort (Table 18.5). The crash has been linked to the disease botulism, which is easily contracted at the landfill sites where Herring Gulls fed in winter at that time – it has also been linked to the

Sites (from NE to South)	1954	1965	1969	1984	1987	2000	2015–18
Inishowen Head			110				
Kinnagoe			100				
Stookaruddan (area)		300	90				
Inishtrahull		3,000+	1,700			20	6
Garvan Isles					531		
Malin Head			280				21
Glashedy Island							20
Dunaff Head			311			20	59
Saldanha Head							11
Tormore (Melmore)					56	73	
Horn Head			500		72		20
Inishdooey							60
Tory	30						15
Umfin							43
Gola			103			65	
Owey							18
Mullaghderg Lough							18
Arranmore							53
Inishkeeragh							26
Illancrone							155
Roaninish			500		250	278	385
O'Boyle's Island				100			
Inishbarnog			100				57
Tormore (Port)			205		39	51	47
Rathlin O'Birne					505		246
Slieve League			185		100		
Tawny					200	83	
Muckross Head					110	7	
Inishduff			300		200		88
Inishgoosk (Lough Derg)			100				
Other Sites			595			182	65
Totals for Full Counts			5,179		2,063	779	1,413

Table 18.5: Herring Gull colonies (individuals)

closure of the landfill sites, reducing the winter survival rate of the birds. While there is no doubt that the numbers have collapsed, the scale of the decline is less certain. Some of the best sites were counted only once, and the 1987 survey counted many of the sites in July, which is not ideal. The huge colony at Inishtrahull was undoubtedly the largest in the county, and its collapse was far beyond what was generally experienced at the smaller colonies.

Numbers in winter are slightly boosted by birds from Great Britain. They can be seen all around the coast, especially at ports and harbours. The record count on Lough Swilly was 2,569 birds on 20 September 2019, and the five-year running mean there has increased from 284 in 2000/1 to 1,224 in 2019/20. Less accurate estimates have been attempted at Killybegs with a figure of 2,000 birds being estimated in 1997, 1,000 in 2000 (considered then a low total for the site), and 3,000 in 2009. Numbers often increase in advance of a storm, which also prompts the return of fishing boats – almost certainly a bigger attraction to gulls than the shelter. The Lough Foyle west bank had a mean of 1,181 from 1996/7 to 2000/1, which would have been influenced by the numbers in attendance at the Culmore landfill site, just across the border. That facility has now closed, so the mean for the west bank dropped to 511 from 2011/12 to 2015/16. Other good sites include Donegal Bay with a peak of 753 for the period 2011 to 2015.

Subspecies
Herring Gull (Scandinavian)
Larus argentatus argentatus | Faoileán scadán (Lochlannach)
SCARCE WINTER VISITOR

The nominate race of Herring Gull breeds in Scandinavia, and winters mainly around the North Sea coasts. Small numbers penetrate west to Ireland, but there is no estimate of how many. The first to be identified in Donegal was on 21 February 1997 at Killybegs (A. McGeehan and M. Garner). A total of sixty-one have now been recorded. Three of these were on Lough Foyle in 1999, as part of a flock of gulls feeding across the border at the Culmore landfill site. One was on Arranmore in 2015, and the rest have all been at Killybegs. The sixty-one birds were spread over twenty-one sightings with the maximum being ten on 20 November 2005. It seems highly likely that the race is under-recorded.

American Herring Gull

Larus smithsonianus | Faoileán scadán Mheiriceánach

RARE VAGRANT

The first of this North American species to be seen in Donegal were two first-winter birds from 3 March to 16 April 1990 (A. McGeehan and S. McKee). Only one had been previously recorded in Ireland, in 1986, but 1990 produced several. At that time the species was still regarded as just a distinctive race of the Herring Gull. It was not until 2002 that it was upgraded to a full species, and as a result is now more sharply on the radar of birdwatchers. The trajectory of these trans-Atlantic vagrants favours more southerly landings than we expect of the gulls from arctic breeding grounds – for which Donegal is better placed. So far only twelve birds have been recorded here, out of 102 for the country as a whole. However, there are a significant number of extra sightings claimed by expert observers visiting Killybegs, which do not appear to have been submitted to IRBC. Details of these are not available, but it is likely that they would more or less double the Donegal total – if submitted and verified.

Eleven of the confirmed sightings have been at Killybegs. As well as the two birds in 1990, there were two in 1998, one in 1999, three in 2004, two in 2007 and one in 2014. The other bird was at Burtonport in 2004. As with most of the rare gulls, they were late-season visitors – only one bird (in December) did not arrive in the mid-February to early April period.

Caspian Gull

Larus cachinnans | Faoileán Chaispeach

RARE VAGRANT

A third-winter bird was seen at Killybegs from 13 to 14 March 1998 (M. Golley and R. Millington). There are no other verified records of this species, which was only recently separated from the Yellow-legged Gull. It hails from the region of its name, but is spreading westwards and now has small, scattered colonies in eastern Europe. These colonies are presumably the source of the birds that are increasingly turning up in western Europe.

Yellow-legged Gull

Larus michahellis michahellis | Faoileán cosbhuí

SCARCE WINTER VISITOR

Like the American Herring Gull, this bird was regarded until fairly recently as a subspecies of Herring Gull, but recognition of its full species status by the various authorities was more staggered. So in 1995 the IRBC decided to adopt a neutral stance on its taxonomy. This remained the Irish position until 2007, when the UK's BOU was added to the list of authorities that had already upgraded it to a full species – and we fell in line. Yellow-legged Gull was first noted in Donegal at Killybegs on 6 December 1998, during that limbo period between 1995 and 2007 (D. Hunter). There have now been twenty-five birds recorded. All are of the nominate race from the Mediterranean region, although one bird at Killybegs from 20 to 22 February 2002 was suspected as being of the race known as Atlantic Gull *L. m. atlantis*, from the Macaronesian islands. The most recent record is of two birds, an adult and an immature, at Killybegs on 15 March 2016. Most birds were seen at Killybegs, but ten were elsewhere around the coast, including one at Blanket Nook. Since the first record, birds have arrived at a rate of one every three or four years, apart from the sixteen birds recorded in the eight years from 2008/9 to 2015/16.

Slaty-backed Gull

Larus schistisagus | Faoileán droim-shlinnliath

RARE VAGRANT

Slaty-backed Gull, Killybegs.

DEREK CHARLES

265

This is one of the most improbable vagrant gulls to have been seen in Donegal, having its breeding and wintering ranges in the western Pacific (Japan to south-west Siberia). A single bird reached Killybegs on 17 February 2015 (M. Callaghan, D. Charles et al.). It was the second Irish record. It could have come via North America, where it is a vagrant, or from the east, as suggested by several records from eastern Europe.

Lesser Black-backed Gull

Larus fuscus | Droimneach beag

SUMMER AND WINTER VISITOR

INCREASING

In the nineteenth century there was doubt that the Lesser Black-backed Gull bred at all in Donegal,[211] and the only record from the first half of the twentieth century is of less than ten pairs on Roaninish in 1948. But that half-century was a time of general increase in Ireland,[212] and in the following half-century the species became well established in the county. Coastal breeders seem to be increasing their hold, but there have been few recent counts, and no consistent monitoring of previously known sites (Table 18.6).

There is now one big inland colony, on Inishgoosk in Lough Derg, which should also be surveyed more frequently.[213] The count of 110 individual birds on 12 March 2003 may not give much idea of what the total number of pairs might eventually have been – work on the colony at Walney Island in Cumbria showed that only half of the breeding Lesser Black-backed Gulls have returned by March, when all the Herring Gulls have already returned.[214]

Migrants arrive here in March and April and can be seen far inland on grass fields, or following the plough.

Until recently, Lesser Black-backed Gull remained as only a summer visitor to Donegal, from wintering in Mediterranean latitudes to breeding in north-west Europe. From the 1990s, the wintering habit has spread north through Ireland, but it only reached Donegal a decade later. The first two wintering individuals were seen on Lough Swilly in 1996/7. The species was not reliably present there until 2005/6, and they finally started to take off in 2013/14. Totals are generally below 100, apart from the exceptional total

Sites (from NE to South)	1948	1969	1978	1992	1993	2000	2003	2006	2015–18
Inishtrahull		7				55			20
Malin Head									14
Dunaff Head									14
Croaghaturr (Rossguill)								2	
Inishdooey						11			20
Umfin					3				
Gola		6			10	30			
Arranmore		8				2			
Inishkeeragh		1							2
Illancrone									29
Roaninish	10								
Rathlin O'Birne		10							11
Inishgoosk (Lough Derg)			900i	132 800i		500	110i		
Other Sites									26
Totals for Full Counts						598			136

Table 18.6: Lesser Black-backed Gull colonies' nests (i denotes individual birds)

of 519 counted in 2019/20 (Figure 18.54). The parallel with a temporary increase in Great Black-backed Gull numbers (Figure 18.50) reinforces the evidence that dredging for Pacific oysters is involved as a direct, or indirect, source of extra food for both species.

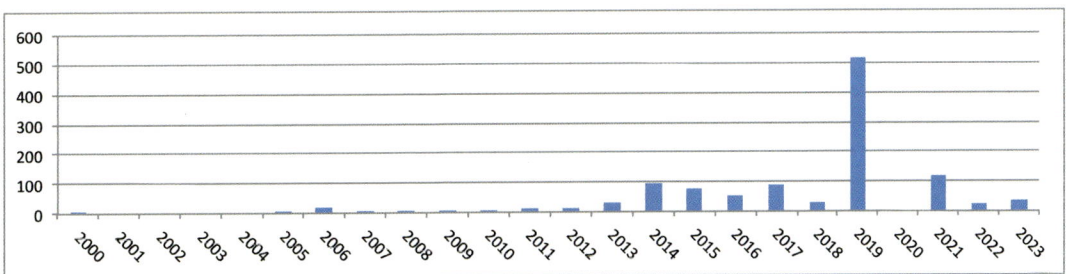

Fig. 18.54: Winter peaks of Lesser Black-backed Gull on Lough Swilly, from 2000/1 to 2023/4

Subspecies

Baltic Gull

Larus fuscus fuscus | Faoileán Baltach

RARE VAGRANT

The first bird of this subspecies, which breeds in northern Scandinavia and winters in Africa and south-west Asia, was recorded on Inishtrahull from 13 to 14 September 1953 (A. Gibbs and I. Nisbet). Both observers had previous experience of the race. Although this record was accepted at the time, the status of the Baltic Gull in Ireland is now regarded as uncertain.[215] Another bird was claimed for Tory on 21 and 24 August 1960 (TBO).

Gull-billed Tern

Gelochelidon nilotica | Geabhróg ghobdhubh

RARE VAGRANT

There are only two records of this rare vagrant from southern Europe. The first was seen from 6 to 19 August 2003 at Blanket Nook (R. Sheppard). The second bird was photographed at Carrickfinn from 27 April to 1 May 2021 (N. Newell et al.). Another bird, seen on a seawatch at Bloody Foreland on 4 August 2012, has not been assessed by IRBC (C. Ingram and R. Sheppard).

Gull-billed Tern, Carrickfinn.

DEREK CHARLES

Caspian Tern

Hydroprogne caspia | Geabhróg Chaispeach

RARE VAGRANT

R.G. Pettitt recorded our only Caspian Tern on 30 September 1959, on Tory. It was the first record of what is still a very rare vagrant to Ireland, and presumed to be from the population breeding in the Baltic area. However, there is also the intriguing thought that a Caspian Tern on Tory would not have been any more out of place if it had arrived, with a good tail-wind, from the North American population.

Sandwich Tern

Thalasseus sandvicensis | Geabhróg scothdhubh

SUMMER VISITOR

INCREASING

Sandwich Tern wasn't recorded breeding in Ireland until 1850. In the nineteenth century, Hart was apparently unaware of it in Donegal,[216] although Ussher and Warren relayed a report from 1862 of birds breeding, and of an addled egg found in 1893. But a search in 1894 found none.[217] A strong colony did exist in the first half of the twentieth century,[218] possibly the same one which re-located in the 1960s from Dunfanaghy New Lake to

Sandwich Terns, Tramore, Gweebarra Bay.

KIM PERIERA

a crannóg in Sessiagh Lough, were they declined to only two pairs in 1995 before abandoning completely.[219] Birds were seen at Vances Point on Lough Foyle with food for young on 30 July 1983, and newly fledged young were seen on 10 August 1987. Nesting may have taken place there, but it is not impossible that these birds belonged to the colony at Inch (see below).

Mulroy Bay had good colonies for many years, moving around adjacent small islands. The most stable one was on Greer's Island, but predation by mink over several years led to the area being abandoned, with a short revival between 2002 and 2005. In general it is believed that the Mulroy Bay birds moved to join those already at Inch Lough (Table 18.7). There was no activity in Mulroy Bay following 2005, until in May 2022 a colony was re-established on the Carlan Islands, with a conservative estimate (viewed from the mainland) of twenty to thirty pairs.

Table 18.7 suggests that the total breeding pairs for the county throughout the second half of the last century was in the 200s, and spread over several colonies. The population has since increased, but is highly concentrated at the single colony on Inch Lough.

The strength of the colony at Inch is due to active management of the small sand bank where the terns nest, firstly by Ken Perry from 1995 to 2005, and subsequently by NPWS. In the early years, first broods were wiped out several times by flooding, so a breeding platform was made which has solved the problem.

Sites	1980	1985/6	1988	1993	1994	2002	2004	2005	2019	2022
Inch Lough		95	73	119	220	219	242	340	388	
Mulroy (Greers Island)*		112	225	117	23	63				
Mulroy (Carlan Isles)				1						25
Sessiagh	4				2					
Dunfanaghy New Lake										
Inishkeeragh		7		29	0					
Illancrone		4								
Roaninish		A few pairs								
Illanfad		50								

Table 18.7: Sandwich Tern nests * Individual birds dispersed across several small sites

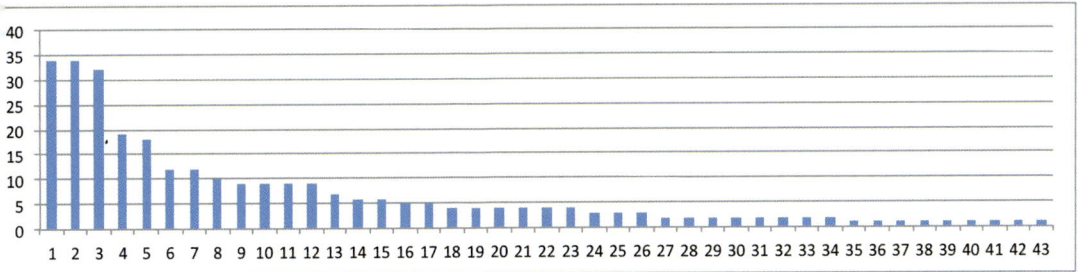

Fig. 18.55: The numbers of Sandwich Tern seen on the top forty-three days of passage migration

Sandwich Terns are among the earliest of our summer migrants to arrive, in their case from wintering quarters around the coasts of West Africa. The first birds usually appear at the end of March, and throughout the summer they can be seen more widely around the coast than other terns. They can also occur in considerable post-breeding numbers well away from breeding colonies, as on the Eddrim estuary (Donegal Bay) on 20 August 1993, when there were about 400 birds.

Of the 280 birds which have appeared migrating south around the headlands in autumn, most have been seen in July and August. These birds have to be from the 'Scottish breeding population, which is about 1,000 pairs. Figure 18.55 shows that most day totals are of a few birds only, and there are also quite a few parties of less than ten. Larger totals, up to a plateau of around thirty birds, usually appear as single flocks, or maybe two or three. Compare this with the Arctic Tern (Figure 18.56).

Little Tern

Sterna albifrons albifrons | Geabhróg bheag

SCARCE SUMMER VISITOR

DECREASING

Ussher and Warren said of Little Terns that 'Donegal is the only county where they are in any numbers' and noted in particular Naran and Roaninish.[220] At the same time, Hart knew of them all around Lough Swilly.[221] Norman and Saunders state that in 1967 there were nine colonies in the county (one doubtful), all but three of them recorded since 1960.[222] Operation Seafarer, spread over 1969–70, covered the terns less thoroughly than other species, and found them at only three sites.[223] But it was my own

Sites	1960s	1969	1970s	1984	1995	1997	2000	2019
Greencastle	10							
Inishowen Head	5							
Doagh (Carrickabracky)	10	5	19	2				
Doagh (Lagacurry)		3						
Tullagh Bay	5							
Tullagh Point	4	3		3				
Rockstown Harbour	3							
Mulroy Bay (north)								
Ballyhoorisky Point			3					
Ballyness Bay	3							
Inishbofin								
Tory	20			8	2	15	10	8
Inishirrer								
Inishmeane							1	
Inishkeeragh		25		5	13			
Roaninish		12						
Rossnowlagh (Lower)		1						
Totals for Full Counts	60	49		18	15			

Table 18.8: Little Tern colonies (pairs)

experience during the 1960s that there were a few pairs present at most beaches around Inishowen, at a time when there were also some strong colonies on the offshore islands. The recorded total for the decade was about 100 pairs (Table 18.8). The more dedicated All-Ireland tern survey in 1984 only recorded eighteen pairs, including ten pairs in two colonies on the mainland.[224] That was the last time any Little Terns were found nesting on the mainland, and by 1995 the total was down to only thirteen pairs.[225]

Two colonies now remain, on Tory and Inishkeeragh, but there may also be a few pairs lingering elsewhere among the west coast islands. Either way, the future for Little Terns in Donegal is threatened. Disturbance is a major problem on Tory, and Inishbofin and Gola lost all their terns following the restoration of human settlement.

There are only about 300 pairs of Little Terns breeding in Scotland, and some of those migrate down the North Sea coast. So it is not surprising that they are rarely encountered here on autumn seawatches. Fourteen were spread over four sessions, and there was one that had fifteen birds. The largest flock seen was at Bloody Foreland on 10 July 2007, when a single flock of nineteen birds paused to feed inshore. It must have represented a significant percentage of the entire west Scotland breeding population. Those birds would have been en route to their wintering grounds, which are mainly in the African equatorial area along the Gulf of Guinea.

Roseate Tern

Sterna dougalli dougalli | Geabhróg rósach

RARE SUMMER VISITOR

The first record of Roseate Terns in Donegal goes back to 1939, and is from Bernard Tucker, one of Britain's most eminent ornithologists. The birds were 'evidently breeding in a small tern colony on the coast of western Donegal … as I saw at least three together it is probable there are two pairs'. Two of the adults were obviously associated with a fledged young bird.[226] In 1968, there was a pair of breeding Roseate Terns at the Common and Arctic Terns colony in Greencastle harbour (J. McLaughlin). At the same time, when surveying for Operation Seafarer, Andrew Ferguson found two pairs on Inishkeeragh and one on Inishfree Lower. The 1995 tern survey found three pairs still nesting at Inishkeeragh.[227]

There are also several non-breeding records. Four birds were seen near Malin Head on 4 October 1963 (O. Merne) and single birds have been seen at Blanket Nook on 12 June 2012 (D. Charles), and on Tory on 16 May 2019 (C. Ingram, T. Campbell).

The prospect of a breeding colony becoming established in Donegal has always been remote, but the diaspora from the huge, and still growing, colony on Rockabill in County Dublin keeps hope alive.

Common Tern

Sterna hirundo hirundo | Geabhróg

SUMMER VISITOR

DECREASING

Sites	1948	1960s	1969–70	1974	1983–5	1992	1994–5	2001–2	2016
Greencastle*		50	29		50				
Inishowen Head**			28						
Kinnagoe Bay		10							
Inishtrahull			32						50
Doagh Isle		10		20	1				
Mulroy Bay (Greers)			2			14	16	14	
Mulroy (Carlan Isles)						60	97		
Island Roy					2				
Sessiagh Lough					3	5			
Dunfanaghy New Lake	15				2				
Inishbofin					25				
Inishdooey	30								
Tory									1
Inishirrer							2	14	
Inishmeane								1	
Umfin			4				2		
Bo Island			4				6		
Inishinny	30								
Inishfree Lower			3						
Mullaghderg Lough					11	2	10		
Rutland			6						
Lough Meela	25								
Inishkeeragh			90		9		30		
Illancrone					14		29		
Roaninish	50								
Illanfad			10		50				
St Peter's Lough			20						
Lough Golagh					40	9			
Inch Lough			30		45	16	21	89	57
Totals for Full Counts	150		258		252		213	118	108

Table 18.9: Common Tern colonies (pairs) * There were at least three colonies on the 4 km stretch between Greencastle Harbour and Inishowen Head.
** A 2.5 km stretch of cliff coastline.

In the first half of the last century it was thought that this species greatly outnumbered Arctic Terns in Donegal.[228] In the 1960s, and presumably earlier, Common Terns bred around the coasts on small islets and rocks, some of them cut off from the mainland only at high tide. Most of these inshore colonies were around Inishowen and in Mulroy Bay, but along the west coast where there were plenty of offshore islands, it was those that were preferred to the mainland. A number of coastal lakes with small islets, and a few inland ones, also held colonies. There is a remarkable agreement in the total number of pairs found on the three surveys of 1969, 1984 and 1995, although the occupied sites were different each time (Table 18.9).[229] It would be nice to conclude that this shows overall stability across the period, but that would be unwise, as it is not always clear whether a site with no birds recorded was actually visited. What is clear is that since at least 1995 there has been a general decline, with the loss of many small colonies.

Common Terns are rarely identified on seawatches, with sightings at only thirteen events, and with a maximum of thirteen birds. This is rather surprising, given their widespread breeding distribution in northern Europe and their migration to the coasts of west and southern Africa. It is possible that passing terns are sometimes too readily assumed to be Arctic.

Arctic Tern

Sterna paradisaea | Geabhróg Artach

SUMMER VISITOR
STABLE

In the 1800s Arctic Tern was the dominant species around Ireland, and Roaninish had what Ussher and Warren believed to be the largest colony in the country.[230] There were other colonies on Inishbarnog and Inishduff. The species was overtaken by the Common Tern in the first half of the twentieth century.[231] Roaninish was reduced to twenty pairs in 1947, and sixty in 1960. As with the other

Arctic Tern, Tory.

MICHAEL BELL

Sites	Pre-1960s	1960s	1969–70	1970s	1980	1984	1994–5	1999–2002	2016
Greencastle		4	3			6			
Inishowen Head			20						
Kinnagoe Bay						2			
Inishtrahull	30	25		25			1		15
Doagh (Carrickabracky)		10	12	5		7			
Doagh (Lagacurry)			3						
Tullagh Point		10							
Rockstown Harbour		12							
Ballyhoorisky Point	50								
Mulroy Bay (north)						20			
Inishbofin						120	44		7
Inishdooey							28	54	
Tory								4	111
Inishirrer							24		
Gola							2		
Lough Meela		10							
Go Island						1			
Bo Island							2		
Inishinny			3						
Mullaghderg Lough		1							
Inishkeeragh						77	189		
Illancrone						132	35		
Illanfad									
Roaninish		60			120	46			175
Inishkeel	40	12							
Dawros Head			16						
Inishbarnog			2						
Totals for Full Counts		144	59			411	325	58	308

Table 18.10: Arctic Tern colonies (pairs)

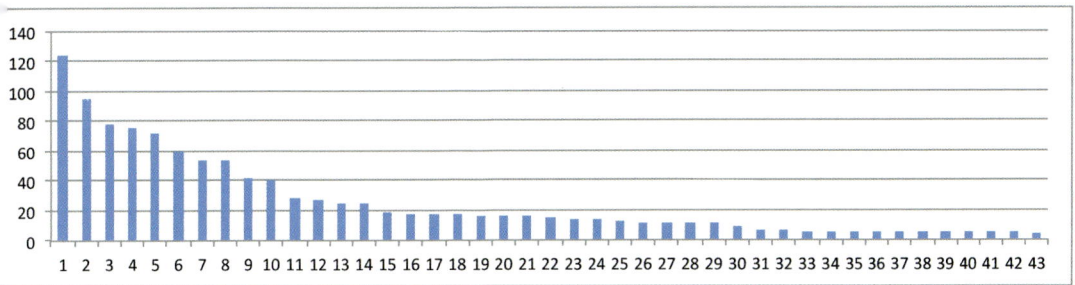

Fig. 18.56: The numbers of Arctic Tern seen on the top forty-three days of passage migration

tern species, great efforts were made to survey all sites on the two national tern surveys of 1984 and 1995, and on the four national seabird censuses, of 1969–70, 1985–7, 1999–2002 and 2015–19 (Table 18.10). For Arctic Tern, the most comprehensive of these surveys were those of 1984 and 1995.[232]

The Arctic Tern's presence at inshore colonies, mainly around Inishowen and Fanad, where the Common Tern bred freely, was always very limited, at least in the latter half of the twentieth century. The mainland colonies have now gone, a fate shared with the Common Tern and probably linked to disturbance, and to catastrophic declines in Arctic Tern colonies elsewhere due to climate-related changes in the availability of fish.

At their remaining sites on the remote offshore islands, numbers of Arctic Terns can still eclipse even the best of the Common Tern colonies. Table 18.10 shows that Roaninish is again thriving. Tory had 200 birds in suitable breeding habitat on 6 June 2003, but with no evidence of nesting at the time. In 2016 and in some subsequent years there has been a good colony. The other big colonies, Inishkeeragh and Illancrone, need to be re-surveyed. The human population of Inishkeeragh had been re-settled on the mainland in 1955, after which several species of tern colonised. A few houses are now being restored, and at only twenty hectares in area it would be hard to see humans and terns co-existing on the island for very long.

Arctic Tern is the most frequently seen tern species on seawatches – on sixty-five events (41 per cent of the total). September is the peak month for passage. These birds come from Scotland and beyond, on their journey to the seas around Antarctica where they can enjoy a second summer in the year – something that no other species does. Figure 18.56 shows the numbers seen on the top forty-three of the sixty-five timed seawatch events, to compare with the forty-three Sandwich Tern days. The flock size distribution increases smoothly with no apparent upper limit, unlike Sandwich Tern (Figure 18.55).

White-winged Tern

Chlidonias leucopterus | Geabhróg bháneiteach

RARE VAGRANT

There have been four sightings (four to six birds) of this marsh tern from Russia and the fringes of eastern Europe. The first was an adult seen at Donegal Bay on 15 July 1964 (E. and R. Millar). After a long gap one turned up at Blanket Nook on 1 June 2009 (D. Brennan). Two were then seen on Tory on 15 May 2014 (A. Meenan). One was also seen at Inch from 15 to 25 May (C. Ingram et al.) which, depending on timing, could possibly have been one of the birds seen on Tory. However, it has not yet been assessed by the IRBC. The most recent record was an adult at Inch on 10 August 2018 (T. Campbell et al.). The birds seen in May and June had probably lost their way en route from their wintering grounds in sub-Saharan Africa to their normal breeding range. The July bird could have been a wandering failed breeder.

Black Tern

Chlidonias niger niger | Geabhróg dhubh

SCARCE PASSAGE MIGRANT

This species of marsh tern is a common summer migrant from African coasts to western Europe, although breeding areas are well scattered and do not include Britain or Ireland. It is the species of marsh tern most likely to reach Donegal. The first two unspecified sightings were in the nineteenth century.[233] The next were two birds at Dunfanaghy on 26 May 1955 (F. Egginton).

There have been at least sixty-five birds seen so far, on thirty-two occasions. This average of two birds per occurrence is skewed by a notable influx at Inch in 2010, starting with two birds on 8 September. The following day there were fourteen, the peak of eighteen came on the tenth, and numbers then faded until the last bird was seen on 25 September. Even apart from that influx, most of our Black Terns, totalling thirty-seven, have been seen at Inch and elsewhere on Lough Swilly – that includes one bird in 1959 at the brickworks pond in Burnfoot, which is on the edge of Inch Levels. There have been two multiple sightings at Malin Head, with a gap of forty years – four birds on 29 September 1975 and six on 24 August 2015. Tory has had two, and likewise Dunfanaghy and Burtonport. Singles have been recorded at north Fanad and Mountcharles.

Most birds occur in autumn, but there have been seven in spring, starting with the pair at Dunfanaghy New Lake in 1955. There was a single at Tory on 30 April 1962. Blanket Nook also had singles, on 22 June 1983, 7 June 2022, and from 1 to 6 July 2022. Inch had one on 30 May 1983.

Great Skua

Stercorarius skua | Meirleach mór

UNCOMMON PASSAGE MIGRANT AND RARE BREEDER

STABLE AS MIGRANT, INCREASING AS BREEDER

The first record of the Great Skua in Donegal was on the relatively recent date of 7 September 1953, at Malin Head (A. Gibbs et al.). Globally, the species was much less numerous in the 1950s, but even so, birds must have been passing the county throughout the nineteenth century on their autumn journey from Scotland and Iceland to the seas off Iberia. It took the development of decent optics to reveal that they are present here, but it is the huge increase in their population over the last half century that has made them now such an everyday experience on seawatching sessions. So familiar have they become that the more interesting Shetland name of Bonxie has virtually replaced the clumsier Great Skua among regular seawatchers.

In strong onshore winds between late July and late October Great Skuas can usually be seen. In fact they have occurred on 89 per cent of all sea-watches, with the higher rates in October, and numbers in July very small. The best counts were 203 past Rocky Point on 24 September 1991 and 177 (44 per hour) at Bloody Foreland on 22 October 2015. However, the maximum hourly rate over the twenty-two Donegal counts during 2022 and 2023 has been only two birds. This striking collapse from what would normally be expected is almost certainly down to the impact of avian

Great Skua,
16 km
north-west
of Tory.

JAMIE BLISS

influenza, which has taken a heavy toll on the Great Skua's main breeding colonies in Scotland and the Faroe Islands. The next two or three seasons will probably confirm this.

Around the turn of the twentieth century, human predation nearly exterminated Great Skuas from their breeding base in the northern isles of Scotland. When that ceased, they rebounded, and spread slowly west and south around the Scottish islands. Until very recently no one could have predicted that they would continue to expand south to Ireland – but they did, and in 2008 the first pair bred in Donegal. There are now at least seven pairs on the offshore islands. Their future will depend on secure breeding locations, a continuing supply of breeding seabirds to rob or kill, and the impact of avian influenza.

Pomarine Skua (Pomarine Jaeger)

Stercorarius pomarinus | Meirleach Pomairíneach

UNCOMMON PASSAGE MIGRANT

STABLE

Breeding no closer than the Russian tundra, and wintering off West Africa, the Pomarine Skua was first recorded in Donegal in the nineteenth century, when there were three records, including a dead bird found at Kiltooris in May. In the mid-twentieth century there were still only a total of four records for the county.[234] Like the Great Skua (above), Pomarine Skuas are now known to be regular visitors on migration. They mainly pass our headlands from August to November. Out of 160 seawatching sessions between

1992 and 2020 Pomarines were present on sixty-eight. The birds logged totalled 672, with the maximum rate of passage being 22 per hour on 26 September 2002 at Bloody Foreland. On that occasion the sixty-five seen on the three-hour event included a flock of forty birds and another of twelve (so an hourly rate of at least forty could equally be claimed). Although less numerous than either Great or Arctic Skuas, Pomarine is the only skua species likely to be seen in a small flock.

There have been eight spring sightings in recent years, totalling fourteen birds. These have all been in the second half of May. With all the skuas, but particularly the Long-tailed and Pomarine, spring passage is more direct, not weaving around the headlands as in autumn. If Pomarines do connect regularly with Donegal in spring, we would expect them to only touch the county at extreme western locations – Rathlin O'Birne perhaps, rather than Rocky Point, and with Arranmore Lighthouse and Tory being the other two obvious locations. There is not enough coverage in the south-west to have produced any evidence of this, but the two north-west sites do get some attention in May, so the relative lack of skua records in spring is surprising. Perhaps it is just that the right weather conditions are too infrequent at these locations in May.

Arctic Skua (Parasitic Jaeger)

Stercorarius parasiticus | Meirleach Artach

UNCOMMON PASSAGE MIGRANT

STABLE

Unlike the two previous species, the Arctic Skua has long been known in Donegal, one of the majority of maritime counties which have had repeated occurrences.[235] Now that it is better known, it is clear that birds can be seen from August to November, but the larger numbers are from September and October. After the Great Skua, this is the most frequently seen member of the family, and the one most likely to be seen away from the exposed seawatching sites. Birds are occasionally spotted inshore, at harbours or in places like Lough Swilly. Some that are moving west along the Northern Ireland coast have been known to evade the rough passage around Malin Head by taking the overland route through Lough Foyle to Donegal Bay. A total of 838 birds have been logged on timed seawatches, compared with

672 Pomarines, and Arctic is the species much more likely to be seen, on 73 seawatching events as opposed to 43 for the Pomarine. But the peak hourly rate is only seven, compared with twenty-two (or more) for the larger bird.

There have been thirteen spring sightings in recent years, totalling twenty-three birds. There is a record of one found dead at Moville on 12 June 1909.

Long-tailed Skua (Long-tailed Jaeger)

Stercorarius longicaudus longicaudus | Meirleach earrfhada

SCARCE PASSAGE MIGRANT

The presence of Long-tailed Skua in the county goes back to Thompson, who noted a single record of four birds at Ards in November 1816 or 1817, with one of them shot (J.V. Stewart).[236] Ussher and Warren reported one shot from a flock on 17 May 1860.[237] While doubt had been cast on this record by Ruttledge in 1966, along with a claim of from fifty to seventy on the River Shannon near Athlone two days previously, he reviewed them in 1982 and found both acceptable.[238]

Until recently, Long-tailed Skua has been treated nationally as a rare species, but is now more frequently recorded. In Donegal there have been 121. Of these, eighty-four were seen on seawatches, a lower percentage than one might expect. The discrepancy highlights the fact that, like Arctic Skuas, Long-tailed Skuas are occasionally seen inshore, perhaps taking shelter from storms. Despite their relative rarity, when they are seen on a timed seawatch – 40 times out of 233 – there is often more than one bird, with the maximum being eight seen on a five-and-a-half-hour session at Bloody Foreland on 24 August 2024. Of these forty seawatches, seventeen were in August, thirteen in September and ten in October – an earlier season than the other skua species. There has been one sighting in November, and a spring bird in May. It is interesting to note that two of the largest counts were on the same day, 7 September 2011, when Melmore Head had five and Bloody Foreland seven. Of course, birds passing Melmore could be at Bloody Foreland in less than an hour, and it is quite possible that the Melmore five were among the Bloody Foreland seven.

Hutchinson, writing in 1989 before the blossoming of seawatching, noted that spring passage in Ireland accounts for more birds than autumn.[239]

He notes two Donegal records for the period 1966 to 1986, and of course there is the remarkable record from 1860 (above). In more recent times the autumn sightings have eclipsed those in spring, of which there have only been two. These were on 26 May 2011 at Bloody Foreland and on 8 June of the same year at Malin Head. However, there is very little systematic seawatching in spring, so the recent dearth of records is not significant. For more details on spring migration, see under Pomarine Skua (above).

Little Auk

Alle alle alle | Falcóg bheag

SCARCE WINTER VISITOR AND PASSAGE MIGRANT
STABLE

The normal winter range of this high-arctic breeding species is to the north and east of Great Britain. Ussher and Warren report a claim that Little Auks were not infrequent in Donegal Bay, but there must be a suspicion that recently fledged Razorbills or Guillemots could have been misidentified.[240] One killed at Arranmore lighthouse in 1913 was collected by Barrington. Kennedy et al. note that during the winter of 1948/9 they were present close to shore at Tory.

There were fifteen birds noted in the early 1960s when Malin Head and Tory Bird Observatories were functioning. There were then no records until 1982. That year had a very large daily count of twenty-five at Rossbeg, on 11 April, and since then there has been a steady trickle of records. The total recorded to the end of 2021 was 162 birds.

Then a wreck took place between 11 and 23 January 2022, when at least 114 birds were recorded. Apart from one at Inishfad in Donegal Bay, all were at or around Tory, at Magheraroarty, or seen from the ferry between the two. Six were found dead and five were found alive but with no resistance. One on the water made no attempt to escape from a Great Black-backed Gull, so it is quite likely that many others were in the same condition. What is curious about this wreck is that it took place in a period of remarkably calm weather. The consensus seems to be that during a period of very stormy weather further north, the birds had been unable to feed and were starving. The move south was probably an attempt to reach calmer waters where they could feed.

Little Auks normally only come inshore in the worst weather. Leaving aside this recent wreck, rather more birds have been seen on late autumn passage than in winter – seventy-five in October and early November, as opposed to forty-eight from late November to early February. But as Ireland is well south of the normal wintering range of this high-arctic breeding species, it has traditionally been regarded as a scarce and irregular winter visitor. In autumn, Little Auks are not expected, but when winter seawatches do happen, that changes, as in late November to early December 2021 when Little Auks were seen on all four seawatches, totalling twenty-six birds. The higher number of autumnal records is easily accounted for by the lack of seawatching in the winter months.

Autumnal migrants, from October to December, are always travelling south – but to where? If proceeding past Ireland, they will end up even further beyond their normal wintering range. Many fewer birds have been seen on the return trip, with the total for March and April being thirty.

Arranmore has had fifty-two birds, thanks to its location, to good coverage by Andrew McMillan, and to nineteen seen on 23 October 1999. Bloody Foreland has had thirty-four, which includes another nineteen birds seen on 20 November 2021. Otherwise, the records are spread around all coasts. Severe storms have delivered thirteen dead or dying birds, but the only inland Little Auk was alive on Trumman Lough, near Laghy, on 18 January 2018.

For the record, but not included with the above, thirty-six birds were recorded on the Rockall Bank on 28 March 2016, 245 nautical miles west-north-west of Tory.

Guillemot (Common Murre)

Uria aalge albionis | Foracha

LOCALLY COMMON RESIDENT

DECREASING

At the turn of the twentieth century, Ussher and Warren stated that Guillemots were considerably less numerous at Horn Head than Razorbills and Puffins, but that it was still the greatest breeding place for them in the north of Ireland. The lofty perpendicular stack at the east of Tory was covered with birds. The next great colony at that time was the Tormore stack, near Port, in the south-west.[241]

Sites (from NE to S)	1954	1960s	1980	1987	1993	2000	2015
Glengad Cliffs							39
Stookaruddan (area)			1,000+		1,104		
Malin Head							330
Horn Head		10,000*	5,500	4,806		6,548	3,855
Tory	165	146		635		53	755
Arranmore (Torneady)						6	
Roaninish				15			
Tormore (area)				50		507	214
Tawny		7				40	39
Totals for Full Counts				5,506		7,154	5,232

Table 18.11 Guillemot colonies (individuals) *10,000 ±5,000

Large changes have taken place since then, but quantifying them is not easy. By the mid-century, Tormore was unchanged, the 'immense numbers' on Tory were reduced to 'very few'. The importance of Horn Head remained, but the Guillemot's strength was 'nearer 2.5% than 5% that of the Razorbill'.[242]

The most useful counts at Horn Head have been those from 1980 onwards, but they fail to establish a clear trend (Table 18.11). The colony at Tormore, which had never been counted, collapsed to a fairly low level, but even that was better than Stookaruddan, which was completely abandoned. With the continuing overfishing of the oceans in general, and increasing disruption of local fish stocks due to climate change and pollution, there can be little confidence that the future is bright.

Roaninish with a maximum altitude of 9 m seems an unlikely breeding site, but its isolation, and the presence of many seabirds of various species, may have been sufficient inducement for a few pairs of Guillemots to breed on its low cliffs.

Migration is on a large scale, but generally unquantified, as the auks streaming past the headlands in autumn are usually recorded as unidentified auks (Razorbills or Guillemots). The few that pass close enough for easy identification hint at parity between the two species, but it can be little more than a hint. The largest numbers of unidentified auks are in October and November (at recorded rates of 50 to 150 per hour). Higher rates are certainly frequent, but at such times, auk monitoring is usually abandoned in favour of recording the wider range of other species on offer. The highest recorded rate is 358 per hour (over 3.5 hours) at Bloody Foreland on 3 October 2021. Other strong movements noted in July are assumed to be the early departure from the breeding cliffs of birds without parental responsibilities.

Guillemots breeding on the coasts of northern Europe generally winter within the region, and no further south than the seas off north-west Africa. Some birds remain in Donegal's inshore waters throughout the winter, but numbers are small.

Razorbill

Alca torda islandica | Crosán

LOCALLY COMMON RESIDENT
INCREASING
RED-LISTED BOCCI-4

For Kennedy et al. (1954) Razorbills at Horn Head had 'probably one of the most extensive colonies on our coasts'. As also for the Guillemot, Tormore remained heavily populated but the numbers on Tory had decreased. The first attempt to count the Horn Head colony in the mid-1960s arrived at the remarkable total of 35,000 to 55,000 (Table 18.12), which has never been approached since.[243] The strong colony at Stookaruddan has completely gone. But for the sites covered since 1987 by all three surveys, including Horn Head, the much lower population seems to be fairly stable. The recent

Razorbill, Tory.

PETER PHILLIPS

Sites (from NE to S)	1954	1960s	1980	1987	1993	2000	2015
Glengad Cliffs							12
Stookaruddan (area)		1,000+		55			
Malin Head							175
Dunaff Head							4
Horn Head		45,000*	12,412	5,628		5,814	5,999
Tory	754			614		44	756
Arranmore (Torneady)							40
Tormore (area)				398		480	444
Twany							
Totals for Full Counts				6,695		6,338	7,430

Table 18.12: Razorbill colonies (individuals). * ± 10,000

NPWS survey of 2015 was the most comprehensive, which would account for the apparent rise in the county total.

Razorbill winter presence and passage is not well recorded, for the same reasons as the Guillemot – the two species not being normally distinguished on seawatches, and often ignored when too many other species demand attention.

The Razorbill is Red-listed in BirdWatch Ireland's 'Birds of Conservation Concern in Ireland 4: 2020–2026' on the grounds that its global conservation status is of concern.[244]

Great Auk

Pinguinus impennis | Falcóg mhór

EXTINCT

There are no historical records of living Great Auk from Donegal. However, remains have been found in the kitchen middens excavated from sand dunes.[245] These date from the Mesolithic period of 3,000 to 7,000 BC. This allows us to record the globally extinct Great Auk as having been a native species during the time of human occupation.

Black Guillemot

Cepphus grylle arcticus | Foracha dhubh

RESIDENT

Map 18.3:
Black Guillemot
numbers 1999–2002

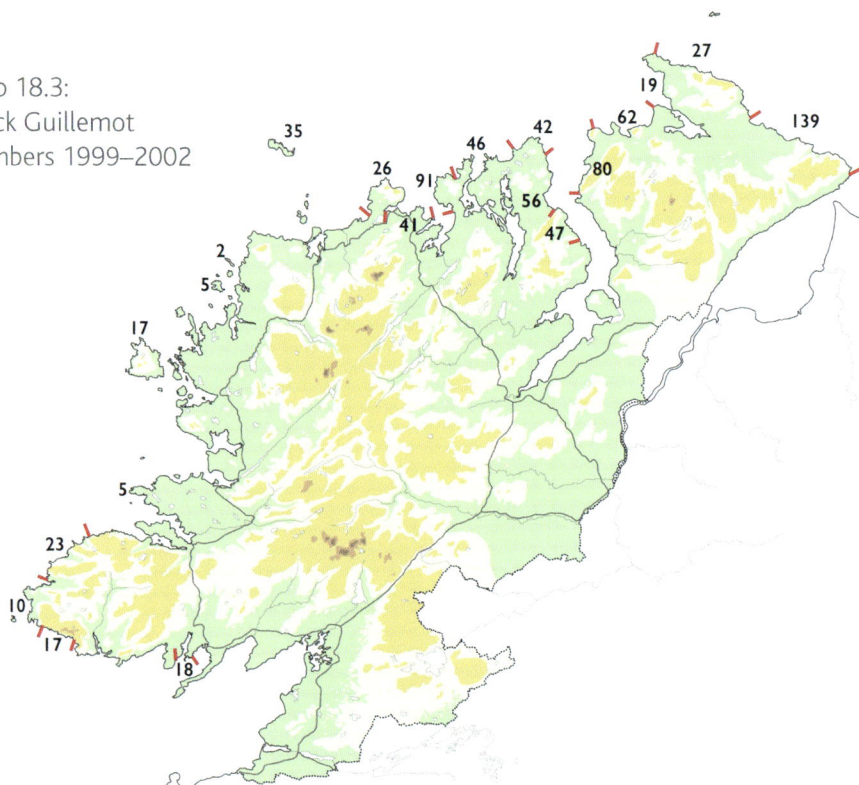

Black Guillemots, Moville.

This atypical auk doesn't mix with the other colonial cliff-dwellers, but it does accept some company at its favoured breeding sites – call them loose colonies, or aggregations of isolated pairs. Like the Puffin, it nests in holes, but chooses rock cavities rather than soil. There has been one dedicated survey, which attempted to cover most of the coast between 1999 and 2002. The total number of birds counted was 766. North Inishowen had 327, Fanad had 142 and Rossguill 137. This puts 79 per cent of the county total in the north-east. However, the west coast, from Horn Head south to Ardara, was not fully covered, and that could account for the bias. However, counts at a few of the western islands and headlands and from the south-western promontory show that the west coast probably does have many fewer birds than the north coast (Map 18.3). There is no ready explanation for this distribution.

Black Guillemots are resident, and can be seen in winter all around the coast. A pattern of roosting at night in more sheltered waters has been recorded elsewhere.[246] There is no evidence of any migration.

Puffin (Atlantic Puffin)

Fratercula arctica | Puifín

LOCALLY COMMON SUMMER VISITOR
DECREASING
RED-LISTED BOCCI-4

Everyone's favourite bird has always been very locally distributed in Donegal, and sadly, none of the sites lend themselves to close encounters. Puffins were certainly more numerous here in the past. Ussher and Warren wrote:

> On the north [coast] … by far the largest colony, possibly the largest in Ireland, is on Horn Head, where the slopes and broken parts are tenanted for miles by these birds. There are haunts on Tory Island and the north side of Aranmore, and a more considerable colony occupies the lofty Tormore, all on the Donegal coast.[247]

Kennedy et al. believed that Horn Head fell short of the whirring multitudes that breed on the north Mayo cliffs, but they include mention of Tormore, which in mid-twentieth century was the second most important colony in the county.

Sites (from NE to S)	1900	1954	1969	1980	1987	1995	1999	2024
Horn Head	Very large	Large		344	93		189†	
Tory	Breeding	1,700*	711		650		68†	2,000††
Arranmore (north)	Breeding							
Tormore (area)		**	Low thousands			13	200	

Table 18.13: Puffin colonies (individuals, unless specified otherwise)
* Pairs – 1,000 on Nougherwole, 400 on Tormore and 300 elsewhere
** Larger colony than Tory
†Burrows
††Rough estimate of 1,000 birds on Nougherwole and 1,000 elsewhere

The first serious attempt to census a colony was by Philip Redman in 1954, when he found Tory to have 1,700 pairs.[248] Casual observations since then, like 400 birds present on 2 April 2021, are unreliable, but the most recent one is remarkably similar to the 1954 census (Table 18.13).

Puffins pass the seawatching headlands in early autumn in small num-

Puffins, Tory.

ANGELA GILLIGAN

bers. These may be local birds from colonies like Horn Head, or perhaps failed breeders dispersing from more northerly latitudes. There are few winter records of this species in inshore waters.

The Puffin is Red-listed in BirdWatch Ireland's 'Birds of Conservation Concern in Ireland 4: 2020–2026' as its global conservation status is of concern.[249]

Red-throated Diver (Red-throated Loon)

Gavia stellata | Lóma rua

WINTER VISITOR AND RARE BREEDER
STABLE

Two pairs of Red-throated Divers were first discovered breeding in Donegal (and Ireland) in 1884. They survived the removal by egg collectors of two clutches per year (three in one year) until 1896.[250] How many years of failed or no attempts is not clear, but at any rate, breeding continues to the present, mostly in small mountain lakes in the west of the county where they are clearly opting for privacy. There would appear to be no shortage of suitable sites, and in 1997, 175 lakes in eighteen hectads (10 km²) were inspected. The estimated population found was only six to eight pairs, declining to three to four pairs in 2002.[251] Since then, NPWS have continued monitoring, and they have tried conservation measures such as trapping mink at threatened sites. The latest information is that in 2018 breeding was confirmed at three sites, was probable at four more, and possible at two. That was a good year! Existing sites, and the many possible alternatives, are being crowded with peat-cutting, afforestation and general development. And of course, with Ireland being at the extreme southern limit of the Red-throated Diver's circumpolar breeding range, climate change may well have the final say. It is hard to see how the species can survive in the long term, but every effort should be made to make sure that it does.

So despite the proven tenacity of the species, its hold continues to look very precarious, and its absence from the current Red-list of Birds of Conservation Concern in Ireland only reveals a loophole in the assessment criteria.

In winter, Red-throated Divers from more northerly breeding populations can be found all around the coast, although numbers are highest in sheltered bays. The peak count on Lough Swilly was fifty birds on 11 January 2003. The highest five-year mean of thirty (for the years 2015/16 to 2019/20) is more recent. Donegal Bay's mean peak for 2011–15 was twenty-one, but it occasionally holds much larger numbers, for example ninety-five at Rossnowlagh on 8 April 1987, and sixty-five on 1 December 1997. Lough Foyle's mean peak for the same period was nineteen, and the maximum recorded there was fifty-six on 18 January 1997 (probably only a small fraction of what was present in the whole lough).

Red-throated Diver, Lough Swilly.

Prior to migration north in spring, birds can sometimes gather together in small groups, with some of them in breeding plumage, as at Maghery on 27 April 1995, when there were sixteen birds. Autumn migration is regularly noted in small numbers at seawatching points.

The Nationally Important figure (1% of the national population) is 20.

Qualifying Sites: Lough Swilly, Donegal Bay

Black-throated Diver (Black-throated Loon)

Gavia arctica arctica | Lóma Artach

UNCOMMON WINTER VISITOR

By far the most abundant of the divers globally, and the most southerly, but despite that, this species is the least common of the three regular diver species to visit Donegal. It was first noticed in the nineteenth century, but remained rare until the 1980s. Up to eleven spent the winter in Donegal Bay in 1983/4, at the same time as the north coast of County Clare first emerged as the only Irish haunt attracting double-digit numbers. Donegal Bay does not achieve this on a routine basis, but has peaked at forty, on 28 February 1993. No other site has more than a few birds at any one time, but sightings are quite widespread, including six birds seen passing Bloody Foreland on migration, and four passing Malin Head.

The total number of records in the county is about 750, of which 78 per cent have been in Donegal Bay. There is a build-up in numbers through the winter to a peak in March. Figure 18.57 shows the monthly occurrence of all birds.

A few birds have remained into summer and/or been in summer plumage. Fourteen of these summer birds were in circumstances that were

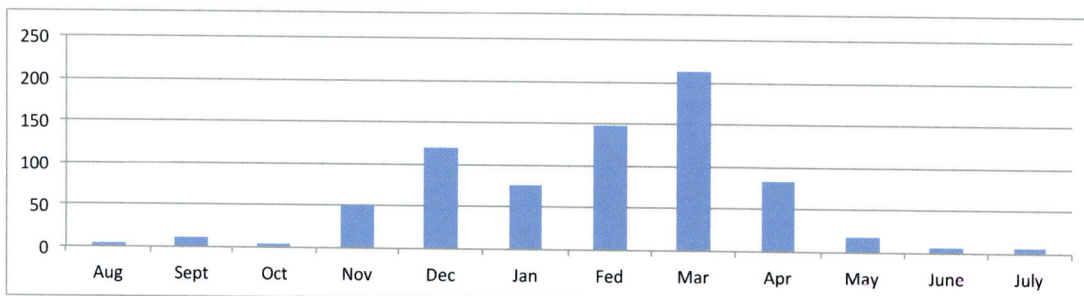

Fig. 18.57: Monthly occurrence of all Black-throated Divers (first sightings)

unremarkable, but there have been three others that are certainly of note. One spent almost two weeks on Lough Beagh in early July 1990, and two were on the same lake for one day in late June 2018. These three birds were all in breeding plumage.

Great Northern Diver (Common Loon)

Gavia immer | Lóma mór

WINTER VISITOR AND PASSAGE MIGRANT

STABLE

Our Great Northern Divers come from Canada and Greenland, although birds from the small population that breeds in Iceland are presumably included with them. They are present all around the coast in winter. A few birds will almost always be found in any sandy bay, feeding in the surf. The best site is Donegal Bay, where they can be seen everywhere from the coastal shallows to mid-way between Inishfad and the Doorin peninsula. The highest count there was 316 birds on 28 February 1996. Other large counts were 132 on 24 April 1997 (two days prior to that, at Arranmore, there were 94) and most recently 102 on 7 May 2021. Lough Swilly in winter has a recent five-year mean peak of twenty-eight, to 2019/20, and the highest count was fifty-six on 10 February 2018.

The counts from late April and May can be regarded as spring gatherings, prior to return migration. As with Red-throated Diver, birds in summer

Great Northern Diver, Sheep Haven Bay.

DEREK BRENNAN

plumage at this time are not unusual, but one that I encountered inland at Lough Inshagh on 23 July 1974 was different. It approached me and called vigorously. Unfortunately, I couldn't find any sign of a mate or a nest, but the breeding intent was clear.

Autumn passage is mostly in October and November. Peak counts have been sixty-nine in three hours on 7 November 2021, and fifty-five in four hours on 22 October 2015, both at Bloody Foreland. An unprecedented movement of forty-nine birds flying east was recorded over four hours on 24 October 2010, in winds varying from west to north-east.

The Internationally Important figure (1% of the flyway population) is 50.

Qualifying Sites: Donegal Bay

The Nationally Important figure (1% of the national population) is 20.

Qualifying Sites: Lough Swilly

White-billed Diver (Yellow-billed Loon)

Gavia adamsii | Lóma gobgheal

SCARCE VAGRANT

There have been a minimum of thirteen White-billed Divers recorded in Donegal. The first was seen from the Tory ferry on 6 May 2000 (B. Robson et al.). It was the seventh Irish record. Since then a bird in moulting plumage was seen at Inishowen Head on 19 April 2014, a breeding plumage adult at Culdaff on 10 August 2018, and a bird passing Malin Head on 17 April 2019.

In addition, there have been twenty-nine sightings in the waters around Tory Island since 2016. Discounting what were probably repeat sightings, that still leaves peaks of three birds between 9 and 13 May 2016, two between 24 March and 10 May 2017, three between 28 April and 2 June 2018, one on 24 April 2019, one on 13 May 2021 and another on 3 May 2022. These annual appearances and some extended stays suggest that a tiny population could be adopting Tory as its base in spring, prior to their departure for breeding in northern Russia. If this is sustained, it would qualify the White-billed Diver as a rare passage migrant rather than a vagrant.

Wilson's Petrel (Wilson's Storm-petrel)

Oceanites oceanicus | Guairdeall Wilson

RARE PASSAGE MIGRANT

Wilson's Storm-petrel is an abundant species which breeds in Antarctica and disperses north to mid-latitudes, as far as the Bay of Biscay. The two subspecies *exasperatus* and *oceanicus* are equally likely to occur in Irish waters (Hobbs, 2022) but no sub-specific identification has been made. The first bird encountered in Donegal waters was on a boat trip to 56 km west of Burtonport on 15 August 1987 (R. Sheppard et al.). Five were seen on a second trip to the same area, on 6 August 1988, and one at 13 km north-west of Tory on 20 September 2007. The first land-based sightings were from Fanad and Melmore Heads on the same day, 28 August 2010, followed by another at Melmore the following day. Another burst of records started on 8 September 2017 when two birds were seen from Bloody Foreland and one from Arranmore, followed by another at Bloody Foreland three days later. So the total now stands at six or seven land-based sightings and seven from boats. There can be little doubt that a number of birds regularly reach the seas west of Donegal, and come within reach of land when conditions are right.

Albatross species (unidentified)

Thalassarche sp. | Albatras

RARE VAGRANT

There have been four unidentified albatrosses seen from Donegal, but none have been specifically identified. It is generally assumed that they would have been Black-browed Albatross *Thalassarche melanophris*, but Yellow-nosed *Thalassarche chlororhynchos* and other 'Mollymawks' *(Thalassarche spp.)* are also possible. All of these records await verification. Three were recorded at Malin Head. They were on 19 September 1966 (MHBO), 10 October 2018 (J. Clark and K. Perry) and 22 August 2024 (R. McLaughlin). The other bird was seen at Bloody Foreland on 9 September 2020 (J. Byrne).

Storm Petrel (European Storm-petrel)

Hydrobates pelagicus | Guairdeall

SUMMER VISITOR AND PASSAGE MIGRANT
STABLE

There is a long history of Storm Petrels breeding in Donegal, but there have been few serious attempts to estimate the numbers, never mind count them (Table 18.14).

Storm Petrel, 56 km west of Burtonport.

There are three large colonies. Rathlin O'Birne and Roaninish have past estimates of 1,000 to 1,500 nests, and despite the lack of any recent figures on that scale, it is quite possible that they still do. Roaninish has been frequently surveyed, with variable results, but some consistency in the 1950s – 255 in 1953, 315 in 1955 and 300–500 in 1957.[252] The other large colony has only just been discovered on Inishtrahull. Being an island that was inhabited in the first half of the twentieth century, and having had a naturalist lighthouse keeper shortly after that, there would surely have been some evidence if petrels had been breeding there at the time. Having now established their presence, the Inishtrahull Bird Observatory has started a nest box scheme, which seems to be successful, and with the Marine Protected Areas Management and Monitoring project (MarPAMM) has surveyed the colony and come up with an estimate of 1,500 pairs.[253]

Inishduff has a moderate colony, and there are smaller ones on Inishdooey / Inishbeg, and on Inishkeeragh. One or two pairs have been found on a number of other islands.

On Tory, the Storm Petrel is probably the most intriguing seabird. There were several reports through the second half of the nineteenth century of breeding petrels, from lighthouse keepers and others. Ussher and Warren specified breeding in the grassy slope at the top of a 300 ft (90 m) cliff – which can only be Tormore.[254] Considerable effort has gone into proving their continuing breeding presence on the island. Robert Ruttledge visited in the 1940s, without success. In 1955, Philip Redman failed to find birds nesting on the face of Tormore, despite three nocturnal visits – although fresh remains of a bird were found.[255] Operation Seafarer in 1969–70 recorded a

Sites	1947–8	1950s	1969	1970s	1985–7	1996	2001	2018–21
Inishtrahull								1,500
Tory	small colony	none	small colony		<100			
Inishbeg	nest found							
Inishdooey								40
Umfin		1+	1–2					
Gola (Torglass)			1–2					
Inishkeeragh			?	75			0	
Roaninish	1,000	255 to 500	250 to 300	1,000 to 10,000	unchanged	400	491	
Rathlin O'Birne	large colony		hundreds	hundreds	500 to 1,000	500 to 1,000	159	208
Inishduff			25		225			

Table 18.14: Storm Petrel colonies (nests or pairs)

breeding colony of unknown size.[256] The Seabird Register included on their map <100 pairs in late 1970s to 1980s, but no detail is given, and there had been no report during the time of their 1985–7 survey.[257]

There have been many June sightings from the Tory ferry, with a high total of 200 on 15 June 1993. These birds may be local breeders on Tory, Inishdooey or Inishbeg, but the mobility of the species leaves many wider possibilities still open. There was a Storm Petrel ringing project at Tullagh Point where for many years amplified sound recordings were used to lure birds at night from an apparently empty ocean, nowhere near any known colonies. As evidence of their mobility, one that Boyd Bryce trapped and ringed on 28 July 1999 was recovered, on the same day, at Sanda Island off the Kintyre peninsula in Scotland, a distance of about 125 km.[258]

Occasional pelagic boat excursions in August, at distances up to 50 km offshore, reliably draw in numbers up to 100, with an estimate of 500 on 15 August 2003. There was an exceptional movement of 4,000 birds passing Arranmore on 2 September 1997. Leaving aside that Arranmore event, and the sightings from boats, there have been 739 birds recorded as migrants. Monthly totals are 222 in July (including a single count of 115 birds, at 29 per hour, passing Bloody Foreland on 17 July 2011), 272 in August, 103 in September and 42 in October. May sightings total sixty-six, but that includes fifty-nine seen from Bloody Foreland on 18 May 2011.

Leach's Petrel (Leach's Storm-petrel)

Oceanodroma leucorhous leucorhous | Guairdeall gabhlach

PASSAGE MIGRANT

STABLE

RED-LISTED BOCCI-4

This more northerly breeding species than the Storm Petrel was first recorded breeding in Ireland in the 1890s, in County Kerry. After many years of searching it was finally established, in County Mayo in 1982, that a few do still breed in Ireland in some of the more remote and inaccessible Storm Petrel colonies. The first hint of breeding in Donegal was a probable bird heard at night on Roaninish on 31 July and 1 August 1952 (D. Wilson). Then nine birds with brood patches were trapped on Rathlin O'Birne in July 1986, and a further fifteen in 1987.[259] Although these records together provide more than a hint of breeding, it is still not proof, as the birds could have been simply paying a visit from a Mayo colony – or even a Scottish one.

On average (excluding exceptional Storm Petrel movements), rather more Leach's Petrels are seen on migration than Storm Petrels, with all records totalling 1,327. Counts in excess of twenty are not unusual. The highest were 110 at Malin Head on 9 September 1984, 89 on 12 September 1997 at Rocky Point, and 68 on 9 September 2017 at Arranmore. The migration season is later, and much more narrowly defined than Storm Petrel's, with 134 birds in August, 1,039 in September and 128 in October.

South-westerly storms can drive Leach's Petrel ashore or inland, where they are usually found, at best, exhausted. All such recorded in Donegal have been as the result of two exceptional storms, one on 27 September 1891 and the other on 27 October 1952. Following the first of these, a bird was found dead near Buncrana and one was found on the outskirts of a wood near Ramelton. On 9, 10 and 11 October individuals were observed in different parts of the parish of Glencolumbkille. At Inver, about the middle of October, one was shot, one picked up dead and two were seen.[260]

Following the south-westerly gales of 24 October 1952 (and subsequent days), an analysis of all reports from Britain and Ireland noted many dead on Tory Island on 27 October, and another freshly dead on 2 November. Dozens were seen alive, and one dead, at the mouth of the Gweebarra estuary on 30 October. One was at Bunbeg on 5 November, and one near Annagary on 19 November. No dates were given for one at Dunkineely and one at Ballindrait. It was calculated that there were not less than 6,700 casualties in Britain.[261] There is a pattern of birds being found some time after these 'wrecks', but as a Leach's Petrel would probably not survive inland for long, the more likely explanation is that birds already weakened by storms are more easily driven ashore on later dates.

The only other inland record was a bird picked up at night south of Carndonagh on 28 September 1998. It was ringed and released the next morning.[262]

There are only two winter records. One at Mountcharles Pier on 1 December 2007, and one in Lough Swilly on 22 November 2015.

Leach's Petrel is Red-listed in BirdWatch Ireland's 'Birds of Conservation Concern in Ireland 4: 2020–2026' on the grounds that its global conservation status is of concern.[263]

Fulmar (Northern Fulmar)

Fulmarus glacialis glacialis | Fulmaire

BREEDING SUMMER VISITOR AND PASSAGE MIGRANT
DECREASING AS A BREEDER, STABLE ON MIGRATION

Fulmars in Britain and Ireland had long been confined to St Kilda, the remote group of cliff-bound islets west of the Outer Hebrides, until in the early years of the last century they started to spread. They reached Donegal and Mayo in 1907, prospecting the cliffs for breeding opportunities. They started breeding here in 1912 (a year after Mayo). By mid-century they were known at Malin Head, Breaghy Head, Horn Head, Tory, Arranmore and Dunmore Head.[264] Since then they have filled in almost all of the gaps.

The geographical spread of birds at 2018 (Map 18.4) reveals only two significant gaps, the north Fanad and Rossguill peninsulas on the north coast, and Crohy Head in the middle of the west coast. These are gaps in coverage – not in the presence of Fulmar.

Map 18.4:
Fulmar breeding
distribution 2018–19

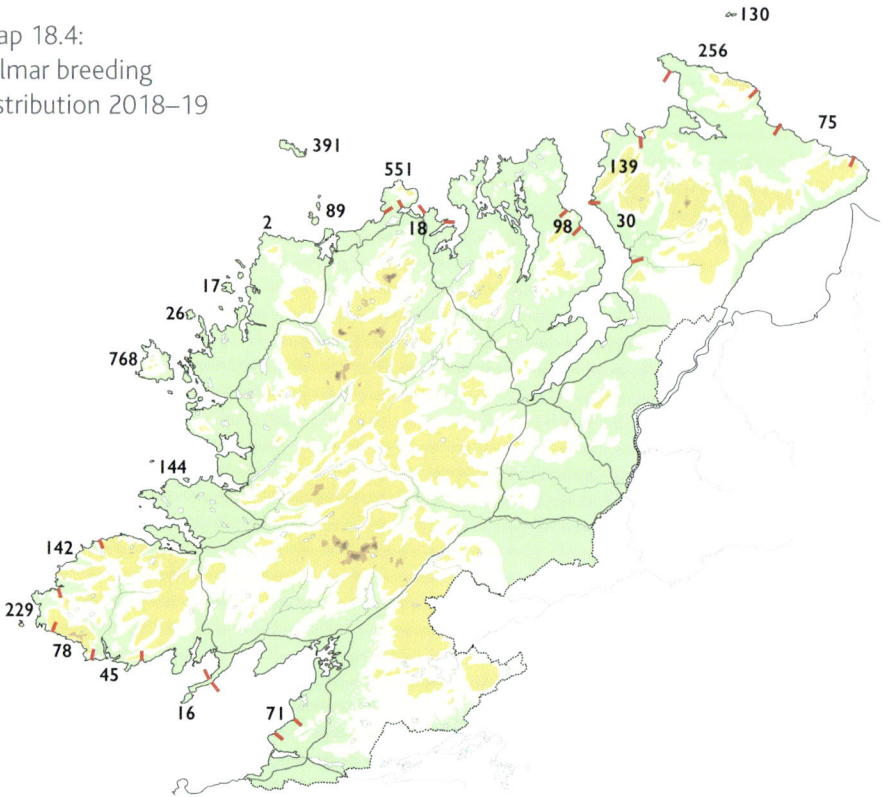

Fulmars are very easily censused, so although the spread of colonies has not declined since those gaps were filled, the table of main sites demonstrates a dramatic decline in numbers over the last twenty years (Table 18.15). The reasons for this are not yet clear, and while most of the larger Irish colonies are also declining, some are increasing.[265]

Fulmars are an almost constant presence throughout the early autumn at seawatching sites, but for that very reason they are often not counted. Numbers rise through July to a peak in August, then fade through September, with very few being seen after that. The highest recorded rates have been 308 per hour on 1 September 2019, 303 on 8 August 2016 and 230 on 18 August 2019, all at Bloody Foreland. The average of the top twenty-five timed counts is ninety-seven per hour.

There is a dark colour phase known as Blue Fulmar which appears more frequently at colonies in the Arctic. It has once been recorded on breeding cliffs in Donegal, when two birds appeared to be paired with two normal birds at a colony on Arranmore, on 1 July 1966 (A. Ferguson). Apart from

Site	1969	1987	1999	2018
Saldanha Head			250	98
Horn Head	1,200	843	1,122	551
Tory Island		246	641	391
Arranmore	786		1,535	768
Tormore (area)		155	534	142
Muckross (area)			173	45
St John's Point	35		23	16
Coolmore (area)	49		90	71
Totals for Full Counts			4,368	2,082

Table 18.15: Fulmar numbers on selected stretches of coast

that, it occurs occasionally on migration, with forty-three birds seen on thirty-one occasions. The highest number seen was five, at Malin Head on 23 September 1984. Three sightings for which Donegal can also lay claim were two over the Rockall Bank (on 13 March 2013 and 26 March 2016) and one 160 nautical miles west-north-west of Tory (on 30 March 2015). The second and third of those sightings were of the so-called 'Double Dark' morph from the high arctic. Taking all forty-three sightings together, every month except January and February is represented. The most favoured months are August with six sightings and March with five.

Gadfly Petrel sp. (Zino's/Fea's/Desertas Petrel)

Pterodroma madeira/feae/deserta | Spéiceas Guairdeall creabhair (Guairdeall Zino/Fea/Desertas)

RARE PASSAGE MIGRANT

The first sighting in Donegal waters of one of these very rare gadfly petrels was from a boat 56 km north-west of Arranmore on 18 August 2001 (B. Robson et al.). This was outside the 30 km limit of Inshore Waters under the remit of the IRBC, but inside the 370 km Exclusive Economic Zone, so the record is recorded as 'at sea', but excluded from the statistics.[266] Four land-based sightings have followed. The first was at Melmore Head on 29 August 2002 (E. Randall) followed by one at Malin Head on 3 August 2014

(R. McLaughlin). Then one was seen from Fanad Head on 4 September 2020 (C. Ingram), and Eric Randall had a second one at Melmore Head on 15 August 2024. The last two have yet to be assessed by IRBC.

Fea's Petrel breeds on the Cape Verde islands off North Africa and has a breeding population of less than 2,000 pairs. Desertas and Zino's both breed only on the Maderian islands and have less than 1,000, and less than 100 pairs, respectively. All three species were until recently thought to be one, but even now that their true relationships are better understood, specific identification away from their breeding grounds is rarely successful. Desertas Petrel is classed as Vulnerable by BirdLife International, while Zino's and Fea's are both Endangered.

Cory's Shearwater

Calonectris borealis | Cánóg Cory

PASSAGE MIGRANT

Cory's is one of two large shearwaters that regularly move along the Donegal coast. It would usually be seen in association with the Great Shearwater, but is the less frequent and numerous of the two, and peaks earlier in the season. Being a breeding species in the warm temperate zone of the eastern Atlantic, Donegal is at the northern limit of its dispersal, so most of the sightings have been in single figures, but there are a few exceptions. The

Cory's Shearwaters, 15 km north-west of Tory.

ROBERT VAUGHAN

first Irish record was as recent as 1958, not long before the first Donegal bird passed Malin Head on 11 April 1965 – still the only spring record. This was followed on 1 October of the same year when Brendan Doherty and Joe Donaldson recorded thirty-nine birds passing Inishtrahull – the only year when there has been recorded seawatching from the island. It was a very late date for Cory's, although that was not known at the time. The first large event was when Martin Garner and I connected with 488 birds passing Fanad Head on 3 August 2005, at a rate of 163 per hour. The rate was still rising at the end of the three-hour watch, despite a falling wind speed.

Although most sightings are still in single figures, the frequency of these has increased greatly in the last two years, with many more larger movements bringing the total seen in the county to over 2,000. August 2024, at the time of writing, is remarkable in almost constant south-west–west winds, and in having Cory's present on all but one of twenty-four timed seawatches. These included the next three highest rates of passage – eighty-four per hour on 29 August, sixty-six on 17 August and fifty-four on 22 August, all at Rocky Point. But all the main seawatching stations, apart from Arranmore, were involved. One previous notable event was at Arranmore on 9 September 2023 when fifty-two birds were counted in one hour.

Passage is highly concentrated in August, with 1,137 birds. July has had 135, September 109, October 40, and there have been single birds in November and April.

Sooty Shearwater

Ardenna grisea | Cánóg dhorcha

COMMON PASSAGE MIGRANT
STABLE

Sooty Shearwaters reach Irish waters in late summer, having travelled from the south Atlantic – most likely the Falkland Islands. Like the Great Shearwater, they arrive here having crossed the North Atlantic from the east coasts of the USA and Canada, but along a more northerly trajectory.

The only nineteenth-century record is of one bird believed to be Sooty Shearwater seen by Barrington 30 km north-west of Donegal on 18 June 1896 – a suspiciously early date, suggestive perhaps of a misidentified

Balearic Shearwater. Confusion over the identification of Sooty Shearwater persisted until the 1950s when observations from Malin Head and Inishtrahull first showed the species to be a regular passage migrant.[267]

Although there are occasionally good passages in August, and even one or two in July, Sooty numbers peak in September, the month with all the best counts. October and November are as July and August. Passage of around 20 per hour is normal, with rates averaging 70 for the best 50 timed counts, and 111 for the best 25. Exceptional rates have been 833 per hour at Melmore Head on 15 September 2013, and 407 at Malin Head on 10 September 1978. Late counts of note were 116 per hour on 22 October 2013 and 54 per hour on 15 November 2023, both at Bloody Foreland.

Great Shearwater

Ardenna gravis | Cánóg mhór

PASSAGE MIGRANT

Unlike Cory's and Sooty Shearwaters, this species has long been known in Donegal waters, from the first record of three seen by R. Ball from a boat off Bundoran on 16 July 1840.[268] Although often associated with Cory's, the Great Shearwaters arrive here by a very different route, and mostly later, having started at their South Atlantic breeding colonies on the Tristan da Cunha islands and travelling via the east coast of North America. Although very much less frequent than Sooty Shearwaters, Great Shearwater appearances can be much more impressive events. The biggest numbers were 3,000 at Tory on 9 September 2007 and 1,200 off Arranmore on 5 October 1993. A loitering flock of 1,060 birds at Rocky Point on 29 September 1993 was exceptional. On 24 September 1999 there were 704 at Rocky Point and 250 at Tory, and 221 passed Bloody Foreland on 18 August 2024 at 62 per hour. A boat trip out of Magheraroarty on 27 August 2006 encountered about 250.

Of 112 sightings, 56 in August had 1,271 birds, 39 in September had 5,865 birds and 9 in October had 1,244 birds. The also-rans were four sightings in July with twenty-two birds, one sighting in May with three birds (in 1890), and one bird each in March, April and June. So the grand total of birds seen is 8,408, although that includes some very large totals that could only have been rough estimates.

Manx Shearwater

Puffinus puffinus | Cánóg dhubh

RARE BREEDER AND ABUNDANT PASSAGE MIGRANT

STABLE

Ussher and Warren and Kennedy et al. both say that Manx Shearwaters breed on Arranmore, but there has been no evidence produced since then. Kennedy et al. report that a few were said to nest on Tory, but despite considerable efforts in 1955, Philip Redman failed to confirm the claim, and also reported that it must have been a misprint, as Major Ruttledge said that it has *never* been known to breed.[269] A nest was found at Malin Head in 1965, at a time when birds were present in the waters around Inishtrahull.[270] Finally, in 2021, about thirty pairs have been proved breeding on Inishtrahull.[271] It is still possible that they breed elsewhere in the county, but nocturnal surveying on very inaccessible islets is probably the only way that it will be established. It is likely that most birds recorded in June are on feeding trips from the large breeding colonies in Scotland or the south-west of Ireland.

Manx Shearwaters pass Donegal in large numbers in early autumn. It is curious that in 1954 Redman failed to detect this passage from Tory, despite continuous seawatching throughout the period. 'The dispersal of this species from the large colonies in the Inner Hebrides does not appear to take in the north Irish coast.'[272] The current evidence leaves no doubt that they *do* pass around the north and west of Ireland on their journey south from Scotland, to winter quarters off South America. This mostly occurs from late July to early September. Most of the higher rates of passage (200–1,500 per hour) are in Force 6 north-west winds. Rates of 3,000 or more are usually ignored or only roughly estimated. They tend to be in south-west winds, or on one occasion in a Force 6–7 northerly, at Bloody Foreland on 13 September 2001. Fanad Head had a rate of 1,303 per hour at the unexpected date of 29 June 2020 (Figure 18.58).

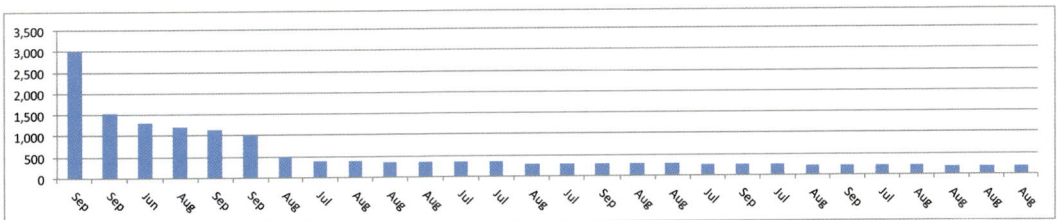

Fig. 18.58: Manx Shearwater highest hourly rates of passage (with the month of occurrence)

Balearic Shearwater

Puffinus mauretanicus | Cánóg Bhailéarach

SCARCE PASSAGE MIGRANT

Part of the population of this rare west Mediterranean species moves north to the Irish south coast in late summer, and a few of those birds continue as far as Donegal. The first record is of the only spring sighting so far, at Tory on 19 April 1963 (TBO). There have been sixty-seven birds recorded passing in the autumn months, almost all of them singly, and with a maximum of three. Four birds were in July, thirty-one in August, twenty-two in September and two in October.

Balearic Shearwater is Red-listed in BirdWatch Ireland's 'Birds of Conservation Concern in Ireland 4: 2020–2026' as the IUCN has assessed it as Critically Endangered Globally – the highest possible threat level.[273]

White Stork

Ciconia ciconia ciconia | Storc bán

RARE VAGRANT

White Storks rarely overshoot this far north on the return from wintering in sub-Saharan Africa to their breeding grounds in continental Europe. There have been three occurrences. The first was at Drumnaraw, south of Creeslough, flying north-east along the ridgeline on 15 April 1997 (R. Sheppard). The second was at Glenveagh on 7 May 1980 (L. McDaid, C. MacLochlainn et al.). The third was an approachable bird, ringed and probably an escape, which has not been assessed by the IRBC. It was present in Gweedore from 7 to 9 May 2011, and two days later in Inishowen (A. Doherty et al.).

Gannet (Northern Gannet)

Morus bassanus | Gainéad

ABUNDANT PASSAGE MIGRANT
STABLE

Gannets can be seen around the coast at all times, but are rare in winter, when they are mostly dispersed throughout the north Atlantic. Birds in

Gannet, Muckross Head.

KIM PERIERA

summer are likely to be on long feeding trips from the nearest large colony, which is on the Scottish island of Ailsa Craig in the Firth of Forth. Individual birds are occasionally seen perched at mixed seabird colonies, but there has never been any suspicion that they had breeding ambitions for Donegal.

From July to early November numbers are consistently high at seawatching points, but, like Manx Shearwater, Fulmar and the auks, they often go uncounted on the days when bird numbers and diversity are high. In winds below Force 4 or 5 they will usually be travelling in both directions. Movement south in stronger onshore winds is typically at 200 to 400 per hour, with a recorded peak of 828 at Bloody Foreland on 24 September 2015. A few storm-driven juvenile birds have been found inland, dead or stranded.

Cormorant (Great Cormorant)

Phalacrocorax carbo carbo | Broigheall

RESIDENT AND WINTER VISITOR

STABLE RESIDENT POPULATION, WINTER VISITORS INCREASING

Towards the end of the nineteenth century, Hart reported fifty pairs of Cormorants nesting on Breaghy Head – the largest assemblage he recalled. He also observed that they used burnt heather-sticks with moss for lining their nests.[274] At that time there was a night-time roost of Cormorants on an islet on Lough Fern, used by birds which commuted to feed in Lough

Swilly.[275] It still exists, and is currently used by at least sixty birds – but there have been no reliable counts in recent years.

Table 18.16 fails to demonstrate any consistent trend in either individual breeding colonies or in the overall population. Macdonald concluded that there was an increase of 42 per cent nationally between the two big surveys of 1969–70 and 1985–7. For Donegal the increase was only from 290 to 294, with no details given of individual colonies.[276] Table 18.16 includes more sites and a slightly wider time-frame, and suggests a fall in the Donegal population over that same period – from 490 to 366. The high peak, in 1999–2002, is accounted for by better coverage of the islands, and what appears to be the discovery at Balbane, in north-east Inishowen, of numbers equivalent to those displaced from Stookaruddan. Between the comprehensive surveys of 1990–2002 and 2015–16, there appears to be a drop in overall numbers of about 30 per cent.

Cormorants are widespread in winter all around the coast, and in estuaries. Inland, they do turn up in lakes, and people can often be surprised by one or two Cormorants flying off from relatively small or tree-enclosed rivers. However, the winter population trend is probably more effectively monitored on a few key sites, like Lough Swilly.

The annual peaks at Lough Swilly (Figure 18.59) show a minor peak around the turn of the millennium, and then a major surge in numbers since 2014. This coincides with the replacement of the native oyster species by the much larger Pacific oyster, and the development of commercial dredging for them in the parts of Lough Swilly where the Cormorants feed – although how the almost totally fish-eating Cormorants could be benefiting from this ecological upheaval is not clear. The recent peak in the numbers wintering in Lough Swilly was 1,214 on 15 January 2022, which crosses the Internationally Important threshold for the first time. No other site in

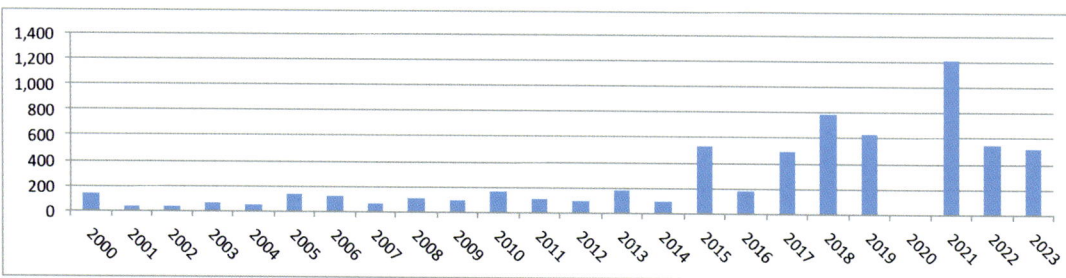

Fig. 18.59: Cormorant peaks on Lough Swilly, from 2000/1 to 2023/4

Sites (from NE to South)	1954	1965–70	1985–7	1999–2002	2010–11	2015–16
Balbane Head				225		210
Kinnagoe Bay		65				11
Kinnagoe – Tremone			62			
Stookaruddan group		200				
Malin Head						17
Garvan Isles			48	50		
Glashedy Island			4	5		20
Dunaff Head		6			12	47
Urris cliffs			30		17	15
Saldanha Head				40		62
Tranarossan Isles		50	17	69	67	7
Clonmass Isle (Sheep Haven)				39		
Horn Head		20				
Dunfanaghy New Lake						1
Dooros Point (Falcarragh)			3			
Inishdooey			10	9		3
Inishbeg			2			
Tory	178					
Bloody Foreland			12	2		
Gola and Umfin Islands		35	45	40		
Owey Island				54		
Arranmore (Torneady)			50	11		
Arranmore (Illanaran)		16	20			
Roaninish				22	27	36
Inishbarnog				36		
Rathlin O'Birne		3				1
Slieve League		95	63	60	20	
Muckross Head					29	
Inishduff						35
Inishgoosk (Lough Derg)						1
Totals for Full Counts		490	366	662	172	466

Table 18.16: Counts of breeding Cormorants (nests or pairs)

Ireland appears to have reached this level.[277] It is worth noting that most of this spectacular increase has happened since the last survey of breeding colonies. It will be interesting to find out if these new wintering birds are locally bred, or immigrants.

Three other sites are important: River Foyle, with a peak of 209 birds between 1994–5 and 2000–1, Donegal Bay (49) and Ballyness Bay (35).[278]

The Internationally Important figure (1% of the flyway population) is 1,200.

The Nationally Important figure (1% of the national population) is 110.

Qualifying Sites: Lough Swilly

Shag (European Shag)

Gulosus aristotelis aristotelis | Seaga

RESIDENT

DECREASING

Shags are marine birds, but firmly attached to land by the need to dry their wings. They breed colonially on cliffs, usually low down, and in caves where these are available. Operation Seafarer gave 850 pairs for the county.[279] This is consistent with the 926 in Table 18.17, which includes some extra sites counted within a year of the main survey. Despite the erratic coverage since then, and the variability at individual sites, the overall total for the county shows a definite downward trend.

In winter, Shags are found dispersed around the coast. They were not counted systematically until 2010, when they were added to the list of species to be counted for I-WeBS – although I-WeBS sites don't hold many Shags. Donegal Bay had a five-year mean of sixty-two, to 2015/16. Lough Swilly, for the five years to 2019/20, has an equivalent figure of only ten. The very dynamic conditions at the entrances to some of the large enclosed bays, like Ballyness and Mulroy, attract large numbers – Trawenagh and Trawbreaga bays are similar. The only one of these that has been systematically counted is Ballyness Bay. It had a three-year mean peak of 115 for the seasons up to 1997/8.

Sites (from North to South)	1965–70	1980	1987	1997–2000	2015–18
Kinnagoe Bay	12				
Kinnagoe – Tremone	13				
Glengad Head					17
Stookaruddan group	120		214		16
Inishtrahull	350			187	90
Garvan Isles			183	183	11
Malin Head	48				95
Glashedy Island	20				
Dunaff Head	61			63	
Urris Cliffs					
Saldanha Head	9				7
Tranarossan Isles	23				
Breaghy Head (Sheep Haven)	5			10	10
Clonmass Isle (Sheep Haven)					13
Horn Head	50	69	191	111	92
Inishdooey				90	74
Tory	4		9	27	77
Inishsirrer				30	3
Gola & Unfim Islands	27			40	38
Owey Island	5			61	23
Arranmore	55			21	90
Roaninish			146		5
Dunmore Head	36			43	26
Port (Tormore)	36		11		
Port (Glen Head)			18		7
Rathlin O'Birne	2		10		5
Slieve League			2		
Inishduff	50		116		81
(five others)					17
Totals for Full Counts	926		900	866	797

Table 18.17: Counts of breeding Shags (nests or pairs)

Glossy Ibis, Rossbeg.

DEREK BRENNAN

Glossy Ibis

Plegadis falcinellus | Íbis niamhrach

RARE VAGRANT

The first record of Glossy Ibis was after 1900, and the second in 1945/6 when there was an influx nationally.[280] The species has been increasing in recent years, and its tendency to disperse widely from its breeding range in southern Europe has brought many more to Ireland, including Donegal. Two were in Dunfanaghy from 13 to 20 December 2013, and a readable colour ring allowed one of them to be identified as from a Spanish breeding population (R. Sheppard). A third was found dead on Tory on 18 December. One was seen on Inch Levels on 1 January 2014. And one was at Malin Town on 20 October 2016. Finally, seven birds were reported on 24 December 2021 near Frosses. One or more were subsequently seen in various locations in south-west Donegal up to 3 January 2022, when one was at Rossbeg. All seen since the 2013 birds have yet to be verified by the IRBC.

Spoonbill (Eurasian Spoonbill)

Platalea leucordia leucordia | Leitheadach

RARE VAGRANT

With a scattered breeding distribution in Europe, and dispersive non-breeding movements, five Spoonbills have reached Donegal over widely spaced intervals. A long period of decline in Europe has been followed recently by a revival that has reached as far as Britain. Unfortunately, that has not, so far, been sufficient to produce more records for Donegal.

There are two very old records. The first was a bird killed in the winter of 1837/8.[281] The second was killed at Dunfanaghy in 1850/1.[282] In 1910 another bird was shot, at Inch on 11 November.[283] Modern records start on 26 October 1965 with an immature bird seen arriving at Malin Head from the north, and departing towards the south (MHBO). A bird was at Blanket Nook and Inch from 1 May into June 1983 (C. Brewster et al.).

Bittern (Eurasian Bittern)

Botaurus stellaris stellaris | Bonnán

FORMER RESIDENT, NOW RARE WINTER VISITOR

The first record of Bitterns breeding in Ireland was in 1840. Ussher and Warren quote W. Sinclair for evidence of breeding in Donegal – 'the locality was an extensive swamp between Strabane and Londonderry, and he said that the bird was not uncommon then and bred in Donegal'.[284] Extensive reedbeds still line banks of the River Foyle on both sides, but are only fragments of what would have been present at that time. Ussher and Warren also reported seven records from Donegal since 1840. Although fairly frequent countrywide as a winter vagrant, there were no further records from Donegal until two recent sightings. The first was at Malin Head on 30 December 2012 (O. McLaughlin). The second was a long-staying bird at Inch from 30 December 2012 to 2 March 2013 (D. Charles et al.).

The large increase in the resident population in Great Britain (the fruits of dedicated habitat creation and management) is likely to have been the source of the last two birds to have reached Donegal.

American Bittern

Botaurus lentiginosus | Bonnán Mheiriceánach

RARE VAGRANT

There is a single record of this species, which usually arrives in Ireland in late autumn and winter. One was found alive in a field near Malin Beg on 21 October 1974. It died, and its skin is preserved in the National Museum.

Little Bittern

Ixobrychus minutus minutes | Bonnán beag

RARE VAGRANT

There is a single record of this European species, which tends to disperse widely in winter, although mainly wintering in tropical Africa. One was recorded from Owey Island on 9 February 1908.[285]

Black-crowned Night Heron

Nycticorax nycticorax nycticorax | Corr oíche

RARE VAGRANT

Donegal's claim for Night Heron rests on a record from March 1834 – the first in Ireland. It is a common European species which disperses post-breeding, as well as migrating to tropical Africa. Donegal appears to be too far north to benefit more frequently from that dispersal.

Cattle Egret (Western Cattle Egret)

Bubulcus ibis | Éigrit eallaigh

RARE VAGRANT

The remarkable global spread of the Cattle Egret is well known. In Europe, the last half century has taken it from a limited core range in Iberia to the early stages of colonising England. Not surprisingly, both individuals and small flocks are now appearing in Ireland, and Donegal has had four recent records. The first was at Inch on 6 January 2008 (D. Breen). The second was a bird that settled in with a free-range herd of pigs at St Johnstown from 11 December 2012 to 21 March 2013 (J. Hamilton, A. Speer et al.). It was probably the same bird which was present at Porthall, about 5 km to the south, from 5 to 8 July. The third bird first appeared at Malin Town on 28 October 2016 and stayed until 16 November (R. McLaughlin et al.). The most recent visitor was at Inver from 4 to 26 October 2017 (F. McDaid, M. Cunningham) where it consorted with a flock of sheep.

Grey Heron

Ardea cinerea cinerea | Corr réisc

RESIDENT AND WINTER VISITOR
STABLE

The familiar Heron, as it is still called by most birdwatchers, is also still called the Crane by many non-birdwatchers, causing much confusion. It is widespread throughout the county at all times of the year. As a breeding species it was colonial in the nineteenth century when the big estates still provided suitable copses of large trees. Ussher and Warren gave a list of nineteen colonies, many of them now gone. The bigger ones were at Fortstewart (thirty nests), Clonleigh (sixteen) and Ards (large).[286] Small colonies of a few nests still exist close to good feeding areas, as at Inch, Blanket Nook, Inner Donegal Bay, Kincrum and elsewhere, but mostly now the Grey Heron nests alone. There have been no recent surveys, other than the breeding atlases, that would give any indication of numbers. Evidence of breeding in the first atlas period of 1968–72 was found in twenty-nine out of the seventy-five 10 km². This was down to sixteen occupied squares in 1988–91. By 2008–11 there had been a further decline, to thirteen squares. Differences in methodology and effort make those comparisons somewhat less than definitive. And the continuing ubiquitous presence of the bird suggests that the real breeding population is more likely to be at least stable, and larger than indicated by the number of occupied ten km squares recorded in any of the atlases.

There is some immigration from the continent and Great Britain in winter, but also two instances when Danny O'Sullivan, the lightkeeper on Inishtrahull, noted birds arriving in late summer, and departing towards Scotland.

At Lough Swilly the five-year running mean to 2019/20 was seventy-four, and the absolute peak was ninety-three on 8 October 2016 – both of them increases from the 2011–15 period when there was a mean of forty-nine and a peak of fifty-nine. Equivalent 2011–15 figures for Donegal Bay are thirty-eight and sixty – for Lough Foyle they are twenty-nine and fifty-one. Lesser sites between 1994/5 and 2000/1 were River Foyle with a peak of thirty-six, Culdaff with twenty-one and Trawbreaga Bay with nineteen.

The Internationally Important figure (1% of the flyway population) is 5,000.

The Nationally Important figure (1% of the national population) is 25.

Qualifying Sites: Lough Swilly, Donegal Bay, Lough Foyle

Little Egret and Grey Heron, Inch.

Great White Egret (Great Egret)

Ardea alba alba | Éigrit mhór

VAGRANT

INCREASING

The Great White Egret is one of several large wading birds which are spreading from their core ranges, in this case eastern Europe, possibly in response to climate change. Good habitat management at key sites has undoubtedly helped to consolidate the movement. With records in Ireland increasing rapidly, it was inevitable that some would reach Donegal, and so far it seems that at least ten, and possibly as many as fifteen, have done so.

The first was seen at Blanket Nook on 10 October 2013 (D. McLaughlin and R. Wheeldon). The second was at Inch Lough from 22 to 23 October 2019. A bird was reported to have frequented Letterkenny golf course from 28 January to 2 February, but was not seen by any birdwatcher. There were several sightings in 2021. The first was at Lough Fern on 31 May. A bird was then seen at Inch on 10 June. There was another in the Sheskinmore area on 13 and 14 June, and finally a sighting at Blanket Nook on 30 June. How many were involved is hard to say, but IRBC records it as three. In 2022 the first bird was seen at Blanket Nook on 1 May followed by several sightings of singles in the general area and one of two birds at a roost, with three ultimately being seen together at Blanket Nook on 14 October. Records continued with a bird at Inch from 3 to 12 April 2023, and another on 2 September at Blanket Nook. Finally, from 17 to 27 May 2024 three separate individuals were identified at Blanket Nook and Inch (including one with a red leg ring, and one with a distinctively coloured bill), but never more than one at a time.

It is worth noting that the first record of the species in Ireland was a bird seen in County Sligo from 22 May to 1 June 1984. At around the same time (probably June) an unidentified egret (thought to be not a Little Egret, still very rare) was reported from Teelin Harbour, in south-west Donegal, but was not re-found by the few birdwatchers who tried to confirm it.

Little Egret

Egretta garzetta garzetta | Éigrit bheag

WINTER VISITOR AND COLONISING RESIDENT

INCREASING

So far, this is the most celebrated of the many large wading species (mainly herons and their allies) that are moving north in response to climate change. It was first seen in Ireland in 1940, and not again until 1956, after which it was of annual occurrence with numbers rising, until by the 1980s it was well established. Breeding in Ireland first occurred in 1997,[287] and it now has breeding and wintering populations that invite comparison with those of the Grey Heron.

The first one to be seen in Donegal arrived with the first national influx. It was near Dunfanaghy, on 2 June 1957 (W. Wheeler and W. Carlyle). Another early record was a bird at Inch on 20 May 1970. It was not until 1998 that Dunfanaghy had its second bird, and then one or two regularly from 2003. Four or five birds could be seen at Donegal Bay from about 2007, rising to nine by 2014 and twenty by 2017. The Malin peninsula is another area with multiple records in recent years. The first bird on a Lough Swilly I-WeBS count was in 2006/7. Growth then was exponential – low single figures for the first ten years, accelerating to fourteen over the next four years, then rocketing to twenty-nine in 2021/2, thirty-four in 2022/3, and on 4/5November 2023 to fifty-nine, which, amazingly, surpassed the fifty-three Grey Herons recorded on that same count (Figure 18.60).

Apart from these favoured areas, individuals had been turning up throughout the establishment phase at widely dispersed locations around the coast, and can now be seen in most muddy estuaries.

A pair with five juveniles seen at Malin on 2 August 2017 was the first evidence of breeding. In 2020 there were reports of two pairs breeding in the Donegal Town area, and one at Ramelton.

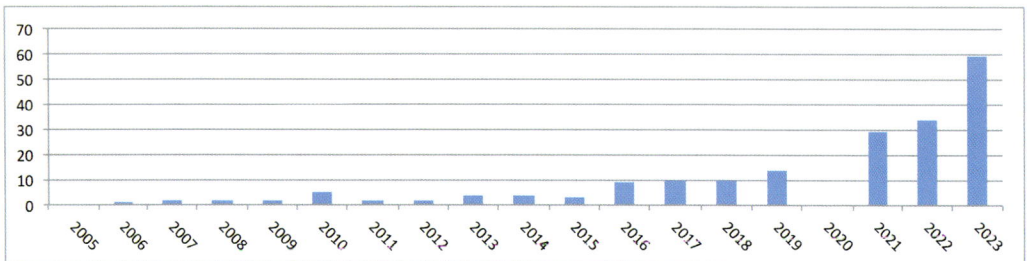

Fig. 18.60: Little Egret winter peaks on Lough Swilly 2005/6 to 2023/4

Osprey (Western Osprey)

Pandion haliaetus haliaetus | Coirneach

SCARCE PASSAGE MIGRANT

It is rather remarkable that there were no records of Osprey in Donegal from the nineteenth century, as was acknowledged at the time by Ussher and Warren.[288] The first one attributed to the county was a specimen labelled 'Obtained at Cultra, 29 October 1907'.[289] Unfortunately, Cultra is in County Down, so we must advance to 31 August 1974 for the first reliable record, when a bird was seen at Lough Eske (N. Murphy et al.). The second bird is particularly interesting, as it was found dead on Slieve League on 1 October 1978, having been ringed in 1976 as a nestling in Finland. A gap of fifteen years without any records followed, to 1991, but that was the start of a steady trickle of twenty-three birds leading up to the present. This increase can probably be explained by a few birds from the growing summer population in Scotland traversing Ireland en route to and from Africa.

Nine of these recent Ospreys have been on Lough Swilly (six at Inch, two at Blanket Nook and one at Dunree), and there have also been three at Glenveagh, one in the Barnesmore Gap and one at Malin Beg – all reasonable locations to expect migrant Ospreys. The remainder are more randomly distributed. The monthly distribution highlights the northward passage in May (Figure 18.61).

There can be little doubt that Donegal has the potential to sustain a small breeding population – although some human assistance might be preferable to waiting for this to happen naturally. For example, the two birds recorded at Inch Lough, on 31 May 2016, is the only record of more than one bird. But if a nesting platform had been in place, it could have been enough to trigger the change we are all waiting for.

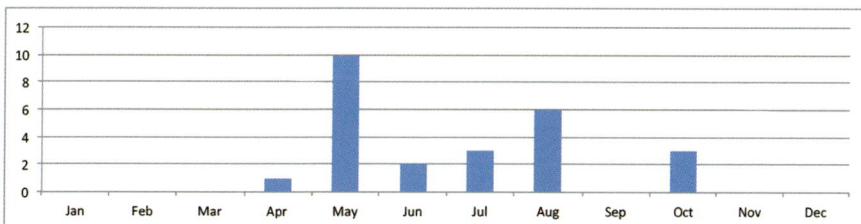

Fig. 18.61: Osprey sightings by month

Honey Buzzard (European Honey Buzzard)

Pernis apivoris | Clamhán riabhach

RARE VAGRANT

There can hardly be a more fortuitous record than Donegal's only Honey Buzzard, which might never have been noticed had it not been wearing a satellite tag. It had been tagged as a nestling in the Scottish Highlands, and those following its progress online noted that it had reached Ireland, and Donegal, at Ards, on 29 September 2001. John Coveney, Kieran Fahey and others followed the trail and managed to see it. By 8 October it had moved to Ardara, and then departed the county. But sadly, it is thought to have perished in the seas south of Ireland. However, its sibling, travelling via England and Spain, arrived safely in North Africa.[290]

Egyptian Vulture

Neophron percnopterus | Bultúr Éigipteach

VAGRANT

The only Egyptian Vulture to have been seen in Ireland is also the first individual seen in the UK for 150 years – recorded in the Isles of Scilly a month before it re-appeared in Donegal. It was seen at Dunfanaghy on 14 July 2021 (Shane Farrell and Robert Vaughan) and stayed for another day, to be seen by a horde of admirers.

Egyptian Vultures are summer visitors from sub-Saharan Africa to the Mediterranean region, and like most large, soaring raptors, they avoid long sea crossings

Egyptian Vulture, Dunfanaghy.

ROBERT VAUGHAN

where possible. They normally reach south-west Europe via Gibraltar, and south-east Europe via the Bosphorus in Turkey, so the English Channel and the further crossing to Ireland are considerable barriers. Having reached Ireland, this bird continued to explore the island up to at least 6 May 2022, although how it found enough to eat is not clear.

Golden Eagle

Aquila chrysaetos | Iolar fírean

EXTINCT RESIDENT
RE-INTRODUCTION PROJECT UNDERWAY
RED-LISTED BOCCI-4

The story of native Golden Eagles in Donegal is one of nest destruction, shooting and egg collecting. In 1832, William Thompson saw two nests on Horn Head, and was informed by the gamekeeper that he had killed only one in the last four years.[291] By 1850 there were less than sixty pairs left in Ireland, and by 1900 only fourteen.[292] Writing in 1891, Hart knew of only one remaining pair, on the west coast, but recalled them on Errigal, Lough Salt Mountain, Poisoned Glen and Fanad.[293] In general, Ussher and Warren concurred, also noting that the area around Errigal and Muckish had been the principal stronghold, with more than one pair in 1866. They also reported that in 1898 a pair had returned to this area.[294]

The last pair to breed in the county was in Glenveagh in 1910, two years before the loss of Ireland's last pair, in County Mayo, but the final demise of the species was long-drawn-out. There are accounts of a bird shot in the spring of 1915, and of its mate trapped eleven years later. This is presumably the bird reported to have been trapped and destroyed near Ardara on 2 April 1926, which was considered to be the last bird in the county.[295]

The *Derry Journal* on 30 August 1937 reported that all efforts to dislodge a huge eagle from its nest above the caves at Maghera had so far failed. 'The tenants in the district are wroth with it, as moorfowl are now realising from 5s to 7s 6d [5 to 7.5 shillings] per brace.' This appears to be a single bird and probably too long after the last breeding pair to have been a survivor. Although the location might suggest a White-tailed Eagle, by 1937 the nearest possible origin for one would have been Norway. So this bird is most likely to have been a wandering Golden Eagle from Scotland.

After that, the only appearances of a wild bird was of one seen passing Tory to the south-west on 12 September 1963 (TBO), and a juvenile in Glenveagh in July 1984.

At the millennium, it was decided to attempt a re-introduction of Golden Eagles to Ireland. The Golden Eagle Trust was established to run the project in cooperation with NPWS, and Glenveagh was chosen as the best centre. Fifty surplus chicks were sourced in Scotland over six years, raised to fledging stage in various suitable locations, and released. These birds eventually matured, set up territories and started to breed. The outcome has been sixteen Irish-bred birds fledging successfully up to 2018 – one in 2007, two in 2009, three in 2010, two in 2013, one in 2014, one in 2016, three in 2017 and three in 2018.[296] Although a careful survey prior to the start of the project established that enough food was available in the Donegal mountains, it is now thought that this may no longer be the case. We can only hope that whatever is slowing the establishment of a self-sustaining population in Donegal and beyond, it will be rectified.

The Golden Eagle is Red-listed in BirdWatch Ireland's 'Birds of Conservation Concern in Ireland 4: 2020–2026' on the grounds of its historical decline to extinction.[297]

Sparrowhawk

Accipiter nisus nisus | Spioróg

RESIDENT

STABLE

Traditionally, this has been our commonest raptor, although it may now have been overtaken by the Buzzard. But being less conspicuous than the Buzzard, it is more difficult to get a good assessment. The Sparrowhawk is a woodland species, but in Donegal, with very little woodland, it can make do with hedgerows, scrub and anything that will provide enough cover for hunting and nesting. It is largely absent from the upland areas of the centre and south-west, and from the islands. The atlas projects have shown it to be stable overall.[298]

Sparrowhawk, Dunkineely.

Tory has a regular trickle of birds between September and October, and just a few from March to April, and in winter.

Marsh Harrier (Western Marsh Harrier)

Circus aeruginosus aeruginosus | Cromán móna

FORMER BREEDING SUMMER VISITOR, NOW RARE VAGRANT

INCREASING

The Marsh Harrier was once common throughout Ireland. Ussher and Warren give more limited evidence for its presence in County Donegal.

> Mr. William Sinclair states that it used to breed in small numbers in Donegal, but that he had not seen one there since about 1870. The late Sir Victor Brooke, in 1890, mentioned a former breeding-haunt in the south of Donegal.[299]

There have been eight (or nine) birds seen since then, all in relatively modern times. The first was a female at Blanket Nook on 24 June 1983 (R. Sheppard). A bird was reported from Sheskinmore Lough on 25 May 1995, and a female was also there on 3 May 1997. Four or five birds have been at Inch. First there was a female on 14 May 1999 and another from 20 to 26 August 2012. Then on 15 to 18 April 2022 another bird was seen (an immature), and again (or a separate bird) on 3 June. And there was a female on 3 April 2023. The most recent sighting was a female or immature bird at Inner Donegal Bay on 26 March 2024.

All but one of the modern sightings have been in spring/summer – the exception being the single bird seen in August.

Hen Harrier

Circus cyaneus | Cromán na gCearc

RARE RESIDENT

STABLE

Ussher and Warren recorded the Hen Harrier as common in Inishowen in the 1860s, but rarely breeding north of Pettigo.[300] They also say that it possibly nested on Lord Leitrim's estate in 1893 – this would most likely have been on Fanad. Not much has changed since then. The first national Hen Harrier survey found two to three pairs on the Pettigo plateau, and two to five pairs in Inishowen.[301] By the second survey, no birds could be found in Inishowen, but the south-east still held two to three pairs, although the area surveyed was somewhat wider than previously.[302] That toehold in the south-east survived and built up to four to five pairs in 2010 and eight to twelve pairs in 2015.[303] That peak seems not to have been sustained, but the toehold remains.

This is a small nucleus in Donegal, but it is far removed from the main populated areas in the south-west of Ireland, with very little in between. So it is highly vulnerable, and its loss would be incalculable in terms of securing the future of this charismatic species on the island of Ireland.

Although they do find conditions ideal for nesting in the cover of conifer re-stock sites, forestry has greatly reduced the area of their favoured hunting habitat – upland grassland. And Donegal's isolated population has not been eligible for the conservation measures that are helping to turn the tide of decline in the core areas in south-west Ireland (hence it not being red-listed).

Winter sightings are comparable in number to those in summer. Many of these are around the coast, as the Hen Harriers' dependence on species like the Meadow Pipit forces them to descend from their breeding habitats in the uplands, to where most upland songbirds will be found in the winter months. Some Hen Harriers do turn up at known breeding sites in winter, presumably maintaining contact with their territories. What is not clear is whether the birds we see elsewhere in winter are all part of this resident population, or do they also include immigrants from Scotland. There have been ringing recoveries in Ireland of Scottish-bred birds, but there is no convincing evidence of significant immigration to Donegal. Small winter roosts have been found in both Inishowen and in the south-west of the county.

Pallid Harrier

Circus macrourus | Cromán bánlíoch

RARE VAGRANT

Only one of these summer visitors to eastern Europe, from African wintering grounds, has reached Donegal. It was a male seen at Lough Eske on 17 April 2024 (R. Vaughan), not yet validated by IRBC. More can be hoped for, as the species is spreading westwards.

Red Kite

Milvus milvus milvus | Cúr rua

RARE PASSAGE MIGRANT /WINTER VISITOR
RE-INTRODUCTION PROJECT UNDERWAY IN IRELAND
RED-LISTED BOCCI-4

Red Kites had been common and widespread in Ireland, but were exterminated quite early on in the historical period – mostly before the Victorian bird collectors would have shot and preserved many specimens. Bones have not been found, and the only textual references to them are in the *Ordnance Survey Memoirs*. Few of the parish accounts in these memoirs make any reference to birds, but Kites appear in three, always along with hawks or eagles – the Townawilly Mountains (Barnesmore), Killymard parish (west Lough Eske) and the Lough Derg area.[304] There is also an interesting comment referring to Crockglass, just inland from Quigley's Point on the shores of Lough Foyle. 'I saw at Crockglass a bird hovering over and descending in the hills which I took to be an eagle. A country lad called it a kite, but I was informed by others that eagles from Binalan are frequent on this hill.'[305] Why would a local lad not claim the bird in view to be the more exalted species, if it wasn't that he knew the difference?

Between six and eight birds have turned up in recent times. The first three or four of these are likely to have been moving south from Scotland, where there is a re-introduction project using birds from the migratory population in Sweden. The first one was at Muff on 31 January 1998, and what was probably the same bird about 2 km to the south, on the border with Northern Ireland. Then two were seen in 2009 in the Buncrana area – one of them from 2 January and then two on 16 January (E. Johnston et al.). One of these birds had a tag that was believed to indicate an Aberdeen release site.

The next to be seen was in the Urris Hills of Inishowen on 17 May 2018, followed by one on 25 July at Falcarragh which may, or may not, have been the same bird. One was then seen on Tory, on 12 November 2018. A few days later, on 19 November, one appeared on the Fanad peninsula and moved around various local venues until last seen on 21 December. In 2023, an injured bird escaped from captivity near Glenties. Shortly after that, on 29 March, one was seen in the town. Sightings in Kilmacrenan on

27 April and Manorcunningham on 9 June are likely to have been the same bird. This wandering behaviour suggested by all the birds since 2018 is more consistent with the dispersal of Irish-bred birds from non-migratory native Welsh stock (now well recovered from near-extinction), than with the southerly migration of Scottish birds. The re-introduction projects in Wicklow, Fingal and Down have been highly successful, and the number of Irish-bred birds is growing rapidly. So there seems to be every reason to expect an increasing number of Red Kites making their way to Donegal, and eventually breeding.

The Red Kite is Red-listed in BirdWatch Ireland's 'Birds of Conservation Concern in Ireland 4: 2020–2026' on the grounds that its global conservation status is of Concern.[306]

White-tailed Eagle

Haliaetus albicilla albicilla | Iolar mara

EXTINCT RESIDENT, NOW SCARCE WINTER VISITOR AND RARE BREEDER
INCREASING TREND FROM NATIONAL RE-INTRODUCTION PROJECT
RED-LISTED BOCCI-4

There is an early reference, in 1739, by Rev. W. Henry to this species being abundant in the north-west of the country, referring specifically to Tory Island.[307] In the early nineteenth century this was the commoner of the two eagle species breeding in Donegal. It was mainly coastal, and concentrated near good seabird colonies. Reports of thirty or forty birds in the air at Horn Head seem hardly credible now, but are in keeping with the known behaviour of other *Haliaetus* species when food supplies are high. Up to four pairs bred on Horn Head, but they had all gone by about 1860. They bred on Arranmore until about 1880. Other locations included Malin Head, Teelin (near Slieve League), Tory and Owey Island.[308]

The re-introduction project in south-west Ireland, from 2007, has produced ripples in Donegal. The first bird seen in the county was a tagged individual, and remained from 17 to 28 April 2009 at the Pollan Reservoir in central Inishowen (D. McLaughlin et al.). The same location was visited by a second bird two years later, on 12 to 13 June 2011. A third one was seen in Inishowen on 23 June 2012. In 2016, there were a series of sightings of what could have all been the same bird. The first on Lough Swilly was present from 19 March to 2 April 2016, then Tory had one on 6 to 7 June, and Durnesh Lough on 22 September. A tag was noticed on the Tory bird, which probably indicates that there were at least two birds in 2016. Three scattered sightings followed – at Fintragh in the south-west on 27 October 2017, Inch Lough on 17 April 2018 and the Lough Derg area on 8 November 2018.

By 2022, visits in all seasons had become routine, with three or four birds in Inishowen, one finding Blanket Nook much to its liking from May through to September (returning in June 2023), and two or three other widely spread locations also being visited. The total number of birds involved is very uncertain, but is certainly in excess of a dozen. Three pairs have held territories, one of them in a failed attempt to breed, and one successfully rearing a single fledgling in 2023. It seems quite likely that White-tailed Eagles from the re-introduction project in County Kerry will eventually re-colonise Donegal.

And looking further ahead, birds from the highly successful re-introduction project started in Scotland in 1975 will surely reach Ireland eventually, and merge with Irish birds – most likely in Donegal. Contact has almost certainly already happened, as tagged birds from Ireland have included Scotland in their wanderings.

On grounds of historical decline, White-tailed Eagle is Red-listed in BirdWatch Ireland's 'Birds of Conservation Concern in Ireland 4: 2020–2026'.[309]

Rough-legged Buzzard

Buteo lagopus lagopus | Clamhánlópach

RARE VAGRANT

There are only two old records of this species, which breeds in Scandinavia and winters to the south in eastern Europe. One was trapped on Horn Head on 26 November 1891 and was in the collection of H. Belcher of Beechwood, Dalkey.[310] The second one was obtained near Ardara on 26 November 1941 and was received by Williams & Son, Dublin.[311]

Buzzard (Common Buzzard)

Buteo buteo buteo | Clamhán

RESIDENT

STABLE TREND

Along with several other raptor species, the Buzzard was exterminated in Ireland in the nineteenth century. It was reported as common in Donegal by J.V. Stewart, in the early to mid century.[312] Hart thought it probably bred in Donegal until 1883, when several were seen, but clearly it would have been declining well before that.[313] It re-colonised Ireland along the coast of County Antrim in the 1950s. A slow spread from there brought it west to Donegal before any move to the south, so there were several sightings here during the first breeding atlas survey in 1968–72.[314] By 1981 it had bred in the border area of the county, and by 1991 there were twenty-six pairs.[315] By the third atlas survey, from 2007 to 2011, its presence was consolidated throughout the lowland east and the highland fringes. The treeless expanses of blanket bog west from the Blue Stack Mountains and Glenveagh to the coast, and in the far south-west, remained largely unoccupied.[316] More recently, Buzzards have become a familiar sight on the sand dune systems around the north and west coasts, where rabbits are usually in abundance.

Buzzard, Tory.

GRACE MEENAN

The only study here to give any indication of density was in 2000/1, when a combination of various national and local groups estimated 112 pairs in the county, mainly in the north-east (east of the central hills and north of the River Finn). Inishowen had most, with seventy-two pairs. There were thirty-one occupied 10 km², with a maximum of seventeen pairs breeding in one square.[317] At that time it was not unusual to see five or more birds (maximum nine) spiralling above the same patch of coveted real estate. For the past five or ten years that has not been the case, and there is always the suspicion that Buzzards are being poisoned or shot. This has occasionally been proven, particularly in the early days of colonisation. However, people seem to have gradually got used to living with a large raptor. That still leaves a probable thinning out of Buzzards in their core east Donegal range to be explained. It may be that their numbers are controlled by the density of their favourite prey species, the rabbit, which is somewhat cyclical.

Barn Owl (Western Barn Owl)

Tyto alba alba | Scréachóg reilige

RARE RESIDENT

DECREASING

RED-LISTED BOCCI-4

Unlike some of the previous species, the Barn Owl was never shot to extinction. What drastically reduced its numbers, and now threatens its survival in Donegal, is a combination of the loss of wet grasslands to hunt in, the loss of suitable farmyard sites to nest in, the reduction in breeding productivity from eating poisoned rodents, and road collisions. But there was also one shot near Newtowncunningham as recently as 29 December 1984.

Barn Owls were scarce but widespread in the agricultural east of the county up to the 1960s. By the end of that decade the first atlas found them confined to Inishowen, the lowlands around Lough Swilly, and the head of Donegal Bay. Even then, the number of occupied 10 km² was very small.[318] By the second atlas (1988–91) they had gone from the Donegal Bay area, and by the third (2007–11) there was only one confirmed pair left in the county.[319]

Reports of a Barn Owl are still received every couple of years, and while many Long-eared Owls are initially reported as Barn Owls, there remains a residue of convincing descriptions. Unfortunately, these are rarely confirmed with a second sighting. It seems possible that a pair will occasionally still breed in the border zone around the north-east of the county, but hard proof is lacking.

The Barn Owl is Red-listed in BirdWatch Ireland's 'Birds of Conservation Concern in Ireland 4: 2020–2026' due to a 50 per cent long-term decline (since 1980) in its population.[320]

Scops Owl (Eurasian Scops Owl)

Otis scops scops | Ulchabhán scopach

RARE VAGRANT

There is one old record of a Scops Owl obtained at Ballyliffin in July 1911. Scops Owl is a common migratory species from Africa to southern Europe, so there is always a chance that one will overshoot on its northward flight.

Long-eared Owl

Asio otus otus | Ceann cait

UNCOMMON RESIDENT
DECREASING

With its fondness for coniferous woodland, it is strange that Long-eared Owls don't seem to have benefited more from the thousands of hectares planted over the last half century. Their main habitats in the past were small copses, especially where they were associated with densely treed hedgerows. These are still in good supply, and are where Long-eared Owls are still most likely to be found. They are even found in scattered localities along the west coast – which is far from being the most wooded part of the county.

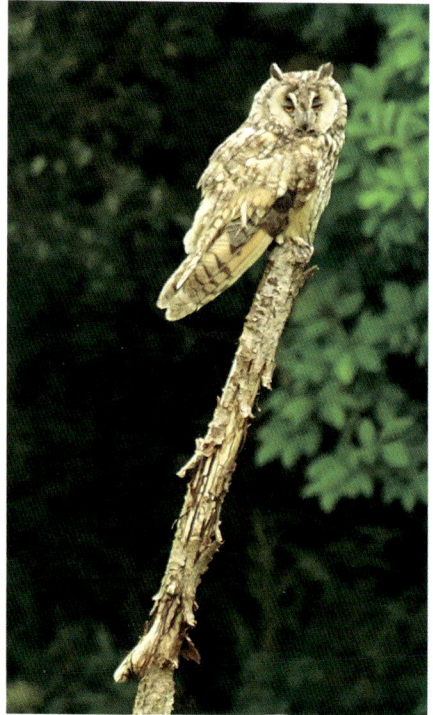

KIM PERIERA

Long-eared Owl, Dunkineely.

The lowlands around Lough Swilly and Inishowen are where most of our Long-eared Owls are located. The more inland parts of east Donegal, which had been well tenanted, have had very few records in the last thirty years. The lack of mature habitat must be a limiting factor, and as with the Barn Owl, the intensification of agriculture undoubtedly limits their hunting opportunities.

Short-eared Owl

Asio flammeus flammeus | Ulchabhán réisc

SCARCE PASSAGE MIGRANT AND WINTER VISITOR

DECREASING

There are late-nineteenth-century records of repeat occurrences of Short-eared Owls at both St John's Point and north Fanad at a time when they were regarded as winter visitors to Ireland.[321] Nowadays the species is more a migrant than a winter visitor.

In modern times (since 1953) a total of thirty birds have been recorded. Of these, eleven were on Tory, but only one of those stayed for the winter – 30 October 2018 to 11 March 2019. There is one record each from Arranmore, Inishtrahull and St John's Point. Lough Swilly has had five sightings since 1982. But there was a period in the 1960s/1970s when several birds were seen at Inch Levels each year, and eight were reported at a roost in 1971 (Perry, 1975). Recorded first sightings have been nine in October, three in May and August, two in November, and one each in January, February, March, July and September – with the one in January involving eight birds. Our birds could come from either Scotland or Scandinavia.

Snowy Owl

Bubo scandiaca | Ulchabhán sneachtúil

SCARCE VAGRANT AND RARE BREEDER

RED-LISTED BOCCI-4

The first alleged Snowy Owl that we know of was shot in November or December 1837, near Killybegs.[322] It had been given to William Thompson's correspondent P.J. Selby. However, it was announced at the Dublin

Snowy Owl,
Malin Head.

RÓNÁN McLAUGHLIN

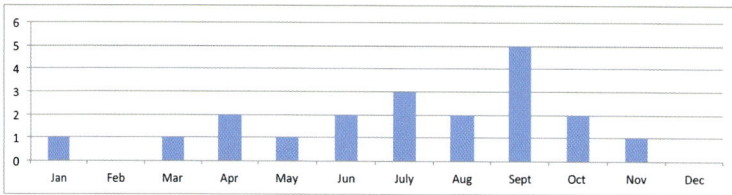

Fig. 18.62: Months of first sightings of twenty Snowy Owls since 1992

Naturalists Field Society as an Eagle Owl, on the basis of the description which has been communicated to them, so it must remain as only alleged. At least four of the seven birds recorded up to 1913 were shot. There have now been a total of seventy-two records, but discounting follow-up sightings, it works out at approximately twenty-seven birds.

Following that period up to 1913, there was nothing until the 1990s, but that decade proved to be a purple patch, with a multitude of sightings. The assumption is that there was only ever a maximum of two birds at any one time, but as many of the sightings were single birds without detailed descriptions or photographs, it is possible that there were more.

The first from this period was a bird that stayed on Tory for three days in November 1992, followed on Arranmore by a large male on 24 June 1993 and an immature on 22 June 1994. A male and female were present in Gweedore from about 12 September to 4 November 1995, and one at the same place on 25 August 1996. From 9 January 1997, when two birds were seen, to at least 26 September 1999, there was a succession of sightings of single birds at various locations in north-west Donegal.

A few yeas later there was a female bird on Arranmore, in August 2004, and a second calendar-year bird at Dooey Point in October 2006. A second calendar-year female stayed on Tory from 12 September to 29 October 2009. An adult male was on Arranmore from 18 July to 10 September 2012, and from 31 May to 6 September 2013, after which, what was probably the same bird appeared in the Dawros peninsula on 1 October. Finally, there was another spate of sightings between Kincaslough and Arranmore from early December 2014 to 19 July 2015, presumably of a single bird.

The old records included three in November, one in December and one in April, very much in keeping with their status as winter vagrants. The modern records peak in September, but in general they are very evenly spread (Figure 18.62). This is not what would be expected for a winter vagrant from Arctic regions. We can only speculate that Snowy Owls, having reached Britain or Ireland in winter, are reluctant to make the return journey. The most recent records show that when not shot, some birds will stay for extended periods. This might be seen as a welcome tendency, but it would be even better if birds marooned here could return to their true home.

However, the culmination of that period of long stays in the 1990s was in 2001, when there was a failed attempt at breeding in the mountains of the north-west. Eggs were laid but were abandoned (NPWS). This is the only known attempted breeding record in Ireland.

The Snowy Owl is Red-listed in BirdWatch Ireland's 'Birds of Conservation Concern in Ireland 4: 2020–2026' on the grounds that the IUCN has classified its global status as Vulnerable.[323]

Hoopoe (Eurasian Hoopoe)

Upupa epops epops | Húpú

RARE VAGRANT

Hoopoe is a summer migrant from Africa to continental Europe, so most birds reaching Ireland are overshooting in spring. Donegal doesn't get the numbers seen in the south-coast counties, but with nine records, we've had enough to keep the Hoopoe on our radar. There are two old records from the nineteenth century, one of them being at Inishtrahull.[324] It was not seen again until one appeared at Tory on 1 May 1960 (TBO). Other singles followed – Tory again, from 28 to 30 September 1964, at Malin Head for about a week from 28 March 1965, at Glencolumbkille on 15 September 1987, at Gortahork from 9 to 10 May 2001, at Crohy Head from 5 to 7 September 2008, and again at Malin Head on 2 May 2015. Of the nine birds seen, three were in autumn.

Roller (European Roller)

Coracias garrulus garrulus | Rollóir

RARE VAGRANT

Southern and eastern Europe is where Rollers breed on return from their winter quarters in southern Africa. Overshooting birds in Ireland are much rarer than Hoopoes, and rarer again in Donegal.

There are two old records, both of them in autumn. The first was a bird shot on Inch Levels on 10 October 1891 by John O'Connell, and the second was a female shot about seven miles from Derry on 27 September 1900.[325] Having retreated from much of its former breeding range in eastern Europe, the prospects for more Rollers reaching Donegal do not look good.

Kingfisher (Common Kingfisher)

Alcedo atthis ispida | Cruidín

SCARCE RESIDENT
STABLE

For such dazzling birds, Kingfishers are surprisingly elusive. They favour lowland streams, which are mostly in the east, but they do also have a minimal presence on the west coast, probably breeding on the Owenea River. They are regularly noticed on the slow, depositing rivers running east from the central highlands to Loughs Foyle and Swilly. These are the Leannan River, the Rivers Swilly, Deele and Finn, and the Swilly Burn. They also use the Mill, Culdaff and Ballyboe Rivers in Inishowen, and the River Eske and Eany Water flowing south into Donegal Bay. Kingfishers are seldom noticed at lakes, which tend not to provide suitable breeding sites, but it is likely that a few do hold breeding pairs, at least occasionally, as indicated by their regular presence on the lagoons of Inch and Blanket Nook.

Crowe et al. give estimates for seven of the main river systems in the country, ranging from about 0.05 to 0.15 territories per km.[326] So here in Donegal, if at the lower end of that scale, we might expect to have one pair requiring at least 15 km of useable habitat. This condition is not often met, so many of the occupied rivers should have only a single pair. But the number of small rivers occupied suggests that we are not always at the lower end of that density scale. Also, the habitat provided by the lagoons and ditches at Inch Lough and Blanket Nook must be at the higher-density level, as they can at times sustain three to four pairs. The total population in the county is unlikely to be much more than twenty pairs.

In winter, some birds disperse to the estuarine reaches of their rivers, and more widely around the coast.

Bee-eater (European Bee-eater)

Merops apiaster | Beachadóir Eorpach

RARE VAGRANT

Like those other flamboyant visitors, Hoopoe and Roller, the Bee-eater is a migrant from southern Africa to southern Europe, and occasionally overshoots on spring migration. There is one old record of a bird shot by J.V. Stewart of Rockhill (near Letterkenny) in 1830 or 1831.[327] A more recent one was reported to me by Helen Campbell from Horn Head. Unfortunately, recalling the date as roughly ten years before our meeting puts it only notionally at May 1975. That was not enough information to justify a submission to the IRBC, but the identity of the bird can hardly be in doubt. The only precise record we have is of a bird photographed on 28 May 2006 at Rathgory, near Portsalon on the Fanad peninsula (T. and J. Bliss).

Wryneck (Eurasian Wryneck)

Jynx torquilla torquilla | Cam-mhuin

RARE VAGRANT

This strange migratory woodpecker remains widespread in continental Europe but is now extinct as a breeding bird in Great Britain. Vagrants are more frequent in Ireland in autumn than in spring,[328] and accordingly, Donegal's birds have all been in autumn – seven or eight in September, and two in October. They have appeared at the main island outposts, and have included two multiple arrivals.

The second Irish Wryneck (the first recorded for Donegal) was a bird shot on Rathlin O'Birne in or around October 1878. The first modern sighting was of a bird on Inishtrahull from 8 to 10 September 1956 (J. Russell). Tory had the next one, on 10 September 1958. We then had to wait until 2002, when between two and four birds came at once – Tory on 9 and 13 September (recorded as two birds, but possibly only one), one on Arranmore, 11 September, and an additional bird that was not assessed by IRBC but is certainly a good record – at

Wryneck, Arranmore.

Malin More on 14 September. There was one on Tory on 27 September 2013, followed by another on Arranmore from 3 to 4 October. The most recent bird was seen on Tory on 19 September 2014.

Great Spotted Woodpecker

Dendrocopus major anglicus | Mórchnagaire breac

RARE BREEDER AND EARLY COLONIST
INCREASING

There is a population of migratory Great Spotted Woodpeckers in Scandinavia which could have delivered one or two vagrants to Donegal over the years, but none have been recorded. However, birds from the normally resident population in Great Britain made the crossing to County Down in 2007, and to Wicklow in 2009. They rapidly consolidated their footholds and started to spread. There were initially

Great Spotted Woodpecker, Tory.

quite a few reports of possible sightings in Donegal, but until very recently these either failed to convince or could not be confirmed. The first credible record for the county was a bird heard calling at Moville on 15 August 2014 (T. Campbell). Then on 18 August 2015, one was photographed on Tory, where there were only a few fence posts to make it feel at home.

More hopeful was a bird present throughout May 2020 in the native deciduous woodland nature reserve at Ardnamona, in south Donegal. In the following spring of 2021, drumming was heard, and a nearby bird table was regularly visited. This culminated in the first proof of breeding, when a juvenile bird died after crashing into a window (G. McGettigan).

The Ballybofey area is also proving productive. A woodpecker was heard calling at Lough Alaan woods on 25 March 2020, and another was seen and heard in the Meenglass area from 5 to 6 August 2021. Finally, a pair bred in coniferous forest south of Killygordon, close to the Tyrone border, in May 2022. And another pair was present in mixed woodland at Mullaghagarry, and drumming was heard.

New areas continue to be occupied. In 2024, there are now at least two breeding pairs in the extensive conifer forests that line the border with County Tyrone between Lough Mourne and Lough Derg, and there is a nesting pair in the deciduous Eany Water woodlands of south Donegal. If, as it appears so far, they are willing to use both coniferous and deciduous woodland, their future looks bright.

Green Woodpecker (European Green Woodpecker)

Picus viridis viridis | Cnagaire glas

POSSIBLE VAGRANT

There were five Irish records of Green Woodpecker prior to 1896, but none since. The third of these was reported in a series of publications right up to the present as having been shot at Rathmullan, County Donegal in July 1854.[329] But there is confusion about which side of the Donegal/Derry border the deed was done.

The first reference, in the *Proceedings of the Dublin University Zoological Association*, states that 'Dr. Ball exhibited a fine specimen of the green woodpecker ... forwarded to him by Thomas Batt, Esq., of Rathmullan, Derry'.[330] Note that Rathmullan is in Donegal, not Derry.

At a later meeting, on 4 March 1894, the report on a talk on rare birds in Ireland from February 1853 to February 1854 quotes Mr E.P. Wright as stating that 'A specimen of this very rare Irish bird was forwarded to Dr. Ball by Thomas Batt Esq., shot on 12 January, in Derry'.[331]

By the time Ussher and Warren had their say, the details had morphed to '... a third was exhibited before the Dublin University Zoological Association by Dr. Ball on 21 January 1854, having been forwarded to him by Mr. Thomas Ball [sic] of Ruthmullen [sic], Co. Donegal'.[332] This statement is clearly correct, as far as it goes, except for the spellings of Batt and Rathmullan.

The next relaying of the tale, by Kennedy et al., is a partial correction. It ignores any possibility that the bird was shot elsewhere and so accepts Donegal as the county of origin. It simply states '... another at Rathmullen, Co. Donegal in January 1854'.[333] This usually authoritative reference is the source of the continuing claim that the bird was shot in Donegal.

If the original error that Rathmullan was in County Derry is sufficient evidence that the bird was shot in Donegal, then the Green Woodpecker's place on the county list is valid. However, Mr Wright is the only witness who stated clearly, without any reference to where Mr Batt lived, where the bird was shot – in Derry. So both Donegal and Derry have grounds for claiming this bird, but the stronger claim would seem to be for Derry. Either way, the record should be withheld from the county list until such time as more evidence (if any exists) can settle the matter.

Kestrel (Common Kestrel)

Falco tinnunculus tinnunculus | Pocaire gaoithe

RESIDENT
DECREASING
RED-LISTED BOCCI-4

The Kestrel is best known as the raptor that hovers over motorway verges and rough grass margins in agricultural land. That image is hard to square with its status in Donegal (quite apart from the absence of motorways). Kestrels have never been abundant in Donegal. Breeding sites on ivy-covered ruins and well-vegetated cliffs are an attraction, but the food supply would always have been the primary determinant of where they occurred. So

Kestrels should have been mainly expected in the lowland farming parts of the county, dividing up the habitats there with the Sparrowhawk. Support for this was provided by the first two atlas surveys (1968–72 and 1988–91), which show gaps in the mountainous and bogland areas of the west.[334]

However, by the third survey (2007–11), those upland gaps had been plugged, and a new one had opened up in the lowland east.[335] This suggests that better coverage in the final atlas revealed the presence of birds in the uplands, and that a declining presence had rendered the species hard to find in the lowlands. These findings chime with the common experience of Kestrels now mainly encountered on hill ground and its margins – the habitats where one might expect Merlins and Peregrines. They have also been recorded on a number of islands – including Arranmore, Tory, Inishbofin and Owey. The shift in agriculture practice in the lowland east, from mixed arable and grazing to intensive grassland, could account for this, but a detailed investigation would be needed to confirm it.

The Kestrel is Red-listed in BirdWatch Ireland's 'Birds of Conservation Concern in Ireland 4: 2020–2026' due to a 53 per cent short-term (twenty-five years) decline in its population.[336]

Red-footed Falcon

Falco vespertinus | Fabhcún cosdearg

RARE VAGRANT

Red-footed Falcon is a summer migrant from Africa to eastern Europe. There have been three records.

A first-summer male bird was seen at Malin Head on 27 June 2004 (D. Radford). The second was a female at Glencolumbkille on 25 September 2006 (A. McGeehan). That second one was only the second autumn bird to be seen in Ireland – the thirty-five Irish records to date have been mostly spring migrants.[337] The third bird was a male at Ards on 2 and 3 September 2024 (T. Campbell, C. Ingram and R. Vaughan).

Merlin

Falco columbarius aesalon | Meirliún

UNCOMMON RESIDENT

STABLE

This small falcon is a low-density occupant of open hill country. The three breeding atlases suggest a peak in the middle period, but this could be influenced by differences in methodology and effort. Of more value are the focused studies of upland birds, and of the Merlin itself. The Upland Bird Study found two Merlin pairs in 20 km² of Ring Ouzel habitat (seven pairs per 100 km²), and one pair in 88 km² of Golden Plover habitat (slightly more than one pair per 100 km²).[338] Two Merlin study areas, Inishowen in the north-east and the Ardara area in the south-west, covering 162 km², held fourteen nest areas, at an average density of between five and six pairs per 100 km² for each.[339] Extrapolating from these figures to find an estimate for the county has too many pitfalls, but it would seem that there are more Merlins than we might have expected.

Merlins in Ireland have traditionally been deprived of tree-nesting opportunities, so nesting on the ground was normal. The conifer forests restored the opportunities, and all Merlins in the Inishowen and Ardara study areas are now nesting in trees. They do still mainly feed in open country, on birds like Meadow Pipit and Skylark, although early in the breeding season the population of these species is still low, and they will then seek alternative prey over the forest canopy. In winter, they move down to the coast, where there are much easier pickings.

Subspecies

Icelandic Merlin

Falco columbarius subaesalon | Meirliún Íoslannach

PASSAGE MIGRANT AND WINTER VISITOR

Merlins are regular migrants on Tory and at Malin Head, in both spring and autumn, with occasional sightings in winter. While these could be Scottish or Scandinavian birds of the same race as our residents, ringed birds from the north-east have rarely been recovered in Ireland. The vast majority have been from Iceland, where the Merlins are assigned to a separate race, wintering in Britain and Ireland.

VICTOR CASCHERA

Merlin (probably Icelandic), Tory.

Two Donegal birds have been confirmed as Icelandic – an injured one at Malin Head on 24 April 1965 (MHBO), and one on Tory on 25 September 1999 (TBO).

Hobby (Eurasian Hobby)

Falco subbuteo subbuteo | Fabhcún coille

RARE VAGRANT

There have been claims of four or five Hobbys in Donegal. A first-summer bird photographed on Tory on 16 to 17 April 2015 was the first recorded – it was possibly the same bird seen on 14 May, although the gap is large for it not to have been noticed in between. Then a female was at Garrabane Mountain, Inishowen on 20 July 2019. In 2020, there were two birds. The first was at Drumkeen on 14 July (C. Ingram) and the other on Tory on 17 September. It is unfortunate that the Drumkeen bird is the only record that has so far been verified by the IRBC. Given the increasing numbers recorded nationally in recent years, these other claims are not at all surprising, but so far they are still unofficial.

Hobby is a summer migrant to Europe from Africa, with in recent years a much expanded population and breeding range in Great Britain.

Lanner Falcon

Falco biarmicus | Fabhcún Lanner

ESCAPE

This resident of the Mediterranean region has not so far reached Ireland naturally, but is kept by falconers and sometimes escapes. This is no doubt the origin of a bird that was seen at Inch on 18 February (Inishowen Wildlife Club) and 5 March 2017.

Saker Falcon

Falc cherrug | Fabhcún Saker

ESCAPE

As with Lanner Falcon, this native of south-east Europe is only likely to reach us having escaped from a falconer. A bird was seen around Lough Swilly on a number of occasions between 9 October 1994 and 8 December 2002 (R. Sheppard).

Gyrfalcon

Falco rusticolus | Fabhcún mór

RARE WINTER VISITOR

The mainly white, northern populations of this circumpolar species are migratory, and the darker, southern ones are more sedentary. So it is thought that white birds reaching Ireland are probably from northern Greenland, and the fewer dark birds are from Iceland. There have been forty-two to forty-four seen in total. White Gyrfalcons are more likely than most rare birds to be recognised and reported by the general public, which must go some way to explain the symmetry of records – fifteen from the nineteenth century, thirteen from the first half of the twentieth century, and sixteen since then. Most of the early birds were shot, but not all. Dates were not always given precisely, but of twenty-nine birds, the spread was August (1), September (3), October (1), November (4), December (5), January (2), February (3), March (5) and April (5). Records have come from many sites around the coast, and a few inland. Six at Horn Head is the highest total, and four at Tory is also high.

Two of the earlier records are worth reporting verbatim from Ussher and Warren;[340]

There is now in the Natural History Museum, South Kensington, a specimen which was caught alive after it had gorged itself with a rabbit at Moville, in the same part of Donegal, about the 1st November 1877. Another individual was captured alive at Glenmore on the 13th September 1882. The butler of Mr. Dames Longworth, when out fishing, was crossing a deep dyke, when a bird of this species, which he described as quite fatigued, flew up and he knocked it down with his landing net. It was kept alive for nearly five years, and is now preserved at Glynwood, Athlone.

In modern times (since 1953) there was a bird on Tory from late February to early March 1955 (J. Dixon). Of seven birds between 1977 and 1996, two were specified as white, two as grey and one as dark (if the colour was not published it was presumably white). Since 2000, there have been from five to seven birds.

Of these post-2000 sightings, three white birds in south-west Donegal in late winter 2002 are likely to have been the same individual – the first at Muckross from 23 to 24 February, another at Killybegs on 16 March, and one at Sheskinmore on 31 March. Then there was a grey bird, probably Icelandic, at Blanket Nook and about 15 km to the south on Mongorry Hill near Raphoe (but in clear view of Blanket Nook) on 1 and 11 November 2003. Another white bird turned up in the Muckross area on 11 March 2006. On 21 December 2011 a white bird was seen in the act of taking an Oystercatcher at Fintragh. Finally, there was a white bird at Glashagh Bay on 15 April 2019.

Peregrine Falcon

Falco peregrinus peregrinus | Fabhcún gorm

UNCOMMON RESIDENT

STABLE

Peregrines were nearly wiped out in Ireland, and elsewhere, from the mid-1950s to the 1970s. They accumulated agricultural pesticides from seed-eating prey, to the point where their eggshells thinned and broke before

	1981	1991	2002	1981	1991	2002
	Occupied territories			% Breeding successfully		
N Inishowen	9	10	12	56	78	58
Derryveaghs	9	8	7	56	57	60
Blue Stack Mountains	5	7	6	20	57	80
TOTAL	23	25	25			

Table 18.18: Peregrine territories in three surveys

chicks were ready to hatch. Breeding success plummeted. Coverage of 134 sites in 1969 and 1970 (out of an estimated total of 180–200) revealed the low point of twenty-nine pairs nationwide.[341] The offending chemicals – dieldrin, aldrin, heptachlor and DDT – were banned, and follow-up censuses in 1981, 1991 and 2002 showed that the Peregrine in Donegal had recovered to a stable level (Table 18.18).

There are about forty traditional sites in Donegal, mostly on coastal cliffs but also among the inland mountains. The 1981 survey covered the twenty-three sites on the north Inishowen coast, and in the Derryveagh and Blue Stack mountains. They found 90 per cent occupancy on the coast and 100 per cent inland.[342] The other surveys repeated coverage of the same sample areas.[343] While a few sites have been abandoned, quarries have provided new breeding opportunities, and a pair has bred on the cathedral spire in Letterkenny.

In winter, Peregrines roam widely. Some will base themselves for the season among the waterfowl hoards at places like Lough Swilly. But they can often be seen crossing urbanised or agricultural land, and even taking prey wherever the opportunity presents itself.

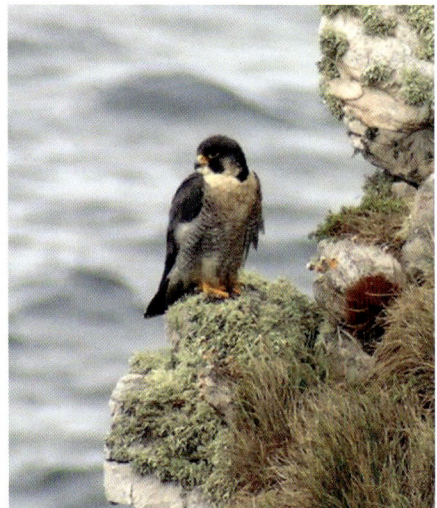

Peregrine Falcon, Tory.

GRACE MEENAN

Red-backed Shrike

Lanius collurio | Scréachán droimrua

RARE VAGRANT

The Red-backed Shrike is a summer visitor from Africa to most of Europe (but not Ireland), although the population which was widespread in England and Wales in the nineteenth century has now faded away. There are no records from the time when that adjacent source was at its full strength – suggesting that birds in Donegal in autumn are more likely to come from Scandinavia. The first of the ten birds to reach us was an adult female on Tory, on 12 September 1959 (J.C. Mortimer). An immature was on Inishtrahull on 6 October 1965, a first-year bird at Rocky Point on 27 August 1994 and a first-winter bird at Glashagh Upper, Gweedore on 2 November 1995. In 2007 there were three first-year birds on Tory, on 1 and 16 September and 2 October. A juvenile appeared on Tory on 13 September 2014. And finally there were two birds in 2019 – a male at Malin Head on 20 April and a juvenile on Tory on 15 October. So all but one of the Donegal birds have arrived in autumn – the lone spring bird was probably an overshoot from continental Europe.

Lesser Grey Shrike

Lanius minor | Mionscréachán liath

RARE VAGRANT

The only Donegal record, and the third for Ireland, was discovered in America! It 'came to light through a fortuitous encounter by a US-based Irish birder with one of the finders who possessed a framed annotated sketch of the bird on his living-room wall'.[344] The bird had been seen, and sketched, at Rossbeg on 28 May 1990 (J. Baird, P. Harrison *per.* E. Masterson). Lesser Grey Shrikes are summer visitors from Africa to south-eastern Europe.

Great Grey Shrike

Lanius excubitor excubitor | Mórscréachán liath

RARE WINTER VISITOR

This very scarce winter visitor to Great Britain from its largely northern European breeding range has turned up in Donegal on three occasions. The first was a bird shot near Dungloe in 1860, obtained by Archdeacon Cox and given to the Dublin Natural History Society.[345] The next was present at Lough Keel from 25 to 27 November 1983 (D. Duggan and R. Sheppard), and seen in flight transferring what was thought to be a Wren, from its bill to its talons. The last one was seen on 2 February 2009 at Ballyshannon (K. and D. Burns), at the end of what was apparently a week-long stay.

Woodchat Shrike

Lanius senator senator | Scréachán coille

RARE VAGRANT

There has been only one record of this southern European migrant from Africa to have reached Donegal. It was present at Dunfanaghy from 13 to 15 July 2020 (W. Farrelly et al.).

Golden Oriole (Eurasian Golden Oriole)

Oriolus oriolus | Oiréal órga

RARE VAGRANT

A summer migrant from tropical Africa to most of continental Europe, the Golden Oriole occasionally overshoots in spring, but rarely as far north as Donegal. There have been five records, only one of them in modern times. The first was near Mountcharles in 1866. The second was on 24 May 1878 near Dawros. The third was found on the shore at Burtonport in May 1891.[346] Then there was one present for two weeks near Naran in June 1902.[347] The final sighting was of one on Inishbofin on 15 April 1987 (N. McGregor).

Jay (Eurasian Jay)

Garrulus glandarius hibernicus | Scréachóg

RESIDENT

INCREASING

Ireland's Jay is an endemic subspecies, slightly darker than the neighbouring birds from Great Britain and western Europe. In the latter half of the nineteenth century its range contracted to southern counties. Whether it had been resident in Donegal prior to that is not known, but there were at least occasional stray birds recorded. After the contraction, it did not reach Donegal again until the period of the second breeding atlas in 1988 to 1991.[348] At that time birds could be found in the zone of small semi-natural woods north of Letterkenny. They initially seemed to be using the new conifer woodlands as stepping stones to spread around the county in search of more suitable habitat. For about the past decade or two they have been present in all types of woodland, and are now even appearing near the west coast, at places like Burtonport.

Jays, Dunkineely.

Magpie (Eurasian Magpie)

Pica pica pica | Snag breac

COMMON RESIDENT

STABLE

From its introduction to Ireland several centuries ago, the Magpie had fully colonised Donegal by the end of the nineteenth century, with only the un-treed parts of the west and most of the islands remaining unpopulated. They are present on Arranmore. Numbers are often thought to be increasing, but prejudice against a bird known to raid the nests of songbirds could cloud judgement on that. While numbers are certainly greater than they were in the 1950s and 1960s, there has been no noticeable increase over the last few decades in the size of winter flocks or roosts.

Chough (Red-billed Chough)

Pyrrhocorax pyrrhocorax pyrrhocorax | Cág cosdearg

UNCOMMON RESIDENT

STABLE

This charismatic crow, with its ringing cry and obvious delight in being airborne, embodies the wind-blown landscapes of sea cliffs and machair. Choughs in Ireland are strictly coastal, although R.J. Ussher found a pair in Donegal 1.5 miles (2.4 km) inland.[349] They breed on the sea cliffs, with a preference for caves. They have had a fairly stable population of breeding pairs here through four national censuses, as shown in Table 18.19.[350] However, as they don't breed until they are two or three years old, a sizeable segment of the population is found in non-breeding flocks. Following that last census, there was a follow-up study on these non-breeding flocks in Donegal in 2004/5.[351] Table 18.19 shows that while the overall number of breeding pairs is holding up, the pool of non-breeding birds is sharply declining. Including these non-breeding birds, the total population in the county has dropped from 366 in 1992 to 326 in 2003 – a decline of 11 per cent. This does not bode well for the results of future breeding censuses.

One site that receives continuous attention is Tory Island. Up to sixty birds were being seen there each year in the early 1960s, with a peak of eighty-nine on 14 August 1965. In 1974 there were nine pairs (the equivalent of

	1962	1982	1992	2003–5
Confirmed pairs	104	109	28	74
Confirmed, probable and possible pairs	107	112	101	129
Birds in non-breeding flocks			164	68
Total population (row two x2, plus row three)			366	326

Table 18.19: Donegal's breeding population of Chough

Trawbreaga/West Malin	100
Horn Head/Tramore	100
Gweedore	32
Arranmore	60
Loughros Point area	60
Glen Head area	40

Table 18.20: Maximum flock sizes in core feeding areas

about thirty to forty birds post-breeding). In the 2003 census, four pairs were breeding. In recent years there have been only two or three pairs, with the maximum number of birds seen being seventeen. Why the decline on Tory should have been so dramatic is not known, but it seems to be greater than at other strongholds.

Choughs are present almost continuously around the coast from Inishowen Head to Muckross Head, but there are particular districts within which flocks gather to feed and roost at all seasons, but mainly in winter. Table 18.20 gives rough estimates of maximum flock sizes in these areas, although the last census would have recorded fewer at most of these areas.

In south-west Ireland, Choughs make much use of improved grassland for feeding, but in Donegal they favour maritime and other unimproved grassland habitats. This gives them some independence from the vagaries of agricultural practice, although they do also make some use of farmland or amenity grassland where the vegetation is kept particularly low.

In November 2006, fourteen Special Protection Areas, under the EU Birds Directive, were established in Ireland specifically for Chough, including 'West Donegal Coast' with fifty-eight breeding pairs, and 'Horn Head to Fanad Head' with thirty-two. These ninety pairs represent 6.6 per cent of the total north-west European Chough population.

KIM PERIERA

Chough, Muckross Head.

Jackdaw (Western Jackdaw)

Coloeus monedula spermologus | Cág

ABUNDANT RESIDENT

STABLE

The Jackdaw is a hole-nesting species, and originally would have mainly nested in woodland. But there are not nearly enough tree holes in Donegal to support the number of Jackdaws, so they also use ivy-covered sea cliffs, old ruins, farmyards and, when not evicted, chimneys. They are somewhat colonial, but isolated pairs can be found in woodlands or in other sub-optimal habitats.[352] They have even been recorded breeding in rabbit burrows. They shun the uplands and are at their most abundant in agricultural land, feeding in flocks on both grassland and tilled ground, usually in the company of Rooks.

Arranmore is the only offshore island on which they breed, although Ussher and Warren said they bred on Tory. Only one of the atlas surveys recorded its presence there in summer, and one in winter, with no evidence of breeding. Colonisation following on from partial winter migration from the mainland is highly likely. Stray birds occasionally visit Tory and Inishtrahull, and presumably other offshore islands where they are not resident.

In winter, Jackdaws gather in large communal roosts with Rooks, sometimes in the majority, but usually not. Numbers can reach into the thousands, and even passing the 10,000 mark, as at Convoy, to which they gravitate along traditional compass lines.

Subspecies

Eastern Jackdaw

Coloeus monedula monedula/sommerringii | Cág oirthearach

UNCOMMON WINTER VISITOR, HAS BRED

Birds showing the pale grey or white collar of one or other of the eastern European races have only recently been recognised as occasional visitors to Ireland. Their status in Donegal is very confused.

Of twelve recorded in the county, the first was at Killybegs on 21 February 1999. In addition to those twelve isolated sightings, an eastern bird first

Jackdaws (probable hybrid western/eastern, and western), Carnowen, near Convoy.

joined a colony of about forty Jackdaws resident in my own yard at Carnowen on 29 December 1999. It paired with one of the local birds. Since then, birds with hybrid-type collars have been frequently present, and apparently pure eastern birds continue to appear from time to time, most recently in February 2021.

Rook

Corvus frugilegus frugilegus | Rúcach

ABUNDANT RESIDENT

STABLE

Corvid roosting assembly, Carnowen, near Convoy.

Rooks are abundant throughout the lowland parts of Donegal, wherever they can find a copse of trees tall enough to suit a nesting colony. On Bird-Watch Ireland's Countryside Bird Survey they are more detected than any other species (at thirty-three birds per transect square). Their population is stable, or increasing. In winter, Rooks associate with Jackdaws searching for invertebrate food in pasture and tilled fields. This is beneficial to farmers, and is only occasionally counter-balanced when they raid a field of ripening grain. In winter, they gather in roosts, sometimes very large ones, along with Jackdaws, when their aerial manoeuvres at dusk can be impressive.

Rooks are absent, apart from occasional stray visits, from the smaller offshore islands, including Tory.

Carrion Crow

Corvus corone corone | Caróg dubh

SCARCE VAGRANT

Carrion Crows are highly sedentary in Great Britain, but have established a toe-hold on the north-east corner of Ireland.[353]

The first to have been recorded in Donegal was at Trawbreaga Bay on 18 October 1961 (O. Merne). There have been eleven sightings in total, involving sixteen birds. They have occurred widely, mainly around the coast. Only Tory with two, and Inch with three, have had more than one sighting. The dates have all been between June and November. Some are presumably overlooked in mistake for juvenile Rooks, but given their non-migratory nature, the total is not likely to be high.

Carrion Crows had long been regarded as a subspecies, along with the Hooded Crow. But the two taxa maintain competitively isolated ranges, with interbreeding only along the zone of contact, so are now regarded as separate species. There has been one sighting of four hybrids at Malin Head, on 19 August 1993.

Hooded Crow

Corvus cornix cornix | Caróg liath

COMMON RESIDENT
DECREASING

The Hooded Crow is now re-garded as a full species (see Car-rion Crow, above). It is common throughout the county, and on the offshore islands.

Hooded Crows are not gre-garious, so do not often occur

Hooded Crow, Lough Swilly.

GERRY STUDD

in flocks. They do gather in small numbers if there is a glut of food, as for example when unauthorised rubbish dumps and authorised landfill sites were widespread, or when there is a lot of sheep carrion on the hills (see also Raven). High numbers recorded were 100 birds at Lough Naminn on 7 April 1994, 40 at Ards Back Strand on 11 November 1994 and 40 on Tory during September 1999. A roost in my own broadleaved plantation near Convoy held a peak of sixty-five birds, on 4 June 2007, when the trees were only seventeen years old. The June date suggests that these would have been non-breeding birds.

The Countryside Bird Survey has shown the Hooded Crow to be in-creasing nationally,[354] but it appears that numbers in Donegal have declined since sheep carrion and refuse dumping have decreased.

The only evidence of migration was when twenty-five birds were seen flying north over Inishtrahull on 7 April 2017.

Raven (Northern Raven)

Corvus corax corax | Fiach dubh

RESIDENT
STABLE

Having been persecuted along with raptors, Ravens have been recovering from a low ebb at the start of the twentieth century, and are now relatively common around the coast and in the uplands. They had been confined to the cliff-bound mountains and coasts, where they often come under

VICTOR CASCHERA

Raven, Tory.

pressure from Peregrines looking to take over a nesting site. The Raven's early breeding season gives it the advantage in this struggle. They were first noted nesting in trees on 20 March 1963, in the Mill River gorge at Buncrana. That practice soon spread to the new conifer forests, so their breeding range is now widespread throughout the uplands.

The Raven has taken advantage, along with the Hooded Crow, of periods when our management of the countryside has supplied more carrion than would be expected from nature alone, as during the 1990s and 2000s. That was a time when, in addition, abundant refuse was also on offer at landfill sites. Eighty were noted at Glenalla landfill site on 6 May 2001, and other poorly managed refuse dumps were attended by numbers from twenty to fifty. These sites are now closed, and sheep numbers have reduced, so flocks are less frequently seen, and presumably the population has contracted to some degree. However, flocking is not new, as there is an old record of fifty-six birds in west Donegal in 1935.[355] Nor has it completely disappeared, as could be seen when a single sheep carcass on the Big Isle polders of Lough Swilly, at the core of a very extensive agricultural landscape, attracted thirty-seven Ravens (with a lone Hooded Crow queueing patiently at a safe distance) on 24 February 2013.

Tory had a maximum count of thirteen birds during September 1999 – not many more than can be accounted for by the usual breeding pair and its family. They are regularly present on other offshore islands.

The Raven has been shown by the Countryside Bird Survey to be decreasing nationally,[356] in contrast to its increasing range.[357]

Waxwing (Bohemian Waxwing)

Bombycilla garrulus garrulus | Síodeiteach

IRREGULAR WINTER VISITOR

When the erratic Waxwing population in Scandinavia peaks, the rowan berries on which they depend in winter can be eaten out. The birds then erupt to the south. Flocks in Great Britain and Ireland are usually concentrated along the east coasts.

The first Donegal record is from 1881 at Dunfanaghy, by H.C. Hart, and there was a second nineteenth-century bird in Donegal Town. Kennedy et al. had details of unprecedented numbers in 1946/7, with Donegal sightings at Portsalon and near Bunbeg, where there were about a dozen. Between 1970 and 1988 there were none recorded in Donegal, but a few can be expected to reach us – on average, about every two or three years – when they are always greeted with delight, and with surprise, as many people are unaware that such exotically clad and confiding birds could possibly be seen in a drab, grey Irish winter. Waxwings tend to come to towns with berried trees along streets or in suburban gardens, but tall hawthorn hedges in the countryside also attract them.

Inishowen is well favoured, with nine records, perhaps because it serves as Donegal's east coast. Letterkenny has had seven records and Tory five. High totals were from Carrick which had at least 120 birds on 26 October 2004, Buncrana had 60 on 30 December 2008 and Letterkenny had 70 on 17 November 2012 and 60 on 4 December 2004.

Coal Tit

Periparus ater hibernicus | Meantán dubh

COMMON RESIDENT

INCREASING

Ireland has its own race of Coal Tit, which is assumed to be the one present in Donegal. It is possible that the British race, which in Ireland is mainly confined to the north-east, is also present, but it has not been confirmed. Coal Tit is a woodland species, favouring conifers more than other tit species, but also widely present in deciduous woods and almost anywhere there is some tree cover. The spread of commercial conifer forests has helped its general status, as has the increase in garden bird feeders. In winter, they join up with roving bands of other, mainly insectiverous, species, but they do not rove far – British ringing recoveries have found that 83 per cent are within 4 km of the place of ringing.[358] There have been relatively few records from Tory, ten birds from 25 to 26 September 1997 being exceptional.

Blue Tit (Eurasian Blue Tit)

Cyanistes caeruleus obscurus | Meantán gorm

ABUNDANT RESIDENT
INCREASING

The Blue Tit is the most abundant and ubiquitous of the tit species, and is more tolerant than its relatives of habitats with minimal tree or scrub cover. Its distribution does not stretch to the offshore islands, apart from Arranmore. BirdWatch Ireland's Countryside Bird Survey shows it to be increasing.[359]

Great Tit

Parus major newtoni | Meantán mór

COMMON RESIDENT
INCREASING

The Great Tit is generally common throughout Donegal, and like its relatives is also absent from the offshore islands, apart from Arranmore. It is more dependent than the Blue Tit on having some mature tree cover. It has benefited from the increase in winter feeding in gardens, but also by the increasing number of trees being planted in towns and villages. BirdWatch Ireland's Countryside Bird Survey shows it to be sharply increasing.[360]

Skylark (Eurasian Skylark)

Alauda arvensis arvensis | Fuiseog

COMMON RESIDENT
DECREASING

Skylark, Cronalaghy, Ballybofey.

Skylarks breed in grassland where they can command their treeless territories from the skies. They find their habitat mainly in the uplands, but also in sand dune systems, and to a lesser extent in agricultural grassland.

BirdWatch Ireland's Countryside Bird Survey shows it to be decreasing. This is probably due to intensification of agricultural grassland, and also to the reduction of grazing in some areas, allowing swards to become too dense.[361] Happily, this trend seems not yet to have reduced numbers in the Donegal uplands. The 2002 Upland Bird Survey found 660 pairs of Skylarks in 88 km^2 of habitat being surveyed for Golden Plover, equivalent to 7.5 pairs per km^2. The density on Ring Ouzel habitat was much lower (0.5 per km^2).[362] There has been no similar work which could test the perception that numbers have declined on machair and dunes. But there can be no doubt that numbers in the intensively farmed lowlands in the east of the county have plummeted.

In winter, the upland birds descend to the coastal dunes and salt marshes, and to arable land. As much of Donegal's arable farmland has been converted to intensive grassland, it is likely that the reduction in numbers found in what remains is because most of our breeding population moves in winter to the mainly arable counties of the south-east of Ireland.

Subspecies

Eastern Skylark

Alauda arvensis intermedia | Fuiseog oirthearach

RARE VAGRANT

The race of Skylark breeding in north-central Siberia has been recorded in Ireland on only three occasions, the third one being in Donegal, at Inishtrahull on 12 March 1915. This bird was presented to the National Museum.[363]

Shore Lark (Horned Lark)

Eremophila alpestris flava | Fuiseog adharcach

RARE VAGRANT

Although its regular wintering range reaches the east coast of England, there has only been a single record in Donegal of this summer visitor to Scandinavia. It was a bird seen on Arranmore Island on 15 December 2009 (G. Griffin), the most recent of Ireland's twenty-one records.[364]

Short-toed Lark (Greater Short-toed Lark)

Calandrella brachydactyla brachydactyla | Fuiseog ladharghearr

RARE VAGRANT

The Short-toed Lark migrates from Africa to breed in southern Europe. There have been six birds in total, all of them in autumn, and all but one in October – overshooting spring migrants would have seemed more likely. The first was on Tory, on 5 September 1963 (TBO). Then an immature was trapped during its stay on Inishtrahull from 17 to 21 October 1965. We had to wait three decades

Short-toed Lark, Tory.

ANTON MEEHAN

for the next one, on 3 October 1994 on Arranmore. Malin More had a long-staying bird from 19 to 28 October 1998. Tory had its second bird on 26 October 2009, and another relatively long-staying bird from 24 to 30 October 2019.

Sand Martin

Riparia riparia riparia | Gabhlán gainimh

SUMMER VISITOR

DECREASING

The Sand Martin is one of the earliest summer visitors to arrive here, usually in late March. Its wintering grounds are in the Sahel zone of tropical West Africa. On arrival, Sand Martins tend to assemble in large numbers at places with an early flush of aerial food, like Inch or Durnesh Loughs, before attempting to re-populate their colonial breeding sites. They nest in vertical sand banks and are quite particular about choosing sand/soil of the right consistency, and other features such as aspect. They find what they want on river banks, coastal dunes and sand quarries. Many of these sites are temporary and are soon lost through erosion or quarrying, so replacements are constantly being sought.

The species is being squeezed by drought in its Sahel wintering grounds, and by an overall loss of suitable nesting sites here – Donegal has suffered more than most counties in that respect. The most recent bird atlas reveals that its distribution here has a large hollowed-out core of upland that is not occupied, although it can at times still be found along some upland streams.[365] The Countryside Bird Survey records both long-term and short-term declines in their national numbers. This seems to also apply to Donegal.

After breeding, Sand Martins again flock to rich feeding sites, especially where they can roost safely in tall marsh vegetation. Estimates are occasionally made at Inch Lough of up to 1,000 mixed hirundines, as on 24 September 2016. Although these congregations do also include lesser numbers of Swallows and House Martins, Sand Martins are usually in the majority. Sand Martins arrive here in spring after surviving the Sahara and Mediterranean crossings, but they return in autumn by an Atlantic coastal route – slower but presumably safer for the inexperienced new generation.[366]

Swallow (Barn Swallow)

Hirundo rustica rustica | Fáinleog

ABUNDANT SUMMER VISITOR

DECREASING

Our Swallows migrate from south-east Africa, on a journey that has always been a source of fascination – now particularly so as conditions anywhere on that long trans-continental flight are liable to be severely altered by climate change.

In the nineteenth century, the first half of April was when Swallows arrived in most of the country, but 'in the bleak counties of Tyrone and Donegal [they] are seldom seen before the latter half of the month'.[367] Hutchinson states that the main passage is in May. Here in 'bleak' Donegal numbers now build up very slowly from the first sightings in early April, with most of their territories reclaimed by late April, but a few of them not until early in May.

Swallow numbers undoubtedly declined in the 1970s and '80s. The intensification of agriculture was accompanied by a clean-up of farmyards, so there has been a growing shortage of mud for Swallows to build their nests, and flies to feed their young. Anecdotal evidence of fewer pairs in out-buildings continues, and is widespread. But nationally, the Countryside Bird Survey shows overall stability since the start of surveying in 1998, so the decline due to intensification may have levelled out, or at least been balanced by a shift in farming away from arable and towards pasture, which better suits the Swallow.[368]

There is one winter record, of a bird seen in Dungloe on 21 January 1980. With climate change, we can expect more late dates, and perhaps more late broods being successful.

House Martin (Common House Martin)

Delichon urbicum urbicum | Gabhlán binne

SUMMER VISITOR

STABLE

House Martins are the last of the three common species of hirundine to arrive in spring, having wintered widely across southern Africa. Their traditional choice of nesting site is on cliffs. Saldanha Head is one such that has recently been used, and perhaps others go unrecorded. But the vast majority now build their mud nests under the eves of houses. So availability of mud, and aerial insects for feeding to chicks, are as important for House Martins as they are for Swallows.

House Martins are widespread around the county, albeit at lower density in the uplands, and often in small, loose colonies. There is little evidence to suggest that they are increasing in Donegal, but they have been shown by the Countryside Bird Survey to be increasing nationally.[369] This is a trend in common with north and west Britain.[370] Paradoxically, it is this north-western zone in which they are at their least abundant, so the trend may possibly be in some way related to climate change.

Like the Swallow, their departure in autumn is protracted, and a few will always hang on longer than they should. One bird seen at Portnoo on 6 December 1968 was not expected to survive.

Red-rumped Swallow

Cecropis daurica rufula | Fáinleog ruaphrompach

RARE VAGRANT

There is only a single Donegal record of this summer visitor from Africa to southern Europe. It was a first-winter bird which lingered on Tory from 27 October to 4 November 2007 (J. Adamson et al.).

Long-tailed Tit

Aegithalos caudatus rosaceus | Meantán earrfhada

RESIDENT

STABLE

Long-tailed Tits are as characteristic of tall hedgerows and scrub as they are of woodland. They are widespread across the county, but least frequent in the hill districts. Their numbers are reduced by hard winter weather, but usually recover quite quickly. Although once a rarity at garden bird feeding stations, Long-tailed Tits are increasingly making use of these in winter. So the overall trend is one of stability.[371]

Long-tailed Tit, Inch.

RICHARD SMITH

Long-tailed Tits spend most of the year in small flocks. These move slowly but steadily through the habitat, with the particularly endearing trait of ignoring any humans standing in their way – as long as the humans are willing to stand still and let it happen.

Wood Warbler

Phylloscopus sibilatrix | Ceolaire coille

RARE SUMMER VISITOR AND PASSAGE MIGRANT
DECREASING
RED-LISTED BOCCI-4

Wood Warbler, Glenveagh.

DERMOT BREEN

Wood Warbler is one of three species of leaf-warblers (*Phylloscopus*) that breed in Donegal – the others being Willow Warbler and Chiffchaff. Unlike the other two, it is extremely rare. It migrates to continental Europe and Great Britain from tropical Africa. In the north-western limits of their normal range they are characteristic of Sessile Oakwoods, and in particular the wet Atlantic variant. As this is the climax vegetation of western Ireland, including

Donegal, it is probably the loss of most of those woods throughout Ireland that has reduced Wood Warbler to its present precarious status.

Wood Warblers are also willing to use non-native beechwood, which, although just as rare in Donegal as old oakwood, produced the first known record. It was of a specimen that was shot by H.C. Hart in the beechwoods at Glenalla in 1878, and sent to the Dublin Museum. The only other old record is also from Hart, who recorded three singing birds at his home in Carraghblagh, in north Fanad, an indication that the species was probably as numerous then as at any time since. Bunlin (Milford) had records of singing birds in the 1950s. In the 1960s a bird was singing at Bogay (Newtown-cunningham) where the habitat was mixed hardwood, including beech.

Between 1980 and 2000 multiple birds could be heard at Ards (two), Clonkillybeg (three) and Glenveagh (three). There were in addition a number of sites with single birds or pairs, including a single bird at Ardnamona, where there had been two in the 1960s. Since 2000, singing birds have only been heard at Ards, Glenveagh and Ardnamona. Habitat reduction and deterioration is almost certainly involved in their apparent decline. But decline of Wood Warblers from an already tenuous presence in the county poses another question. Are our current breeding birds the descendants of a long line of birds returning to Donegal, or are they casual migrants on their way to Scotland, but liking the look of what they are flying over and deciding they don't need to proceed any further? If they are Donegal birds, a break in continuity would likely be permanent – if migrants, there should be a continuous replacement of casual breeders as long as the population in western Great Britain holds up, and the habitat here doesn't deteriorate any further. It has to be said that the Donegal native origin of our few breeding pairs is much more credible.

Migrant Wood Warblers have always been as rare as the breeders. Malin Head had an autumn bird on 31 August 1967, and Tory had one from 19 to 24 September 2011 (or up to three individuals). Spring migrants are slightly more frequent. There was one on Arranmore on 7 May 1995 and one on 3 May 1997. There was also one at Malin Beg on 25 May 1997. Tory had one from 22 to 26 May 2014 (or up to three individuals), and one on 9 May 2017.

The Wood Warbler is Red-listed in BirdWatch Ireland's 'Birds of Conservation Concern in Ireland 4: 2020–2026' due to a 71 per cent long-term (since 1980) decline in its population.[372]

VICTOR CASCHERA

Yellow-browed Warbler, Tory.

Yellow-browed Warbler

Phylloscopus inornatus | Ceolaire buímhalach

UNCOMMON VAGRANT

INCREASING

This remarkable little bird should be migrating in autumn from Siberia to south-east Asia. But a tiny percentage go in reverse, and end up in western Europe. So far, Donegal has had a minimum of seventy-four birds. The first was on Tory on 30 September 1964 (TBO). The next was at Glencolumbkille on 10 October 1988. Since 2010 they have been annual, which ties in with a dramatic increase in the already large numbers that reach Great Britain each year (Figure 18.63).

All have arrived between the last few days in September and first few in November. Tory has had most birds by far, and it has also had the highest number at any one time, with seven on 10 October 2016. The far south-west, from Malin Beg to Glencolumbkille, has also had quite a few, and more would undoubtedly be found there with greater coverage. The only occurrence away from the headlands and islands was one at Letterkenny on 3 October 2013.

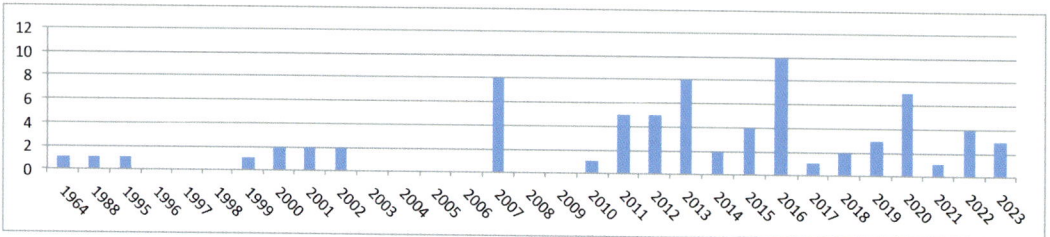

Fig. 18.63: Annual totals of Yellow-browed Warblers since 1964

Pallas's Warbler (Pallas's Leaf Warbler)

Phylloscopus proregulus | Ceolaire Pallas

RARE VAGRANT

Pallas's Warbler is a tiny Goldcrest-sized bird, which like the Yellow-browed Warbler should be migrating to south-east Asia – in this case from south-central Siberia and northern China. Although the numbers reaching Great Britain and southern Ireland are increasing, it is still a very rare sight this far west. There have only been three birds reported so far in Donegal – the first was on Arranmore on 2 October 1993 (A. McMillan), and the third on Tory on 2 November 2023 (A. Meenan). The second, at Malin Beg on 31 October 2016 (J. O'Boyle), is the only verified record so far.

Radde's Warbler

Phylloscopus schwarzi | Ceolaire Radde

RARE VAGRANT

The tenth Irish record of Radde's Warbler was a first-year bird on Tory, on 2 October 2003 (A. Kelly). It is the only record for Donegal. This is another leaf-warbler that oscillates between Siberia and south-east Asia.

Dusky Warbler

Phylloscopus fuscatus | Ceolaire breacdhorcha

PROBABLE RARE VAGRANT

There are *no* confirmed records of Dusky Warbler from Donegal. The species is included here only because there are three unverified records, two of them photographed, that should prove to be acceptable. The first was on Arranmore on 4 October 2013 (R. Vaughan), the second on Tory on 13 May 2017 (B. Carruthers et al.), and the last at Malin Beg on 9 October 2020 (C. Ingram). Like the previous three species of leaf-warblers, Dusky Warblers normally migrate from breeding grounds in Siberia, to south-east Asia.

Willow Warbler

Phylloscopus trochilus trochilus | Ceolaire sailí

SUMMER VISITOR

INCREASING

The delightful babbling song of the Willow Warbler puts the seal on the arrival of summer in what is usually called poor or marginal land – the heavier soils of the south and the scrubby habitats within the upland zone of farming. In fact, anywhere where scrub replaces managed hedges or woodland. So in practice it is found throughout the county, but is less dominant in the intensively farmed east. It can also take advantage of the succession from grassland to closed-canopy conifer forest, but that scrubby stage doesn't last long. In common with other long-distance migrants from Africa, it is doing well in the cooler, wetter north and west of Britain and Ireland. The Countryside Bird Survey shows it to have increased nationally by 95 per cent since 1998, in contrast with a decline in the UK of 9 per cent. It is the second most abundant summer visitor to Ireland after the Swallow, although here in Donegal it probably holds the top spot.

Willow Warblers are frequently seen as migrants on Tory, Malin Head and similar locations in autumn.

Chiffchaff (Common Chiffchaff)

Phylloscopus collybita collybita | Tiuf-teaf

SUMMER VISITOR

INCREASING

The Chiffchaff arrives here in late March, having migrated from the Mediterranean zone. Although needing tall trees, it can make do with a small copse or a well-treed garden, and is common and widespread away from the treeless uplands and the conifer forests. Like the Willow Warbler, it is on an upward trend.[373]

Some Chiffchaffs can hang on through the winter along the warmer coasts of the south of Ireland, or at other locations such as sewage treatment works, where a supply of insects and good cover are always available. Two birds over-wintered on Tory in 2012/13, but so far Donegal seems to be a step too far for repeated over-wintering – although with climate change, is it perhaps only a matter of time before that changes. Chiffchaff are common as migrants in treeless coastal locations like Tory, where they make do with whatever cover is available.

Subspecies

Siberian Chiffchaff

Phylloscopus collybita tristis | Tiuf-teaf Sibéarach

RARE VAGRANT

Some authorities regard this as a separate species, but the IOC treat is as one in a series of three races, distinct from our own nominate form occupying the west of Europe, and a second one in Scandinavia which has not yet been recorded here. We have eight records (five of them verified) of this distinctively pale bird, which, like other vagrant leaf-warblers, tends to arrive very late in the autumn season. The first was on Arranmore on 11 November 2004 (A. McMillan). The second was at Malin Beg from 7 to 11 November 2012 (not yet verified). Another Arranmore bird was seen on 4 November 2013. Then Tory had one on 10 October 2016. In the same year, Malin Head had one from 30 October to 1 November 2016. Arranmore had yet another bird on 5 November 2021. Two records are yet to be processed, both from Tory – the first on 10 November 2021 and the second on 5 October 2022.

Greenish Warbler

Phylloscopus trochiloides viridanus | Ceolaire scothglas

RARE VAGRANT

The Greenish Warbler breeds in eastern Europe and central Russia, wintering in south Asia. Only one has been verified as reaching Donegal; it was on Tory, on 21 September 1998 (J. Dowdall et al.). However, Tory might yet claim a second bird, seen on 10 September 2022 but still to be assessed by IRBC. In a review of Greenish Warbler records, an earlier record of a bird that had been ringed at Malin Head on 18 October 1965 (MHBO) was found to be no longer acceptable.[374]

Arctic Warbler

Phylloscopus borealis borealis | Ceolaire Artach

RARE VAGRANT

Arctic Warblers occupy a breeding zone in north-east Scandinavia and Russia, to the north of Greenish Warbler, and migrate to south-east Asia. There have been three records. The first was trapped on 1 September 1960 on Tory (TBO) and was also the first Irish record. The second bird was at Bloody Foreland, and lingered from 28 to 30 September 2003 (V. Caschera, J. Coveney, P.J. Doyle et al.). Finally, one arrived on Tory on 8 September 2022 – well photographed, but still sub-judice (A. Meenan).

Sedge Warbler

Acrocephalus schoenobaenus | Ceolaire cíbe

SUMMER VISITOR
STABLE

When they arrive here from Africa, Sedge Warblers are commonly found in the margins of wetland sites – scrubby ditches, streams and lakes, reedbeds and marshes. One new habitat that has proved useful is the scrubby stage of growth found in new plantings and re-stock plots of conifer forests. They are otherwise absent from the uplands, and as suitable habitats are scarce throughout the intensively farmed east of the county, the Sedge Warbler is also

Sedge Warbler.

scarce there. Multiple records are few, but include one of thirty-five birds on Arranmore on 16 May 2010. The Countryside Bird Survey shows no clear trend in their numbers.

Paddyfield Warbler

Acrocephalus agricola | Ceolaire gort rise

RARE VAGRANT

There is a single record of this very rare warbler from Central Asia which normally winters in India. One was trapped on Tory on 21 September 1998 (J. Dowdall et al.). It was the third Irish record.

Blyth's Reed Warbler

Acrocephalus dumetorum | Ceolaire Blyth

RARE VAGRANT

Blyth's Reed Warblers normally migrate from south Asia to central Russia, but they are now spreading west into Finland and the Baltic states. So it is not so surprising that the sixteen Irish records have all been since 2005.[375] All the four (or five) birds to reach Donegal so far have been on Tory. The first was on 8 October 2012 (R. Vaughan et al.). The second was on 6 October 2018 (A. Meenan), followed quickly by the third, at Malin More on 28 October 2018 (C. Ingram). What may have been either two birds, or a long-staying single, were reported from 16 to 17 and from 22 to 27 October 2022 (A. Meenan). Only the first two records have so far been verified, but it looks as if Tory is proving to be a hotspot for this species.

ROBERT VAUGHAN

Blyth's Reed Warbler, Tory.

Reed Warbler (Eurasian Reed Warbler)

Acrocephalus scirpaceus scirpaceus | Ceolaire giolcaí

RARE VAGRANT

The Reed Warbler is a common breeding species throughout Europe, wintering in sub-Saharan Africa. Although colonising Ireland as a breeding bird, it still only reaches Donegal as an occasional vagrant. There have been six records so far.

The first was on Tory from 21 to 22 September 1998. There were two at Malin Beg in 2000 – one from 29 September to 3 October, and the other on 1 October (A. McGeehan and D Hunter). Then two more were seen on Tory – on 12 September 2009 and 5 September 2010. Finally, on 17 May 2019, a spring bird was seen on Tory.

It may take some time for the breeding population in Ireland to spread to Donegal, but it probably will happen. The extensive reedbeds lining the River Foyle at St Johnstown might be a good place to look (and listen), or the fens around Durnesh Lough.

Marsh Warbler

Acrocephalus palustris | Ceolaire corraigh

RARE VAGRANT

Marsh Warbler has declined as a breeding species in Great Britain, almost to the point of extinction, but it is generally common in Europe. It is even

Marsh Warbler, Tory.

ANTON MEENAN

expanding in Fennoscandia, which may explain why Donegal has had three individuals very recently. Its winter home is in south-east Africa.

All sightings have been on Tory, and found by Anton Meenan. The first was present from 8 to 10 June 2020, and in the same year another turned up on 20 September. The third was on 31 May 2021. The first and third birds were well photographed and have been verified by IRBC.

Booted Warbler

Iduna caligata | Ceolaire tosaithe

RARE VAGRANT

The first Irish record of a Booted Warbler was a bird seen on Tory, on 27 September 2003 (D. Weir, W. McDowall et al.). These are Russian birds, wintering in India, and like Blyth's Reed Warbler they are expanding westwards into Finland. So it is just possible that their vagrancy to Ireland could increase. But as the subsequent six Irish records have all been on the south coast, another Donegal bird still seems to be only a remote possibility.[376]

Eastern Olivaceous / Sykes's Warbler

Iduna pallida / rama | Ceolaire bánlíoch /Sykes

RARE VAGRANT

UNCONFIRMED

A warbler was trapped on Tory on 29 September 1959 (R.G. Pettitt). A full description was submitted and it was accepted at the time as the first Irish record of Olivaceous Warbler *Hippolais pallida*, with measurements indicating that it was of the south-east European race *elaeica*, which is now re-classified as a full species – Eastern Olivaceous Warbler *Iduna pallida*.

A 2003 re-evaluation by IRBC of what were by then three Irish records of the species found that the Tory record was inconclusive. But they left the door open for its re-evaluation as Sykes's Warbler *Iduna rama* (from eastern Iran to Pakistan). If confirmed as a Sykes's Warbler, it would be accepted as a first record for Ireland – for the second time! So it seems that this bird is a first record of its species for Ireland, but we don't (yet) know what that species is.

Melodious Warbler

Hippolais polyglotta | Ceolaire bin

RARE VAGRANT

This species is a common summer visitor to south-west Europe, from winter quarters in tropical West Africa. Only six have reached Donegal, all but one of them on Tory. The first was on 21 August 1995 (M. O'Clery). The second was at Malin Beg from 24 September to 10 October 1999 and a photo of it was published.[377] The third was on 15 September 2015, and the fourth from 19 to 20 September 2016. The fifth record was on 8 September 2021 and the most recent on 2 September 2023. The second, third and sixth of these records have yet to be processed by IRBC.

Grasshopper Warbler (Common Grasshopper Warbler)

Locustella naevia naevia | Ceolaire casarnaí

SUMMER VISITOR

STABLE

The mechanical buzz of the largely nocturnal Grasshopper Warbler is the best clue to its presence. Unfortunately, its high frequency makes it inaudible to most people after their fifties, so the Grasshopper Warbler is almost certainly more common than reports would suggest. Having arrived here from west Africa, they take up their territories in tall, wet grassland along river floodplains, around lakes or in tall, marshy vegetation wherever it occurs. They are also willing to accept conifer re-stock sites as a good substitute. This expands their distribution to the hilly areas, so they are now found throughout the county. The conifer habitat is short-lived, but new sites become available all the time as forests are felled and re-stocked. Inasmuch as it can be determined with such a secretive bird, the population appears to be stable.[378]

Grasshopper Warbler, Lough Derg.

ROBERT VAUGHAN

Blackcap (Eurasian Blackcap)

Sylvia atricapilla atricapilla | Caipín dubh

BREEDING SUMMER VISITOR AND OCCASIONAL WINTER VISITOR
INCREASING IN SUMMER

The Blackcap is one of the more notable and interesting success stories of recent years. Ussher and Warren reported it from Glenalla and Rathmullan, but without proof of breeding, and also as occurring at light stations.[379] Kennedy et al. reported a bird which struck the Tory lighthouse on 15 October 1928. Otherwise, there was no evidence of its presence until the late 1950s and early '60s, when Tory was functioning as a bird observatory. The first atlas period of 1968 to '72 showed breeding birds to be present in a confined area north-west of Letterkenny,[380] and soon afterwards they were at several locations near the border with Londonderry.[381] The subsequent atlases confirmed consolidation in the eastern half of the county.[382]

Blackcaps like scrubby habitat with trees, but their presence now throughout the whole county, abundantly in much of it, suggests that they are very flexible about the details. This breeding population migrates from the western Mediterranean and tropical west Africa. The Countryside Bird Survey has shown it to have risen from 1998 to 2016 by an extraordinary 1,689 per cent.[383]

The wintering habit has been increasing in Britain and Ireland since the 1960s but is still a rare event in Donegal.[384] However, the Blackcaps seen in Ireland in the winter months are from a separate population, breeding largely in the Low Countries and Germany and migrating south-east, rather than south-west as our summer visitors do. The birds that are detected in winter are usually at garden bird feeders. One was seen on Inch Island on 3 December 1992. They have turned up three times at my own garden feeder near Convoy – 3 December 1993, 1 March 2021, and one which arrived on 17 December 2022 and was still present at the end of the year. Malin Head had one on 2 January 2016, as had Milford on 17 December 2017. Other unreported birds are known to have wintered at a bird table in Kilmacrenan around the turn of the millennium.[385] Migrants pass through Tory regularly in April, May and October, with a few birds most years into November.

Garden Warbler

Sylvia borin borin | Ceolaire garraí

SCARCE MIGRANT AND POSSIBLE RARE BREEDER

Garden Warblers breed throughout Europe, migrating from tropical Africa. Determined by their particular habitat requirements in Ireland, they have a very restricted breeding distribution in the scrub zone around midland lakes.[386] This doesn't stretch to Donegal, but we have had a few hopefuls. The first was a bird that sang for about a week in my own garden at Carnowen, from 28 May 1970. Then two birds were singing in potentially suitable habitat on the shores of Lough Eske on 3 June 1978 (J. Lovatt), and another at the same place nearly two decades later, on 10 June 1996 (R. Sheppard). No proof of breeding was obtained, but if a female had also been present at the Lough Eske site, they would surely have tried. There are other locations in south Donegal which could be suitable – so Garden Warblers have only to explore a little beyond their comfort zone in County Fermanagh.

Three areas have attracted almost all of the migrant Garden Warblers. Of the seventy-two autumn birds, Tory has had thirty-seven, twenty-five have been in the far south-west from Malin Beg to Glencolumbkille, and ten were recorded at the north-east corner of the county, in the Malin Head / Inishtrahull area. One bird each has been seen at Arranmore and Bloody Foreland. All these autumn birds were in September or October, apart from one at Malin Head on 29 August 1963. All five of the spring birds were at Tory, four of them in May, and one on 11 June 2012.

Barred Warbler

Curruca nisoria | Ceolaire barrach

SCARCE VAGRANT

Barred Warblers breed in eastern Europe and Russia, and winter in east Africa. Of the vagrant species visiting Ireland which share this particular geographical range, Barred Warbler is one of the most frequent, and Donegal has had a respectable total of twenty-one birds so far. One was on Inishtrahull, one on Arranmore, two in the far south-west and the rest on Tory. The first record was of a first-winter bird on Tory on 1 September

1960 (TBO), and the most recent was from 29 September to 2 October 2023, also on Tory. Most records have been in September, with only six in October and two in late August.

Lesser Whitethroat

Curruca curruca curruca | Gilphíb bheag

SCARCE VAGRANT

INCREASING

In summer, the Lesser Whitethroat is widespread across Europe, apart from the south-west, and it winters in east Africa. There are a few pairs trying to establish a breeding presence in the far south of Ireland, but most of the strays reaching Donegal would be migrating from Great Britain or, perhaps even more likely, from Scandinavia. There have been fifty birds recorded in the county. The first one was at Inishtrahull, caught at the lighthouse in 1899, and received by Barrington as the second Irish specimen.[387] The next was at Falcarragh on 6 September 1959. The flow of modern records then followed from 1991.

Tory has had the lion's share, while eleven have been in the far south-west, and one each at Inishtrahull, Malin Head, Falcarragh and Arranmore. All records have been in September or October, apart from one in August, one in November and six in May. It is curious that the six May birds included the only two records of more than one bird – but it would be stretching the case to suggest that these were two pairs.

Lesser Whitethroat, Tory.

ANTON MEENAN

Eastern Subalpine Warbler

Curruca cantillans albistriata | Ceolaire oirthearach fo-Alpach

RARE VAGRANT

The Subalpine Warbler has recently been split into three species. With hind-sight, it is clear that the steady flow of vagrants to the Irish south coast sites belong to the Western Subalpine Warbler *Curruca iberiae*, which has a wide breeding range in the western Mediterranean. So it is rather surprising that the only Subalpine Warbler to reach Donegal proved to be the second (or third) Irish record of the Eastern Subalpine Warbler *Curruca cantillans*. Of that newly defined species, our bird was of the subspecies *albistriata* which occupies south-eastern Europe in summer, and winters south of the Sahara in the eastern Sahel zone. It was seen and photographed on Tory on 18 May 2018 (B. McCloskey, G. Murray et al.).

Whitethroat (Common Whitethroat)

Curruca communis communis | Gilphíb

SUMMER VISITOR

STABLE

Whitethroats breed throughout Europe, and winter in the scrubby savannah areas of Africa – our birds migrating to the dry Sahel, south of the Sahara in west Africa. Their numbers in western Europe crashed during the period from 1969 to 1975, at a time of severe drought in the Sahel.[388] Field work for the first breeding atlas (1968 to 1972) straddled the crash, but in Ireland very little atlas work had been done prior to it. Even so, the distribution recorded in Donegal had remained widespread, with only the mountainous core being without birds.[389] The drought in the Sahel was probably less of a problem for Whitethroats than the widening of the Sahara into the north-ern edge of the normally more vegetated Sahel, thus increasing the distance they had to fly on migration without a break. But natural selection favoured the Whitethroats with the longest wings, so populations have recovered, although not yet to the level they had once been.

The most recent atlas shows that more of the uplands are still unoccupied than in the early 1970s.[390] Scrub clearance on marginal and hill farms must have reduced the area of suitable habitat, but in the west there has been

Whitethroat,
Burtonport.

DEREK CHARLES

widespread abandonment of farming, which would have favoured the Whitethroat. Combining these trends may account for the Whitethroat being now mainly coastal, with conifer re-stock sites retaining a small population inland. The Countryside Bird Survey shows a continuing rise in their fortunes, so it is probably only lack of habitat in Donegal determining that they still remain fairly scarce.[391]

Firecrest (Common Firecrest)

Regulus ignicapilla ignicapilla | Lasairchíor

RARE VAGRANT

Along the south coast of Ireland, Firecrests are scarce autumn migrants, with smaller numbers in spring and winter. They had been rare, and their increase mirrored an expansion in breeding range and numbers in Europe, and the colonisation of southern England.[392] That trend has had little (if any) impact at our corner of this island, where we have only had three records. The first was a spring bird at Malin Beg on 25 May 1997 (A. McGeehan). The second was also at Malin Beg, on 27 October 2017 (C. Ingram and T. Campbell). The third record is of a bird on Tory, on 29 March 2024 (A Meenan). Only the first of these birds has so far been verified.

Goldcrest

Regulus regulus regulus | Cíorbhuí

COMMON RESIDENT, PASSAGE MIGRANT AND WINTER VISITOR
STABLE

Unlike most insectivorous birds, the Goldcrest hangs on through the winter as a resident species, when available insect food is at a minimum. It is also our smallest species, and therefore at greater risk from cold nights and long freezes. Goldcrests favour conifers, but for millennia these have not been generally available in Ireland, so they have had to adapt to living in any available woodland, while still being able to take advantage of the spread of the new conifer plantations since the 1950s. So the species has done well here in Donegal, against the odds – suffering population crashes when we have particularly severe winters, as in 1962/3 and 2009/10, but bouncing back again when competition from neighbouring pairs will be low. It is widespread and common throughout the county. There is no evidence of any long-term change in its status here, or nationally.[393]

Some birds travel west from north-east Europe to winter here, and these account for spring and autumn passage noted on the headlands and islands. Numbers are usually small, but variable, and there are occasionally larger numbers that can be described as falls.

Wren

Troglodytes troglodytes indigennus | Dreolín

ABUNDANT RESIDENT

Like the Goldcrest, the Wren is a small-bodied resident species which can suffer from hard winters. But its mouse-like lifestyle in the shelter of ground-level vegetation enables it to survive, even in upland areas, and its use of communal roosts in old nests can make the difference between life and death in freezing night-time temperatures. More than any other species, the Wren merits its reputation as ubiquitous, being present throughout the offshore islands, uplands and lowlands. The Upland Bird Survey showed it to be the most numerous species, after Meadow Pipit, in Ring Ouzel habitat, and the fourth most numerous in Golden Plover habitat.[394] The overall trend nationally shows a moderate increase,[395] but in Donegal it is only the population crashes, and the recovery from them, that so far are detectable.

Wren, Tory.

Wrens are well known for their tendency to evolve into island races as a result of reproductive isolation. St Kilda and Fair Isle are the homes of two well-studied endemic races. There is some initial evidence that the Wrens on Tory may also qualify for sub-specific recognition, and the matter is currently being investigated. The photo of a Wren shown here was taken on Tory – just in case!

Treecreeper (Eurasian Treecreeper)

Certhia familiaris Britannica | Snag

RESIDENT
INCREASING

The Treecreeper is another resident insectivorous species, like the Goldcrest. It survives in all kinds, and sizes, of woodland, as long as there are some mature trees, but in coniferous woodland it also needs some standing deadwood. The third breeding atlas shows gains in distribution towards the edge of the treeless west of the county, presumably helped by the spread of conifer plantations.[396] In winter, a single bird can often be found keeping pace

Treecreeper, Convoy.

379

with roving bands of tits and Goldcrests. The Countryside Bird Survey suggests a moderate increase over the years from 1998 to 2016, but with a fairly low registration on survey plots, this conclusion is rather tentative.[397] In Donegal the gains in distribution indicate at least a moderate increase in numbers over recent decades.

Rosy Starling

Pastor roseus | Druid rósach

SCARCE VAGRANT

Scanning coastal Starling flocks in late summer and autumn could be rewarded with a sighting of one of these brilliant birds. Rose-coloured Starlings, as they were called until very recently, are from the steppes of southwest Asia, and have a tendency to disperse or erupt erratically.

Donegal has had records of twenty-eight birds. Six were in the nineteenth century, starting with one at Woodhill, Ardara in July 1825. That was the third Irish record. The first in the twentieth century was at Inishtrahull on 18 September 1925. There was then another long gap before the first modern record, which was a juvenile bird on Tory, on 9 October 1993 (M. O'Donnell et al.).

The geographical extremities have attracted twelve of the twenty-one modern birds, with five of them on Tory. The remainder have been scattered down the west coast (could it be that the urge to join flocks of the Common Starling can sometimes over-ride the appeal of the usual emergency landfall sites, like Tory or Malin Head?).

Rosy Starling, Tory.

ANTON MEENAN

Most records are of single birds, with the first of the exceptions being two each at Malin Head from 15 to 17 July 2002, and at Sheskinmore on 10 October 2018. Two birds were also seen at the north entrance to Trawenagh Bay, but with two years between each sighting. They were on 1 August 2018 and 19 June 2020. Also in 2020 there was one bird from 5 June and a peak of three on 8 June at Carrickfinn, with singles at Tory on 4 June and from 23 June to 7 July, and at Rossnowlagh from 11 to 19 August. While this might total five birds during 2020, the safer estimate would be three wandering birds. Of the twenty-one modern sightings, eight were first seen in June, four in July, three in August, three in September and three in October.

Starling (Common Starling)

Sturnus vulgaris vulgaris | Druid

RESIDENT AND WINTER VISITOR

DECREASING

It seems hard to believe that Starlings were not always part of the farming landscape in Donegal. But at the end of the nineteenth century, Ussher and Warren reported that they first bred at Killybegs in 1890, and about Glenties in 1893. They were long established on Arranmore, but not on the adjacent mainland. And they were generally more numerous in winter in Ireland.[398]

By the 1950s, Kennedy et al. noted that they were still somewhat scarce on the west Donegal mainland. But by the time of the three breeding atlases they were ubiquitous. However, in recent years the breeding population in farmland has clearly declined, although this has not been quantified. The national trend shows a slight decline.[399]

While there is no doubt that immigrants from the continent hugely increase the numbers in Ireland as a whole, that is certainly not the case in Donegal. The winter atlas shows smaller numbers in the north-west of Ireland than in the east and south.[400] Peak numbers in Donegal are to be seen in the post-breeding flocks around the coast, but even these would normally hold only a few hundred birds. Winter flocks are widespread, but are both sparse and small, and could easily be accounted for by the resident population.

Song Thrush

Turdus philomelos clarkei | Smólach ceoil

COMMON RESIDENT, WINTER VISITOR AND PARTIAL MIGRANT
STABLE TREND

Like the Blackbird, the Song Thrush is a woodland edge species, favouring a mixture of closed and open habitats. As woodland has declined, both species have learned to thrive in the landscapes that have replaced it – the mixture of hedgerows, gardens, lawns and fields. Song Thrushes are found throughout the county, but are less numerous than Blackbirds. They are also more secretive, although when they start to sing in early spring they are hard to miss. The recent breeding atlases and the Countryside Bird Survey all show little long-term change.

Song Thrushes are susceptible to hard weather, which sometimes reduces their population. They do, to some degree, evade the problems of hard weather by retreating to the coast, but remaining within Ireland.[401] But hard weather can just as readily bring birds to Donegal from Scotland – sixty at Glencolumbkille on 23 October 1999 was probably one such event. These immigrants are known to boost winter numbers.[402] But overall, their numbers in Donegal are fairly stable.

Subspecies

Hebridean Song Thrush

Turdus philomelos hebridensis | Smólach Inse Ghall

WINTER VISITOR

The IOC describe this race as inhabiting the west of Scotland and the west of Ireland. In his list of the birds of Ireland, Hobbs says that its status is uncertain. 'It is a possible resident in the north, and a winter visitor to parts of the west and the northwest.'[403] As with the Icelandic Redwing, it is a darker bird than the race generally seen throughout Ireland, and such birds can frequently be seen in Donegal, although identification in the field is less obvious than for the Icelandic Redwing. The only attempt to confirm the subspecies was in 1964, when the Malin Head Bird Observatory recorded several birds with its characteristics: 'On 9th October some at least were probably hebridensis. One or two on 10th had characteristics of this race, as also had one on 18th.'[404]

Mistle Thrush

Turdus viscivorus viscivorus | Liatráisc

RESIDENT

DECREASING

Mistle Thrush, Tory.

The Mistle Thrush is common through-out the county. Like the Magpie, it was not known in Ireland until a time when its arrival would be noted – in the case of Donegal, it was recorded in the west of the county in 1808, by William Sin-clair, only ten years after its first arrival in Ireland.[405]

Unlike the Blackbird and Song Thrush, it is happy to utilise the open expanses of pasture fields in the search for invertebrate food, and not confine itself to the edges. So its presence is often obvious from the vantage of a moving car. In winter, it gathers into loose flocks, which also makes it even more conspicuous. Like its winter relatives, the Redwing and Fieldfare, it is fond of berries, and a single bird will often dominate and defend a good bush of hawthorn, rowan or holly. All this disguises the fact that the Mistle Thrush is considerably less numerous than either of its two resident relatives. The Countryside Bird Survey has also shown it to have declined nationally in number by 17 per cent over the 1998 to 2016 period,[406] and this would accord with the general perception in Donegal.

Redwing

Turdus iliacus iliacus | Deargán sneachta

WINTER VISITOR

STABLE

RED-LISTED BOCCI-4

Like the Fieldfare, with which they often associate, Redwings of this nom-inate race come to Donegal from the northern European part of their ex-tensive range, which reaches as far as eastern Siberia. Flocks in the low hundreds are frequent. It is not unusual for one or two birds to hang on into

383

May, and singing has been noted. However, there has been no indication that breeding has been attempted.

Arrivals in autumn can be impressive, as on Tory between 6 and 12 October 1959 when major movements took place. And on the night of 9/10 October 1962, birds passing in the beam of the lighthouse peaked at 1,500–1,950 per hour (TBO). Four birds were caught and examined during this movement, and all were subspecies *iliacus*. It is normal for flocks here to move on before the winter is much advanced, and remaining numbers are smaller than in the rest of the country.

Like the Fieldfare, they feed on hedgerow berries and on the invertebrates of open pasture, but it is not clear whether their early departure is due to running out of food, or in response to harder weather. As Redwings are known to suffer from hard weather, moving south in Ireland would not necessarily help very much, and any movements south for that reason could well be heading for south-west Europe.

The Redwing is Red-listed in BirdWatch Ireland's 'Birds of Conservation Concern in Ireland 4: 2020–2026' on the grounds that its European conservation status is of concern.[407]

Subspecies
Icelandic Redwing

Turdus iliacus coburni | Deargán sneachta Íoslannach

WINTER VISITOR AND PASSAGE MIGRANT
STABLE

This dark race of the Redwing breeds exclusively in Iceland, and winters in Scotland and Ireland.[408] Hutchinson notes that some birds occur in autumn on the north and west coasts at Tory Island and Malin Head, and Erris Head in Mayo, and that there is probably a small wintering population.[409] This seems to understate the position in Donegal, where it is quite normal to identify wintering birds as of this race. The Tory Bird Report for 1958/9 states that all birds seen in the spring were subspecies *coburni*. At Malin Head on 17 October 1961, 372 Redwing were recorded, and many more were heard passing before dawn. The majority of those examined were *coburni*. One day later, at Malin Head, 30 per cent of more than 400 birds were *coburni* (MHBO).

There are usually a few Redwings which are slow to make the return journey north in spring. One particularly late bird at Dungloe on 4 June 2017 was Icelandic.

Blackbird (Common Blackbird)

Turdus merula merula | Lon dubh

ABUNDANT RESIDENT, WINTER VISITOR AND PASSAGE MIGRANT

STABLE

Blackbirds are an almost constant presence in all but the most treeless bog or upland landscapes. Offshore islands are occupied if there is enough cover, so for example they were not found on Inishmeane but were present on Inishfree Upper.[410] Numbers are boosted in winter by immigrants from across northern Europe. On Tory, Blackbirds are migrants and occasional winter visitors, but as a breeding species the estimate is one pair or less, with no conclusive record of a successful breeding attempt.[411]

The breeding atlases show no change, and the Countryside Bird Survey is in agreement. However, other indicators in Ireland and in Great Britain suggest that the Blackbird is creeping uphill into ground that would normally be used by Ring Ouzel. While this may be associated with land use changes, climate change is probably the chief driver.[412]

Fieldfare

Turdus pilaris | Sacán

WINTER VISITOR

STABLE

Fieldfare desert their breeding grounds in northern and eastern Europe in winter, when flocks descend on most of the continent to the south, including Ireland. They feed in tall hedges on berried trees, like hawthorn and rowan, and on invertebrates in open fields. Large numbers are not as frequent as Redwing, with less than 100 being normal, and the larger flocks are usually in the low hundreds. The winter atlas[413] shows west Donegal and a couple of the wilder parts of Connacht as the only parts of Ireland to be largely without any Fieldfare present, but two decades later, only west Donegal remained unoccupied.[414] However, the number of birds coming here each winter, and their movements when they are here, are very erratic. So it is not totally surprising that one of the largest mixed flocks of Fieldfare and Redwing to be recorded here was one of more than 1,000 birds in the heart of west Donegal, near Burtonport, on 19 November 2016 (T. Gallagher). Fieldfares generally move south after they have used up the berry supply and can often be scarce in mid and late winter.

Ring Ouzel

Turdus torquatus torquatus | Lon creige

SCARCE SUMMER VISITOR

DECREASING

RED-LISTED BOCCI-4

The mountain blackbird was once present as a breeding species in all but five counties in Ireland.[415] In Donegal, Hart regarded them 'as by no means rare'. In his own district of Fanad 'they breed at Knockalla, Glenalla, Auchterlinn and Lough Salt, etc'. And 'in September and October they appear in small flocks, usually less than a dozen, and frequent rocky places about the mountain tops before leaving for the winter'. He cites his regular correspondent, Arthur Brook from Killybegs, who said they were 'very common during the breeding season in all the mountains of S.W. Donegal'.[416] R.J. Ussher found it twice at Doochary, near sea level.[417]

Since then the Ring Ouzel has been in decline throughout Ireland, and by the time of the first breeding atlas it was more or less confined to the remote fastnesses of the north-western coastal counties from Mayo to Donegal,

Ring Ouzel, Derryveagh Mountains.

ROBERT VAUGHAN (under licence)

Map 18.5:
Ring Ouzel locations
occupied since 1963

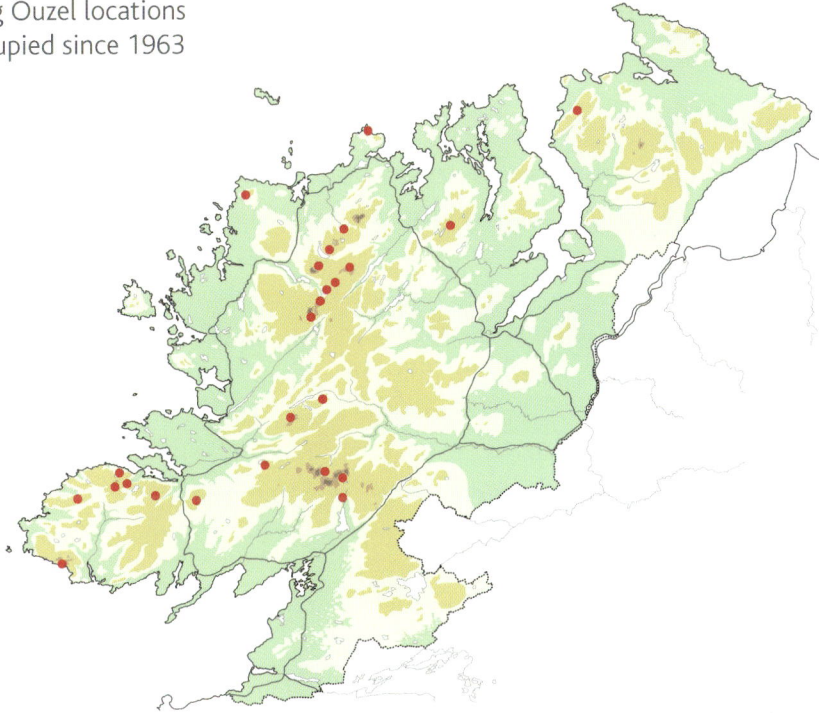

and to the mountains of Down, Wicklow and Waterford in the east.[418] By the third atlas, almost all Irish birds were in Donegal.[419]

There are modern records (from 1963 onwards) of Ring Ouzels present and presumed to be breeding in 15 hectads (10 x 10 km grid squares) (Map 18.5). The Upland Bird Survey looked at thirty-five known sites. The habitat was found to be wet heath, with much exposed siliceous (acidic) rock and cliffs. The altitude that Ring Ouzels were found at was lower than in Great Britain, in some cases quite close to sea level. The survey found ten to fifteen breeding pairs – ten confirmed or probable, and a further five birds that can be considered possible breeders. These were distributed across eight hectads.[420] More recently, a thorough survey found nine occupied tetrads (2 x 2 km squares) in 2021, two in 2022, and in 2024 a single bird in the southern uplands and a successful breeding pair in the northern uplands (NPWS unpublished data).

Ring Ouzels' sites are not particularly threatened by peat cutting, being located on shallow dry heath. But hill walkers and climbers are increasingly penetrating their remote breeding habitats, and conifer plantations have spread over large areas surrounding suitable terrain. If the impacts from these changes can (in principle) be contained, the same cannot be said for those due to climate change. As an upland species, it seems inevitable that Donegal, and Ireland, will cease to be acceptable for the few remaining birds.

But these modern threats do not explain the long-term decline since the mid-1800s (Map 18.5). In a detailed study in County Kerry, Allan Mee points to the need for mosaics of heather and grass, possibly including wet flushes.[421] And Anthony McGeehan argues that increases in predator numbers, and decreases in bilberry, heather and native deciduous trees have all flowed from replacing cattle in the hills with sheep.[422]

The historical decline in breeding numbers qualified Ring Ouzel for the first of all the Irish red lists, by Tony Whilde in 1993,[423] and it has never departed from subsequent lists, in 1999, 2007 and 2013.[424] It is a sad fact that we are witnessing what appears to be the last gasp in Ireland of this once-familiar songbird. If some measures are not taken urgently to conserve its specialised mountain habitats, it will very soon join the Corn Bunting in oblivion.

As a migrant, Ring Ouzels turn up somewhat less than annually, but there can be several at a time. The maximum was six on Inishtrahull on 7 October 1965. Of twenty-seven modern records of migrants, seven have been in spring.

The Ring Ouzel is Red-listed in BirdWatch Ireland's 'Birds of Conservation Concern in Ireland 4: 2020–2026' on grounds of an 80 per cent long-term decline in its population.[425]

Spotted Flycatcher

Musicapa striata striata | Cuilire liath

SUMMER VISITOR

DECREASING

Spotted Flycatchers are among the last of the summer migrants to arrive in Donegal from winter quarters in tropical Africa, usually in early May at about the same time as the Swift. They are woodland birds, but need open spaces for their aerial manoeuvres. They may be plain little birds, but their swashbuckling behaviour more than compensates.

The Spotted Flycatcher is undergoing major declines, more than most other summer migrants. The Countryside Bird Survey records a considerable loss in distribution across the country amounting to 25 per cent over forty-four years from 1972 to 2016,[426] but, perhaps surprisingly, it does not meet the criteria for Red-listing.[427] Between the first and the third breeding atlas there appears to have been a retreat from the coastal and lowland 10 x 10 km squares.[428] This concurs with the experience of it being now easiest to find in the small scraps of tree habitat around abandoned settlements in the upland areas. None of the usual explanations for significant declines have been

Spotted Flycatcher, Carnowen, near Convoy.

proven in Ireland, but a case can be made that they are particularly vulnerable to nest predation.[429] That has certainly been my own experience over many years of monitoring the local nests.

Robin (European Robin)

Erithacus rubecula melophilus | Spideog

ABUNDANT RESIDENT, WINTER VISITOR AND PARTIAL MIGRANT
STABLE

The Robin is one of our best-loved birds, and deservedly so, given its confiding ways, and for lifting our spirits through the winter with its cheerful song. It is almost ubiquitous on the mainland, and is present in the breeding season on Arranmore, Inishfree Upper and Inishmeane. Having been only a regular autumn migrant on Tory, Robins have, over the last five or six years, become more frequent, until breeding took place in 2022. It is absent as a breeding species from other offshore islands.

In hard winter weather some will descend to the coast, where small numbers have been seen to set aside their territorial instincts and feed together peaceably on the tide line. Others, particularly females, will migrate, but

Robin, St John's Point.

KIM PERIERA

they are replaced by winter immigrants, mainly from Scotland. The evidence from the Countryside Bird Survey 1998–2016 is that the Robin's population is stable.[430]

Bluethroat

Luscinia svecica svecica | Gormphíb

RARE VAGRANT

Bluethroats breed throughout continental Europe in two different races, which are only distinguishable in the field in their breeding plumage, when the colour of the spot in the middle of the blue throat is diagnostic. *Luscinia s. svecica* is the red-spotted race. It breeds in Scandinavia, and migrates to North Africa. Fortunately, the only birds recorded in Donegal

Bluethroat, Tory.

PETER PHILLIPS

have both been spring males, so they were easily identified. The first was on Arranmore on 19 May 1996 (A. McMillan et al.), and the second was on Tory from 20 to 22 May 2012 (T. Campbell et al.).

Nightingale (Common Nightingale)

Luscinia megarhynchos megarhynchos | Filiméala

RARE VAGRANT

There has been only a single Donegal record of this summer visitor from Africa to Europe (excluding Scandinavia, Ireland and most of Great Britain). Its song, although perhaps not the secretive bird itself, would be familiar to many Donegal tourists on their summer trips to continental Europe. Our bird was seen on Tory, on 13 May 2019 (J. Adamson et al.).

Red-breasted Flycatcher

Ficedula parva | Cuilire broinnrua

RARE VAGRANT

Red-breasted Flycatcher is an African migrant to breeding grounds in eastern Europe. It is one which comes frequently to Ireland, but only nine have reached Donegal. The first was killed at the light station on Tory on 28 October 1894. It was the third record for Ireland, and its wing and leg were preserved. The next one was also obtained at the Tory light, on 25 October 1948. The first modern record was at Tory, where a bird was trapped on 10 October 1961 (N. Curran). It was followed three days later by one at Malin Head. Glencolumbkille had the fifth record – a first-winter bird on 15 October 1996. The remaining birds were on Tory, the first of which was an adult on 16 October 2012. Then there were three birds in 2014 – one alive and one dead on 18 September, and an adult on 12 October.

Pied Flycatcher (European Pied Flycatcher)

Ficedula hypoleuca hypoleuca | Cuilire alabhreac

SCARCE PASSAGE MIGRANT AND RARE SUMMER VISITOR
STABLE AS A MIGRANT

Pied Flycatcher is a summer visitor from west Africa to virtually all of Europe – apart from Ireland. With its British distribution predominately western, it is unfortunate that there have so far been only three breeding records of Pied Flycatcher in Ireland. But in July 1989, two singing males

ANTON MEENAN

Pied Flycatcher, Tory.

were found in Donegal, one of them at Glenveagh, which is undoubtedly potential breeding habitat. The other one was at Glencolumbkille, where one would only expect migrant birds. In 2023, a singing male held territory at Ardnamona, but with no sign of having attracted a female. Good coverage at Glenveagh has revealed the presence of three other summer migrant species which, along with Pied Flycatcher, are common in the western oakwoods of Great Britain, but extremely rare in Ireland – Wood Warbler, Redstart and Tree Pipit.

Two autumn migrants were obtained at light stations in the first half of the last century, and one at Tory on 13 April 1952. It was then only two years until Philip Redman had the first modern sight record, also on Tory, on 15 October 1954. Autumn passage is now known to be almost regular. There have been sixty-four modern autumn records so far, forty on Tory and nineteen in the far south-west. These are unusually high concentrations, even for those two eminent migration hot spots, and no doubt the balance between them is skewed by the greater observer coverage on Tory. Arranmore and Malin Head have had two each, and a single bird was recorded at Bloody Foreland. Six of the autumn migrants were in August, thirty-three in September and twenty-five in October. Spring passage is rare, with only two May records – one on Tory and one on Arranmore.

Collared Flycatcher – Ireland's only sighting, Tory.

Collared Flycatcher

Ficedula albicollis | Cuilire muinceach

RARE VAGRANT

This eastern European migrant from Africa has only once been recorded anywhere in Ireland. It was a first-summer bird on Tory, on 29 May 2012, and we can thank Robert Vaughan for finding it.

Black Redstart

Phoenicurus ochruros gibraltariensis | Earrdheargán dubh

SCARCE PASSAGE MIGRANT, RARE WINTER AND NON-BREEDING SUMMER VISITOR

The Black Redstart is resident in south-west Europe and is a summer visitor to central and eastern Europe. In Donegal there have been twenty-nine birds seen so far. The most remarkable of these records are the bookends. The first birds were two seen at Malin Beg on 12 July 1948. Despite the date, and the number of birds, there was no suspicion of breeding.[431] The most recent sighting was even more unexpected. A bird was seen on mountain scree in July 2021. This is a natural breeding habitat for Black Redstarts in continental Europe, but is not used at all by the small breeding population in Great Britain, which is virtually confined to urban and industrial sites.

The European populations have very mixed seasonal movements, so it is not unreasonable that the few birds which reach Donegal have appeared

Black Redstart, Tory.

in all seasons. There are ten records in the spring months of April and May, and eight in the autumn months of September and October. Six birds have arrived in November and December, and five in June and July. There have been no records in August or March, and none in the core winter months of January and February.

This seasonal breakdown is tending to shoe-horn birds into brackets that don't necessarily fit. It can be argued that the absence of birds in January, February and March means that there have been no true winter visitors to Donegal – the November and December records being only late autumn migrants. But one December bird was on the 26th, and another stayed from the 12th to the 30th. Of the five birds recorded in these months, these two at least were surely true winter visitors. The June and July records could also be treated as passage migrants – were it not for the fact that the mountain bird was unquestionably a summer visitor.

There have been fourteen birds on Tory, four at Malin Head and three each on Arranmore and the southern coastal strip from Bundoran to Tullaghan.

Redstart (Common Redstart)

Phoenicurus phoenicurus phoenicurus | Earrdheargán

SCARCE PASSAGE MIGRANT AND RARE BREEDER
DECREASING
RED-LISTED BOCCI-4

Redstarts winter in Africa, and in summer occupy all of Europe – apart from Ireland. Our western oakwoods should be perfect habitat, as they are in Scotland and Wales, but the scattered remaining fragments of these in Donegal are probably just not enough to sustain a viable breeding population of Redstarts. Yet Donegal is one of the more favoured counties in Ireland. The first proof of breeding was a nest discovered at Glenveagh in 1968 (R. Forbes). Ardnamona had a breeding pair in the same year. Subsequently, Glenveagh has had a series of records. In 1975, there was a pair, and a bird seen at another site within the wood. In 1981, there was another pair, and in 1984 a singing bird on 28 April and at the same location on 22 May. It would be reasonable to assume that more intensive vigilance in the 1970s and 1980s would have found breeding birds in most years. The lack of any records since then is more puzzling, and may be related to deterioration in the quality of the habitat.

Fifty migrants have been recorded. A high proportion (forty-two) have been on Tory, and five on Arranmore. The thirty-seven autumn birds arrived between late September and late October, and the thirteen spring birds were all in late May. It may be significant that in the six operational years of Tory Bird Observatory, from 1959 to 1964, twenty-five Redstarts were recorded. In the most recent six years from 2015 to 2020, with fairly good coverage, only nine birds were seen on Tory. This hint of declining numbers would tie in with the known decline in the breeding population in the oakwoods of western Great Britain.

The Redstart is Red-listed in BirdWatch Ireland's 'Birds of Conservation Concern in Ireland 4: 2020–2026' due to a 72 per cent long-term (since 1980) decline in its breeding range.[432]

Whinchat

Saxicola rubetra | Caislín aitinn

UNCOMMON SUMMER VISITOR AND PASSAGE MIGRANT
DECREASING
RED-LISTED BOCCI-4

Whinchats spend the winter in tropical Africa, and the summer in Europe. It would appear that they have always been scarce in Donegal. At the end of the nineteenth century Hart describes them as very local, mentioning only two pairs about Trawenagh Bay, and one near Pettigo.[433] They were absent in Fanad, where he lived.

Whinchat, Tory.

Whinchats are found in areas of rough or tussocky grassland with song posts. Habitats matching this description can occur at all elevations, and for various reasons. One classic variant is to be found along river flood-plains, and in Donegal the Gweebarra River below Doochary is one of very few such locations. Whinchats have bred there as widely apart as 1955 and 2016. Mown hay meadows also fit the habitat description, but are now equally rare. Occasional birds at a hay meadow on the Sheskinmore Nature Reserve demonstrate that this habitat could still be useful, if conservation management to favour Whinchats is ever contemplated. At Malin Head they have had a significant breeding presence, with about ten pairs each year during the period from 1993 to 1999, where there had only been two pairs in 1992. One site that has been faithfully occupied for many years is in Glenveagh, north-west of the main lake.

In recent times the most productive habitat has been the large expanses of harvested conifer forests, re-stocked with young trees that have reached the height of typical song posts. The 1990s and 2000s showed that the species was scattered widely in such areas. But these are transient sites, and finding those at the right stage to suit Whinchats involves a lot of time and legwork. One was clearly at the right stage on 28 May 1998, when a 2 km stretch of forest road south of the Mourne Beg River had seven territorial males.

Searching for these sites has been neglected over the last ten or more years, so each year only one or two pairs are found, and these are often in the heathery margins of agricultural land in the central hills – not ideal hab-itat, as they can find themselves in competition with Stonechats. The three breeding atlases show a steady decline from the 1960s.[434]

Autumn and spring migrants are fairly regular, but very localised. The recorded forty-nine birds were at Tory (27), Malin Head and Inishtrahull (14), the group of locations around Glencolumbkille (5), Fanad Head (2) and Arranmore (1). The seasonal spread has been August (3), September (20), October (10), May (15). Wintering birds are rare anywhere in Ireland, but one was seen at Bogay on 1 March 1964.

The Whinchat is Red-listed in BirdWatch Ireland's 'Birds of Conservation Concern in Ireland 4: 2020–2026' due to a 62 per cent short-term decline and an 89 per cent long-term decline in its population, and a 76 per cent long-term decline in its breeding range.[435]

Stonechat (European Stonechat)

Saxicola torquata hibernans | Caislín clochn

RESIDENT

STABLE

Stonechat is one of the most characteristic species on rough ground throughout the county. Generally, it is common in low scrub with gorse or heather, and particularly around the coast, where the winter temperatures are more tolerable. The Upland Bird Survey found it on 45 per cent of all sites chosen as Ring Ouzel habitat, and on 28 per cent of sites suitable for Golden Plover.[436]

Low temperatures reduce numbers of Stonechats dramatically, and they can seem to disappear completely for a few seasons, as after the particularly hard winters of 2009/10 and 2010/11. Given time (and survival of the habitat), they eventually bounce back to previous population levels, as is revealed by the tracking of the Countryside Bird Survey.[437]

Wheatear (Northern Wheatear)

Oenanthe oenanthe oenanthe | Clochrán

SUMMER VISITOR

STABLE

Our Wheatears migrate from the Sahel region south of the Sahara to breed all across Europe. They are one of the first summer migrants to arrive, usually in early March. They like open ground with sparse vegetation, and stony or rocky areas. As these are mainly found around the coast, it is there, and on the offshore islands, that they reach their highest densities. They can also be found at lower density in the uplands, where dry-stone walls provide nesting opportunities. The Upland Bird Survey found Wheatears to be present on 86 per cent of all areas surveyed as Ring Ouzel habitat, at a density of 1.7 pairs per km². On Golden Plover habitat, where unbroken expanses of vegetated ground is more typical, they were less frequent, being present on 58 per cent

Wheatear, Cronalaghy, Ballybofey.

of the surveyed sites, at a density of 0.6 pairs per km² .[438] They are generally absent from the agricultural lowlands. Numbers have been fairly stable over the years.[439]

At migration time, particularly in spring, quite large numbers are sometimes encountered. Counts of over 100 have been made from time to time on Tory, Arranmore and Malin Head. Individuals sometimes turn up in the agricultural east of the county on ploughed fields.

Subspecies

Greenland Wheatear

Oenanthe oenanthe leucorrhoa | Clochrán Graonlainne

PASSAGE MIGRANT

Some migrant wheatears in spring are significantly larger than the local breeders, and the males have more richly coloured underparts. These birds are on a journey to Greenland (or Iceland), for which natural selection has progressively lengthened their wings – a very appropriate measure as this is the longest migration taken by any songbird, and one of the most hazardous. There is no indication of how many Greenland Wheatears pass through Donegal, but they are quite reliably seen anywhere around the coast a bit later than the locally breeding birds. Of 135 wheatears seen on Arranmore Island on 14 May 2012, a good percentage were of this race.

Black / White-crowned Wheatear

Oenanthe leucura / leucopyga | Clochrán dubhchorónach / bánchorónach

RARE VAGRANT

A wheatear seen at Portnoo golf course on 10 June 1964 was one or other of these two species (H. Copeland et al.). There are no other Irish records of either species – specifically identified or not. This is a remarkable record, as both species are non-migratory and sedentary – *leucura* from Iberia and North Africa, and *leucopyga* from North Africa and the Middle East. The record was accepted by the Irish and international authorities at the time.

Dipper (White-throated Dipper)

Cinclus cinclus hibernicus | Gabha dubh

RESIDENT
STABLE

Ireland shares a resident race of the Dipper with the west of Scotland. It can be found in Donegal along fast-flowing, stony and eroding stretches of streams and rivers. This is usually the character of upland streams, yet in their 1992 study of Dippers in six river systems (two in Inishowen and four across the border), Perry and Agnew found that of fifty-four nests, thirty-nine were below 100 m, and all but three were below 150 m.[440] They surveyed the Crana and Mill Rivers in Inishowen on three

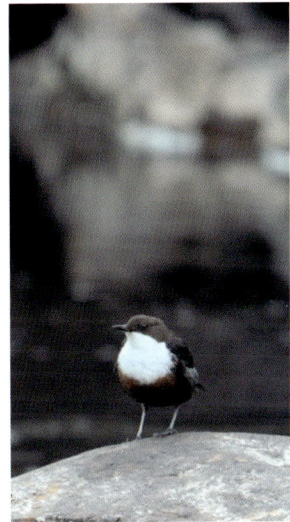

Dipper, Blue Stack Mountains.

JOHN CROMIE

occasions, at ten-year intervals. Dippers on the Crana went from three pairs in 1972 to six in 1992, while the Mill River stayed at two pairs throughout. Territory length was calculated on the River Faughan in County Derry. It went from 2.9 km per pair when there were seven pairs in 1972, to 1.8 km per pair when there were twelve pairs in 1982.[441] These territory sizes are probably comparable to those on the Crana River, but shorter than what is required on the Mill River.

The breeding atlases show widespread presence of Dippers, with odd gaps that may only indicate variable coverage.[442] The winter atlas shows the

Dipper to be generally absent, with significant presence only in Inishowen.[443] This is also likely to reveal only a lack of coverage, as the third atlas shows no difference between summer and winter distributions. It is known that some Dippers go downstream towards the coast in winter, but there is little evidence in Donegal of this being on a significant scale.

There is a single record of a Dipper on Tory. It was a bird in flight, on 13 October 1962 (TBO).

House Sparrow

Passer domesticus domesticus | Gealbhan binne

RESIDENT

STABLE

The familiar House Sparrow is not as abundant as it used to be, but there is no evidence of continuing decline. Efficiencies in the transport of feedstuffs have reduced the easy pickings along roadsides, but House Sparrows are still fairly widespread, although more localised to farmyards or other places where they can find spilt grain. The last breeding atlas shows that almost the whole of Donegal has a very low density of House Sparrows in comparison with the rest of Ireland.[444] The Countryside Bird Survey shows an upward trend nationally.[445]

The species is highly sedentary. Hutchinson states that they withdraw from the mountains of Donegal in winter.[446] In fact there are very few House Sparrows there to withdraw. Where the atlases record them as still present in the core areas of upland habitat in the county, it would only be in a few farmyards in valleys around the fringes of the area.

House Sparrows have reached Tory from time to time, but breeding numbers never reach double figures.

Tree Sparrow (Eurasian Tree Sparrow)

Passer montanus montanus | Gealbhainn crann

RESIDENT

DECREASING

This more smartly unisex-dressed relative of the House Sparrow has had fluctuating fortunes in Donegal, as elsewhere, but here the rhythm has

Tree Sparrows, Tory.

differed somewhat from the other regions. The first record in Donegal was of a pair breeding in a cabin roof on Arranmore in 1886. They were not found in 1896, but a pair was again breeding there in 1939.[447] Since then there have been three records for the island – at Torries in 1993, and at the lighthouse in 2010 and 2011, all of them in summer months, but with no evidence of breeding.

Tree Sparrows first arrived on Tory in the summer of 1960 (TBO), and have been there ever since, with a peak count of 155 on 13 October 1995, and up to 100 estimated during 1998. They are not threatened by competition with House Sparrows, which colonise erratically in very small numbers which do not persist. Tree Sparrow numbers on Tory in recent years have mostly been in single figures. It may be only a coincidence, but the main decline started when the road through the two villages was first tarred. The gritty and weedy margins of the un-tarred road had been prime habitat for foraging Tree Sparrows. A similar situation pertained at Malin Head, where about fifty birds were present in the area in 1961, after which they declined rapidly. At least one pair was reported to have bred in 1965 (MHBO). On the nearby Inishtrahull, two adults were present on 22 July 1963, and one adult with a juvenile on 12 August. The *Irish Bird Report* suggested that breeding may have taken place on the island or on the adjacent mainland. In retrospect, it seems highly probable that breeding did take place on the island. More recently, a Tree Sparrow was seen at Malin Head in July 1999.

Malin Beg and Glencolumbkille each held breeding birds in the 1960s. At the western end of Dunfanaghy New Lake, a small colony present from at least as far back as 1935 was last recorded in 1966.[448] In the 1990s, small inland colonies emerged in old stone structures – ruined castles, outhouses and bridges – mostly in the agricultural lowlands around Lough Swilly, and between there and the River Finn. These have mostly faded away, but the area continues to hold scattered pairs with no apparent continuity of site. The exceptions are around Blanket Nook and Inch Lough, where small colonies and wintering parties persist.

With House Sparrows so few on Tory, it is perhaps not surprising that the two species have interbred. One or two hybrids were seen each year from 1994 to 1997.[449]

Dunnock

Prunella modularis hebridium | Donnóg

COMMON RESIDENT
INCREASING

The Dunnock is a widespread and common species in woodland edge and scrub habitats, which in practice includes hedgerows and gardens. Their range has not changed. Nationally, they have been steadily increasing in abundance,[450] and that is likely to be the case in Donegal.

Dunnock, Farland Bank, Inch.

Western Yellow Wagtail*

Motacilla flava flavissima | Glasóg bhuí

SCARCE VAGRANT

The Yellow Wagtail was always a suite of many races spread across Eurasia, but those in the Far East have now been separated into a different species, so the residue have been given the longer name of Western Yellow Wagtail. The race classified as *Motacilla flava flavissima* is a common summer migrant from Africa to a large part of Europe, including England. Up until about 1940 it was a very local breeding species in Ireland, but only twenty individuals have made their way to Donegal. The first two were in the nineteenth century, for which no details are available.

The eighteen records since then start with one on Tory, on 28 September 1962 (TBO). In total Tory has had twelve birds, three have been at the south-western locations of Rocky Point, Malin More and Glencolumbkille, two were at Blanket Nook and one at Kiltooris Lough. All occurrences have been of single birds, apart from four on Tory on 5 May 2015, and the only other year with more than one bird was 2015. The monthly breakdown is April (2), May (5), June (1), August (1), September (7) and October (2). One of the May sightings was a bird on Tory from 31 May to 1 June 2011, and the other June record was at Blanket Nook on 20 June 2004. Given that Yellow Wagtail was a regular breeding species in Ireland in the past, these spring records, in particular those in June, hint that future breeding attempts are not totally out of the question.

* This very distinctively yellow subspecies always had the English vernacular name of 'Yellow Wagtail', which unfortunately it shared with the full species. Re-naming the full species as 'Western Yellow Wagtail' still leaves the subspecies vernacular every bit as inappropriate as before. A new name for this race would be timely. See below for the names of other subspecies.

Subspecies
Blue-headed Wagtail

Motacilla flava flava | Glasóg cheannghorm

RARE VAGRANT

This neighbouring race of the previous one breeds in north and central Europe. There are three records of it from Tory. The first, on 22 May 1954 (P.S. Redman), was accepted at the time, but not published until recently.[451] It was the first Irish record. The second Donegal bird arrived sixty years later, on 26 May 2014 (D. Brennan), and the third on 30 April 2022 (R. Vaughan).

Subspecies
Grey-headed Wagtail

Motacilla flava thunbergi | Glasóg cheannliath

RARE VAGRANT

There are two records of this Scandinavian race of the Western Yellow Wagtail, both of them still to be verified by the IRBC. One was seen in the Glencolumbkille area from 29 to 30 September 2000, and one was present on Inishtrahull from 11 to 12 June 2022.

Eastern Yellow Wagtail, Tory.

Eastern Yellow Wagtail

Motacilla tschutschensis tschutschensis / plexa | Glasóg oírthearach

RARE VAGRANT

The Eastern Yellow Wagtail has only recently been split from what had been the Yellow Wagtail, which incorporated many races from western Europe all the way east to Alaska (see also Western Yellow Wagtail, above). The eastern races, which now comprise the Eastern Yellow Wagtail, breed in north-east Siberia and Alaska, and winter in south-east Asia. A first-winter bird seen on Tory was the first record of the species for Ireland. It was present from 12 to 28 October 2013 (V. Caschera, J.F. Dowdall, J.E. Fitzharris). DNA analysis of its faecal matter determined that the subspecies concerned was either the nominate race *tschutschensis*, which straddles the Bering Strait, or *plexa*, from further west in north-east Siberia.

Citrine Wagtail

Motacilla citreola | Glasóg chiotrónach

RARE VAGRANT

Essentially a central Asian species, the Citrine Wagtail is spreading into eastern Europe. The three Donegal records so far are probably as much as we could have expected, although one of the beautiful spring males would be a very welcome addition. The first bird was on Arranmore on 4 October 1993 (P. Farrelly and A. McMillan). The second was at Malin More on

24 September 2002 (A. McGeehan). The third was again on Arranmore, on 23 September 2018 (J. O'Neill).

Grey Wagtail

Motacilla cinerea cinerea | Glasóg liath

RESIDENT
DECREASING
RED-LISTED BOCCI-4

The Grey Wagtail is one of the few species we have that are closely associated with rivers. It likes fast-flowing streams, but can be found in both lowland and upland stretches as long as there is a good supply of insects to chase. So it is widespread as a breeding species. The Upland Bird Survey found it to be present in 32 per cent of Ring Ouzel sites sampled, and in 13 per cent of Golden Plover sites.[452]

The Countryside Bird Survey records declines in the breeding population nationally.[453] The first winter atlas survey shows it mainly in the north and north-east, but by the second one it was more generally present, and at a density as high as anywhere in Britain and Ireland.[454]

In more recent years Grey Wagtails have been rarely seen in winter. Throwing some light on this observation is a useful piece of research which found that Irish birds start to breed earlier, lay smaller clutches and produce fewer young than other European populations.[455] This presumably slows their recovery from hard winter losses, to which they are very susceptible – like many species depending on insect food in winter.

The Grey Wagtail is Red-listed in BirdWatch Ireland's 'Birds of Conservation Concern in Ireland 4: 2020–2026' due to a 50 per cent short-term decline in its breeding population.[456]

Pied Wagtail

Motacilla alba yarrellii | Glasóg shráide

RESIDENT
INCREASING

The Pied Wagtail, along with Swallow, Starling and House Sparrow, is most familiar as a farmyard bird. It is also found at lower densities in other man-made habitats, including towns, and in natural habitats such as streams, estuaries and coasts. Apart from the higher uplands and woodland, it is almost ubiquitous, as confirmed by the most recent atlas.[457] It is much more readily found in winter than the Grey Wagtail, which is explained by its ability to recover more quickly from population crashes in cold winters. The Countryside Bird Survey has found it to be increasing nationally.[458] That is likely to be also the case in Donegal, but evidence is lacking.

Subspecies

White Wagtail

Motacilla alba alba | Riabhóg bhán

PASSAGE MIGRANT

White Wagtail, the nominate race of *Motacilla alba*, breeds throughout continental Europe, and passes through Donegal on its way north to Iceland or Scandinavia. Numbers in late April can be substantial at suitable locations – much higher than any aggregation of Pied Wagtails. There is a record of 190 on a wet grass field on Inch Levels on 30 April 2016, and 110 on 19 April 2016 at Blanket Nook, where they are most easily viewed along the embankment wall. They are regular each year at these two sites. That may also be the case at Malin Head, if a sighting of 100 plus on 21 April 2019 is an indication. The seasonal distribution of 179 sightings is shown in Figure 18.64. Of 3,243 birds counted, the first half of April had 18 per

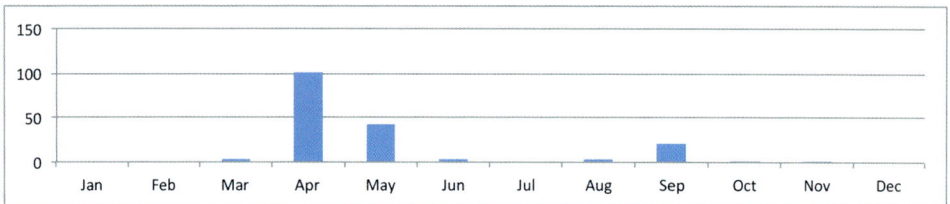

Fig. 18.64: Monthly distribution of 179 White Wagtail sightings

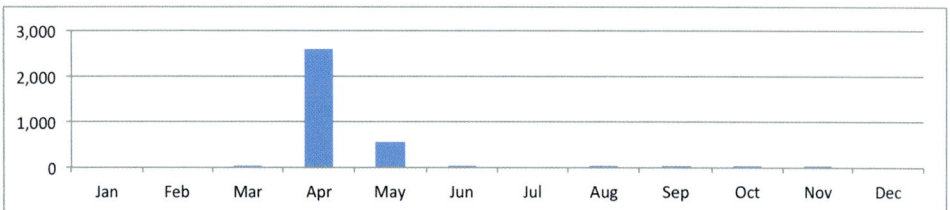

Fig. 18.65: Monthly distribution of 3,243 White Wagtail birds seen

KIM PERIERA

White Wagtail, St John's Point.

cent, 62 per cent were in the second half of April, and the whole of May had 17 per cent (Figure 18.65).

There may well be more birds on the return journey in autumn than suggested by Figures 18.64 and 18.65, as White Wagtails are much more difficult to identify in autumn or in their juvenile plumage than in their sprightly spring dress, and most observers don't try.

June sightings are of birds paired with Pied Wagtails – one at Horn Head on 4 June 1989, and one on Tory, on 4 June 1994. There is no record of any breeding attempt by a pure pair of White Wagtails.

Richard's Pipit

Anthus richardi | Riabhóg Richard

RARE VAGRANT

Richard's Pipit is one of the largest species of its genus, breeding in mid-latitude east Asia, and wintering in south Asia. Why so many reach western Europe is a mystery, and we can only marvel that they even reach Donegal at all. So far there have been ten recorded.

The first record is of a bird seen on Horn Head on 19 August 1955 by the eminent British ornithologist E.M. Nicholson.[459] There was no other acceptable record until the more recent phase of intensive birdwatching, which has produced nine birds, all of them on Tory. The first three were in September 2001 – a first-winter bird on the 23rd, two on the 25th and one on the 26th. In 2012 there was one from 14 to 18 October. In the following year, 2013, there was one from 20 to 24 October. Finally, there were four birds in 2015. Three of these were present from 1 to 6 October, and the fourth was seen on 4 October.

Meadow Pipit

Anthus pratensis | Riabhóg mhóna

ABUNDANT RESIDENT AND PASSAGE MIGRANT
DECREASING AS A RESIDENT
RED-LISTED BOCCI-4

Meadow Pipits are by far the most abundant birds in open landscapes, particularly in the uplands. Machair is an important secondary habitat. They are largely absent from the intensively farmed land in the east, although suitable higher ground is never too far away. The absence of a good insect food supply in winter forces them to descend from the uplands. It must also help them avoid the worst impacts of cold weather, which nonetheless is the major cause of fluctuations in their numbers. They breed on many of the offshore islands, although the estimate for Tory is only about ten pairs.[460]

The only general indication of numbers we have is from the Upland Bird Survey, which looked at habitats suitable for Golden Plover and Ring Ouzel. This survey found nine pairs per km² on Golden Plover habitat, and eight on Ring Ouzel habitat.[461] The national population estimate is 1.0 to 1.7 million individuals,[462] which, if taken as twice the number of breeding pairs, would give seven to twelve pairs per km² averaged over all habitats. These figures are very close to what was found in our hills by Cox, but the Donegal estimate is for what should be prime habitat, while the national figure includes much unsuitable land. Higher figures have been found in other habitats elsewhere – up to sixty pairs per km² in Wexford and Cape Clear[463] – so something is suppressing the Donegal density. Altitude might be a factor, or some, as yet unidentified human-induced habitat degradation. The Countryside Bird Survey has found the species to be decreasing nationally.[464]

Meadow Pipits are the bread and butter for many of our raptor species, especially Merlin and Hen Harrier, which follow them off the hills in winter. They are also the chosen surrogate in Donegal for the Cuckoo.

Meadow Pipit, Doochary.

Irish birds are mostly resident, and there is no hard evidence that winter numbers are boosted by immigrants. But there is considerable passage migration of birds from breeding populations in Scotland and beyond. The observatories at Tory and Malin Head / Inishtrahull in the 1950s and 1960s logged these movements. Numbers frequently reached into the hundreds, with one large movement of 800 birds on Tory on 8 September 1958.

At that time, the Meadow Pipit was split into two races, the nominate *pratensis*, and *theresae* from the west of Ireland.[465] The local *theresae* race was later re-defined as *whistleri*, and the area it occupied was extended to include the west of Scotland and all of Ireland.[466] Currently, the IOC doesn't recognise any sub-specific division of Meadow Pipit. During the years when they functioned, the two observatories each reported daily movements of *theresae*, with a maximum of 250 on 4 September 1960, comparable to the peak count of 300 *pratensis* on the 25th to 27th. Some of the putative *theresae* birds were caught and identified in the hand, which supported the observatory's tentative identification of the larger numbers observed in the field.[467]

The Meadow Pipit is Red-listed in BirdWatch Ireland's 'Birds of Conservation Concern in Ireland 4: 2020–2026' on the grounds that its European conservation status is of concern.[468]

Tree Pipit

Anthus trivialis trivialis | Riabhóg choille

RARE PASSAGE MIGRANT AND SUMMER VISITOR

The Tree Pipit is another of that elite group of migrant African songbirds that are part of the bird community of the western oakwoods in Great Britain but are largely absent from Ireland. Tree Pipit is the rarest of the group here, with no proven breeding records so far, nor even in Ireland as a whole. However, in Donegal there is a record of a bird singing at Glenveagh from 27 to 31 May 1988 (D. Duggan et al.), something that is perhaps overlooked among the abundance of Meadow Pipits.

The first bird in Donegal was not recorded until 1952, when a migrant was killed at the Tory lighthouse on 13 April.

There have been twenty-seven birds seen, all singles, apart from two doubles in May, one of them on Arranmore, the other on Tory. The sixteen autumn birds were mainly in September, and the eleven spring birds centred on May. Tory has the largest share with twelve birds. The south-west group of sites around Malin More has four, Arranmore three and Magheraroarty two.

Pechora Pipit

Anthus gustavi | Riabhóg Pechora

RARE VAGRANT

This very rare pipit breeds in Siberia, between the tundra and the taiga, and winters in the Philippines and Indonesia. There is a single record in Donegal, of a first-winter bird seen on Tory on 22 September 2001 (J. Dowdall et al.). It was the second Irish record, and to date there have been no others.

Red-throated Pipit

Anthus cervinus | Riabhóg phíbrua

RARE VAGRANT

Red-throated Pipit breeds right across the Eurasian tundra, with western birds wintering in tropical Africa. It is a fairly frequent vagrant to Ireland, but not to Donegal. There have been only two records, both on Tory, and both seen by Robert Vaughan. The first was on 30 September 2015 and the second on 17 September 2017. The second bird has yet to be assessed by IRBC.

Water Pipit

Anthus spinoletta spinoletta | Riabhóg uisce

RARE MIGRANT AND WINTER VISITOR

Once regarded as the alpine race of the Rock Pipit, this species spreads out from its breeding bases to spend the winter in wet meadows and on coasts around most of Europe, but relatively few reach Ireland. Donegal has had only four, all of them at Glashagh Bay on the north Fanad coast. The first was present on 16 and 17 March 2008 (W. Farrelly and D. Breen). The second was on 16 April 2019 (R. Vaughan et al.). The third, which could have been a return visit by the previous bird, was present for a long stay during the following winter, from 30 November 2019 to 26 January 2020 (J. Bliss et al.). Finally, the same site held a bird on 7 March 2024 (C. Ingram). The date span makes it certain that more than one bird has been involved. There is no obvious explanation why all our Water Pipits have been drawn to the same part of the north Fanad coast.

Rock Pipit (Eurasian Rock Pipit)

Anthus petrosus petrosus | Riabhóg chladaigh

COMMON RESIDENT

STABLE

All around the coast the Rock Pipit will be found, at least on rocky and stony shores. It will also use sandy beaches where there are zones of rotting seaweed generating plenty of flies. Rock Pipits would normally be within reach of the sea spray, but on Tory they are found throughout the island, where very few pairs of Meadow Pipit breed. There was an estimate of about 150 pairs on Tory in 1954.[469] This compares with only ten pairs of Meadow Pipit (see above). Rock Pipit is likely to be present on all of the offshore islands.

There is no significant difference in its winter status, but there could be local movements within Ireland.

Subspecies

Rock Pipit (Scandinavian)

Anthus petrosus littoralis | Riabhóg chladaigh (Lochlannach)

VAGRANT SPRING MIGRANT, OR RARE WINTER VISITOR

The first record in Ireland of this race was, predictably, on Tory, on 4 April 1961.[470] Being from Scandinavia, it migrates to the North Sea and continental Atlantic coasts for the winter, so it is not unreasonable that some should pass through Ireland, or even winter here. But the race was not detected again until 1997, when one was seen in Donegal Bay on 3 March (R. Sheppard). Seventeen have now been seen, all in spring, with five in March, nine in April and three in May. This suggests that they are spring migrants only, but it is more likely that they will have spent the winter here, and are only detected when they start moulting into their fairly distinctive summer plumage.

Chaffinch (Common Chaffinch)

Fringilla coelebs gengleri | Rí rua

ABUNDANT RESIDENT

STABLE IN SUMMER, DECREASING IN WINTER

The Chaffinch is one of our most abundant and widespread birds. It is a woodland species, but so adaptable that anywhere with a few trees or shrubs will do. This of course rules out the offshore islands apart from Arranmore. Local evidence for any increase is still lacking, but the Countryside Bird Survey reveals a steady increase in numbers, making it Ireland's fifth most numerous breeding species.[471]

Chaffinches gather in flocks in winter, particularly in farmland if there is grain to be gleaned in harvested fields. This race is largely resident in Ireland, but many of our breeding Chaffinches must be moving south in winter (see below).

Subspecies

Chaffinch (Continental)

Fringilla coelebs coelebs | Rí rua (mór-roinneach)

WINTER VISITOR

DECREASING

Wintering flocks of Chaffinches need to be scrutinised closely, as they could well include males with a rosier breast than our local birds. These are birds of the nominate race which occupies most of continental Europe. Those from more northerly latitudes, mainly Scandinavia, migrate south in winter, and many come to Ireland. They were conspicuous in the days when farming allowed more pickings to remain after harvest, and in tillage fields that were not ploughed until spring. At that time, broadly speaking up to the 1970s, mixed flocks of seed-eating birds were frequent throughout the countryside. These flocks, and the continental Chaffinches, are now hard to find.

Brambling

Fringilla montifringilla | Breacán

UNCOMMON WINTER VISITOR
DECREASING

The classic habitat for Brambling is beechwood, where they feed on the carpets of fallen seeds (beech mast). This habitat is not found in Donegal, but here and there are small stands or avenues of ancient beeches left over from the days of the large estates, particularly along the west bank of Lough Swilly, and at Ards Forest Park. They are always worth a look, and in most years, patient searching produces a few Bramblings. They were more easily found prior to the 1970s, when a few would usually be found among each flock of mixed seed-eating birds in the harvested potato, vegetable or grain

Brambling, Dunkineely.

KIM PERIERA

fields. That the Brambling could be more frequent here if conditions are right was clear to see on 6 February 1994 at Speenoge, near Inch Levels, where a field of linseed had been sown, and left unharvested. The estimated flock of about 250 Bramblings were making the most of the opportunity, along with thousands of other birds of various species. How the news got out to all those Bramblings when they were still in more regular haunts somewhere well to the east of Donegal is something to ponder on.

Among all the changes and chances of finding Brambling, perhaps the most predictable is their appearance each year as migrants, mainly on Tory, but that could be just that it has more regular coverage. Sightings there are mainly in October. Numbers are usually very small, but occasionally reach double figures, as on 14 October 2012 when there were twenty, and 12 October 2014 with thirty. Also in 2014, Arranmore had fifteen on 27 October, and Malin Head had eleven on 31 October 2018. Spring migrants are few and far between. Rocky Point had one on 10 May 2003, and Tory had one on 14 April 2019.

Hawfinch

Coccothraustes coccothraustes coccothraustes | Glasán gobmhór

RARE VAGRANT

The Hawfinch is uncommon in Britain and largely sedentary throughout Europe. Movements are minor, and are either dispersal, or partial migration in winter to more southerly parts of its range. Neither Ussher and Warren nor Kennedy et al. make any reference to Hawfinches having occurred in Donegal. But Hart had good evidence: 'One was captured at Lough Eske many years ago in winter, and kept in confinement for some time by Mr J. Young. Mr Stewart shot a pair at Ards early in this century, he believed the first obtained in Ireland.'[472] Stewart's account from 4 December 1828 of the anatomy of one of the birds killed at Ards is quoted at length by Thompson.[473]

Almost two centuries elapsed before there were any more records in Donegal. The first of our five recent birds was in Portnoo between 12 and 15 June 2014 (B. Naughton). The second was on Tory on 5 June 2016 (A. Meenan). The third was at Malin Head on 21 April 2019 (R. McLaughlin). One was at Culdaff, from 15 to 16 April 2021 (P. McBride). The most recent one was on Tory from 21 to 24 October 2023 (A. Meenan et al.).

All of these birds have been in private gardens, and all but one of them arrived in spring or summer. This doesn't fit the pattern of Irish Hawfinches being mainly present in winter, and usually in woodland. In hindsight, the overlooked nineteenth-century records look very convincing, whereas the modern records (the first three of them verified) raise the unsettling thought that they could be escapes from captivity – although none of these birds were reported to have shown signs of a captive past.

Bullfinch

Pyrrhula pyrrhula pileata | Corcrán coille

RESIDENT

INCREASING

This beautiful finch, which is most frequently seen in pairs, is much loved by everyone – except those who value their fruit trees more. The Bullfinch is a woodland species which is well adapted to farmland, gardens and scrub. It seems to be faring quite well in Ireland, in contrast to Britain and the rest of Europe, where it is declining along with many other woodland species.[474] The first two breeding atlases show it to have been absent from the west of the county, but the third one records it in many 10 km^2 in the mid-west, in the heavily populated zone between Gweedore and Ardara.[475] Bullfinches are systematic in their attack on fruit tree buds, but that dedication can be used to retain them around the orchard margins, by planted tempting native alternatives like wild cherry.

Subspecies
Bullfinch (Northern)

Pyrrhula pyrrhula pyrrhula | Corcrán (tuaisceartach)

RARE VAGRANT

The nominate race of the Bullfinch occupies northern Europe, where it is usually sedentary, but undergoes partial migrations south, and very occasional larger irruptions. A female bird showing northern characteristics was recorded on Tory, from 25 to 26 October 2004 (A. and P. Kelly). It was from an exceptional influx of birds, assumed to be of this form, into Britain and Ireland in 2004. This was the first from that influx to be submitted to the IRBC.[476]

Common Rosefinch

Carpodacus erythrinus | Rósghlasán coiteann

SCARCE VAGRANT

STABLE

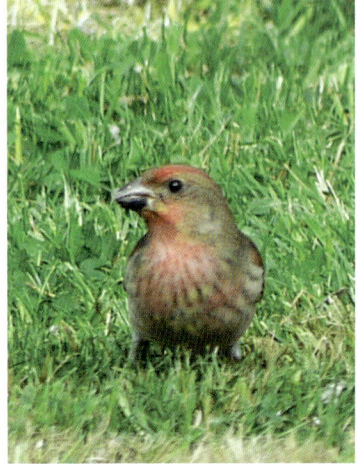
TONY GALLAGHER

Common Rosefinches breed across Russia and Siberia and winter in south Asia. In recent decades they have spread west into eastern Europe and Scandinavia. This should be reflected in an increased number of birds seen here, but that is not at all clear from Figure 18.66. The first Irish record of the Scarlet Grosbeak (as it was called at the time) was a bird seen on Tory on 8 September 1954 (P.S. Redman), and the second was at Inishtrahull on 25 September 1965. There were then no sightings in Donegal until 1992, after which

Common Rosefinch, Meenbannad, near Burtonport.

it started to arrive with some sort of regularity (Figure 18.66). By far the largest share of the Donegal records have been on Tory – twenty-eight out of the total of forty-one. Eight were at the south-western locations around Malin More, three were in the mid-west (Arranmore and Burtonport), and two in the north-east (Inishtrahull and Greencastle).

The seasonality is very much that of an autumn vagrant. Of the forty-one birds seen, two were in August, twenty-five in September and seven in October. In spring, there have been five in May and three in June. The bird that Tony Gallagher photographed in his garden was singing, but neither he, nor it, could have had any expectations of a response.

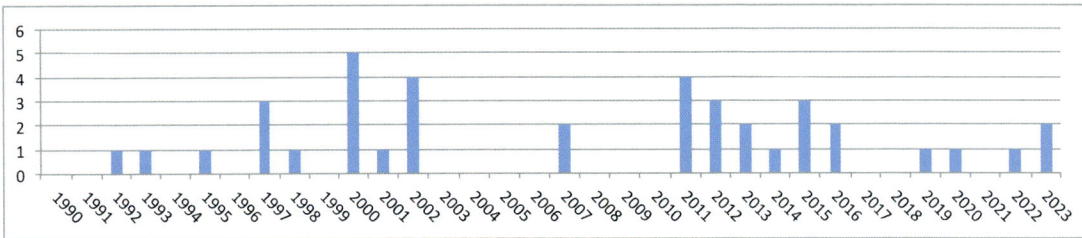

Fig. 18.66: The number of Common Rosefinches each year in Donegal since 1990

Greenfinch (European Greenfinch)

Chloris chloris harrisoni | Glasán darach

RESIDENT

DECREASING

This common finch is widespread across the county, but of the islands only Arranmore is known to be occupied.

Ussher and Warren recorded them as common in every county, although they were not aware of them in the bare coastal districts of north-west Donegal.[477] However, by the 1950s they were noticeable in places like Naran, Bunbeg and Arranmore.[478] The atlases reveal a patchy presence in the more challenging landscapes, but that is no more than can be expected for many species dependent on farmland, woodland, gardens, etc.

Greenfinches can gather in large flocks in winter, feeding on weed seeds and left-over grain in harvested tillage fields. About 1,000 were taking advantage of an un-harvested field of linseed at Speenoge, near Inch Levels, on 6 February 1994. And there were 300 at Raphoe on 3 February 1988. In more recent years there is no doubt that their numbers have dropped. This has been attributed to outbreaks of a disease called finch trichomonosis, caused by the protozoan parasite *Trichomonas gallinae*. The spread of this has been assisted by contact with infected birds and their food droppings at garden feeding stations. The decline has been dramatic, at just over 50 per cent during the period 1998–2016,[479] but they do not quite meet the criteria for the Red List.

Greenfinches are rare on Tory, with one from 5 April to 12 May 2015 being the only documented sighting. This must be taken as evidence that migration is not normal for birds in Ireland and Britain.

Subspecies

Greenfinch (Continental)

Chloris chloris chloris | Glasán darach (mór-roinneach)

WINTER VISITOR

The nominate race of the Greenfinch, which breeds from north Scotland and southern Scandinavia, is more migratory than our resident race, and comes to Ireland as a winter visitor. It has not been identified in the county, but migration of small flocks of Greenfinches was noted at Malin Head in autumn 1964 and 1965, and also in spring 1965.[480] Those birds are more than likely to be continental (see comments on migration in the entry above on our resident Greenfinch race).

Twite

Linaria flavirostris pipilans | Gleoiseach sléibhe

SCARCE RESIDENT AND WINTER VISITOR

DECREASING

RED-LISTED BOCCI-4

Twite is a Tibetan species with an outpost on the fringes of north-west Europe – a unique combination that can only be explained by the intervention of the last glaciation, which presumably split what had been a Eurasian-wide distribution. In Ireland, Twites mostly breed along the brows of coastal cliffs, with a secondary presence in upland moors, usually not far from the coast.[481]

They were common here in the nineteenth century. Ussher and Warren record a flock of fifty which roosted in Hart's plantations in the north of Fanad.[482] Tory had fifteen to twenty breeding pairs in 1954,[483] but that had declined to two pairs in 1962,[484] after which they quickly disappeared. Malin Head had what appeared to be a resident population of fifty to seventy birds in 1965.[485]

By the time of the first breeding atlas, Twites still bred on the coasts and mountains all down the west of Ireland.[486] Birds were present in summer on the north-west Inishowen coast in the early 1980s, and in winter at Muff on the Lough Foyle shore, in Trawbreaga Bay, and on Inch Island. There are records of breeding near the Gweebarra estuary in 1990, and at St John's Point in 1991. By the third atlas they were almost entirely confined nationally to a small corner in south-west Donegal, and another in north-west Mayo.[487]

That conclusion was arrived at in the course of research which produced an estimate of the Irish breeding population at between 54 and 110 pairs. It was also found from ringing studies that the Irish birds are largely sedentary, moving mainly from their breeding areas on heathery slopes, to local foraging areas where they can find wildflower seeds. Places where traditional farming survives, and the associated human settlements,

Twite, Malin Head.

RÓNÁN McLAUGHLIN

generally provide these foraging conditions. In Donegal, Twite breed in the Slievetooey mountains, and forage in the nearby semi-natural grasslands of Maghera, Loughros peninsula and at Sheskinmore.[488]

Management prescriptions which could help to save the species from extinction in Ireland were proposed,[489] and some of these recommendations were incorporated into an action on Twite in the Green, Low Carbon, Agri-Environment Scheme (GLAS). NPWS had seeded a small field at Sheskinmore with a wildflower mix which sustained up to 150 Twite for a number of years in the early 2000s. This hasn't been renewed, but without such seed-rich fields, flocks decline.

Apart from the resident population, evidence of migration was lacking until the Tory Bird Observatory reported that 'almost certainly migrants were passing through, up to 20 being usual' with a peak of seventy on 7 September 1960.[490] That was the case throughout the early 1960s. More recently, Tory has only had birds in low single figures, with ten on 26 March 2017 being exceptional. The experience on Arranmore is similar, but Malin Head has always fared better. In 1965, there was a peak of 200 birds on 15 September, and peaks of thirteen and eighteen birds on the nearby Inishtrahull.[491] Up to fifty birds are still being recorded at Malin Head on the best days in autumn, mainly in October and November, and with up to thirty in April.

In winter, our resident birds tend to visit salt marshes and estuarine areas where they can forage for seeds. Lough Swilly's limited salt marsh habitat has proved attractive to them, but not in recent years. However, birds from

Scotland can still boost wintering numbers in Ireland to about 650 to 1,100 birds. Mostly they will join with the Irish breeding birds around the coast. Malin Head and the west coast are generally favoured.

The Twite is Red-listed in BirdWatch Ireland's 'Birds of Conservation Concern in Ireland 4: 2020–2026' due to a 98 per cent long-term (since 1980) decline in its population, and an 80 per cent long-term decline in its breeding range.[492]

Linnet (Common Linnet)

Linaria cannabina cannabina | Gleoiseach

RESIDENT

DECREASING

Linnets are fairly common and widespread in Donegal. They are to be found most easily on marginal farmland, coastal heath and dunes, where they feed on weed seeds, and use scrub for nesting. They are less frequent on the more intensive farmlands, and in the barren uplands. Ussher and Warren recorded them breeding on Tory and Arranmore.[493]

The atlases do not show significant change in the range of the species, and the Countryside Bird Survey records a stable population. It is suggested that losses from the intensification of farmland have been balanced by gains from increased weed presence on abandoned farms.[494] This argument is not very convincing for Donegal. The last twenty years seems to have witnessed a large drop in numbers, if not in distribution.

In winter, they gather in flocks on farmland wherever they can find a field with a good crop of weeds. A flock on Inch Levels on 29 December 1990 held what was estimated to be about 5,000 birds. Less challenging estimates were made of roughly 2,000 birds in 1993, and 1,500 in 1996. Flocks of thirty to fifty can still turn up almost anywhere, but the medium-sized flocks of 200 to 500 that were frequent up to the 1990s seem to have become very scarce since then, and the large flocks have gone.

Heavy migration occurs at Malin Head. The highest daily total was of 1,750 birds arriving on 17 October 1964.[495] Curiously, numbers at Tory in the early 1960s were mostly in single figures. Sixty birds on 28 September 1999 was exceptional. More recently, numbers are still mostly in single figures, with the peak of thirty on 9 and 14 September 2014. This dominance of Malin Head over Tory for falls of large numbers of passerine immigrants is in contrast to Tory's dominance in vagrants (see also Twite, Continental Greenfinch and Reed Bunting).

Mealy Redpoll (Common Redpoll)

Acanthis flammea flammea | Deargéadan liath
RARE PASSAGE MIGRANT AND WINTER VISITOR

Frequent changes in the classification of the redpolls have probably made observers wary of identifying the rarer forms, particularly away from Tory where rare forms can be anticipated. That has been balanced to some degree by the fact that most of the different forms are (with care) recognisable in the field.

Mealy Redpoll is the nominate subspecies of the Common Redpoll (the full species encompasses Mealy, Greenland and other races). It breeds in Scandinavia and across Russia where it is mainly resident, but partially migrates to the south, and makes occasional mass irruptions which don't quite reach Ireland.

The first record for Donegal, and Ulster, was at Inishtrahull in the first week of January 1898.[496] The total so far is thirty-five – twelve confirmed, and twenty-three presumed, but not yet assessed. Two others have been accepted as either Mealy or Greenland Redpolls. Of the thirty-five total, nineteen were on Tory and five at Malin Head / Inishtrahull. The remaining eleven were widely dispersed, and all of them in winter or spring. Three each were at Letterkenny and Magheraroarty, two at Convoy, and singles at Killybegs, Big Isle and Ballybofey (Figure 18.67).

Subspecies

Greenland Redpoll

Acanthis flammea rostrata | Deargéadan Graonlainne
SCARCE PASSAGE MIGRANT

This large-billed sister-race of the Mealy Redpoll breeds in both Greenland and Iceland. It is the more frequent visitor to Donegal.

The first Greenland Redpoll to be recorded for Donegal was captured on Inishtrahull on 24 September 1913, and lodged in the Patten collection at the Belfast Museum.[497] This was followed by ten on Tory on 12 September 1954.[498]

Mealy Redpoll, Magheraroarty.

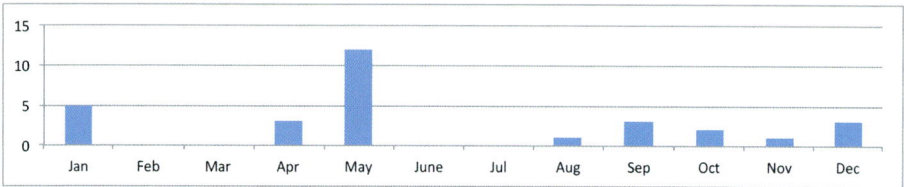

Fig. 18.67: Monthly distribution of all probable Mealy Redpolls

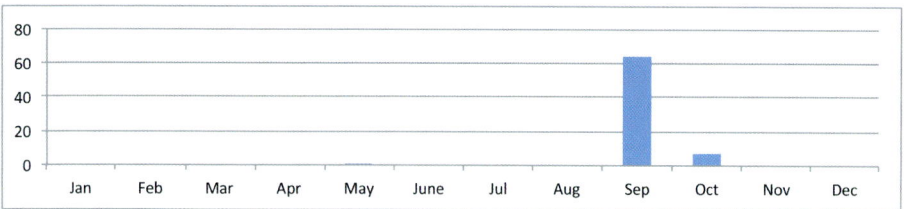

Fig. 18.68: Monthly distribution of all probable Greenland Redpolls

The total number reported to have reached Donegal is seventy-one, yet the IRBC have only sixty-one records for Ireland, including some for other counties.[499] Even if these records refer to events rather than individual birds, the Donegal total clearly includes quite a few that have not been assessed by the IRBC, for example twenty birds at Bunaninver, near Bloody Foreland, on 5 September 1997, and all but two of the ten birds recorded in the Glencolumbkille area between 1995 and 1999. Apart from one bird at Fanad Head, all the rest have been on Tory. These birds are included here only to emphasise what appears to be the emerging picture of a very different pattern of occurrence from Mealy Redpoll (Figures 18.67 and 18.68).

The *rostrata* birds in Iceland are regarded as largely sedentary, whereas those from Greenland move south-east in winter, which would bring them initially to Iceland. But they are also somewhat irruptive, which could account for them reaching Donegal. Only three out of the seventy-one birds recorded have been singles, which also hints at a deliberate move south to Ireland on north-westerly winds, rather than arrival by accident. So the pattern of arrivals here does not look like vagrancy – Greenland Redpoll is probably best treated as a scarce passage migrant. This was first suggested in 1954 by Redman, although he was of the opinion, based on the examination of skins, and consideration of the meteorological conditions at the time of arrivals, that Iceland rather than Greenland is the origin of our birds.

Lesser Redpoll

Acanthis cabaret | Deargéadan beagh

RESIDENT AND PARTIAL MIGRANT
INCREASING

Our resident Redpoll was for long treated as a subspecies of the Common Redpoll (or the Mealy Redpoll, as it was called then). But it gained full species status in the recent taxonomic upheavals. They are confined to Ireland, Britain and central Europe south from southern Scandinavia.

The zone of heathy scrub between marginal farmland and the open hills, where it still exists, is where to expect Lesser Redpolls in summer. The seeds of deciduous trees are what they mostly feed on – birch, alder and willows are their favourites. That habitat is greatly reduced now, so birds have moved to breed more in the younger conifer plantations. But even here, they still depend to some extent on finding remnants of the original habitat for food.

Four pairs were recorded on the 88 km² of Golden Plover habitat covered during the Upland Bird Survey – no doubt only in sheltered corners or on the fringes, where there might be some trees.[500]

In winter, they largely move out from the forests, and even leave the county altogether, unless tempted to stay by garden bird feeders or a good stand of alder. If the latter, they will usually be in mixed flocks with Siskins and Goldfinches.

Redpoll was the most abundant species in younger plantations in a study of the bird communities in plantation forests,[501] so its population and distribution has improved in recent times. This is confirmed by the Countryside Bird Survey, which also records a substantial increase in numbers,[502] so it is probably also the case in Donegal.

Arctic Redpoll

Acanthis hornemanni | Deargéadan Artach

RARE VAGRANT

The Arctic Redpoll replaces the Common Redpoll (Mealy and Greenland races) in the arctic tundra, where it is circumpolar. A North American race, *Acanthis h. hornemanni*, and an old-world race, *Acanthis h. exilipies*, have both been recorded in Ireland, but the two birds to reach Donegal were not identified to subspecies level.

The first record was on Tory, from 10 to 24 September 2000 (A.A.K. Lancaster et al.), and the second, also on Tory, was on 18 September 2001 (J. Fitzharris et al.). These were the second and third Irish records for the species. As the first Irish bird was identified the year before as of the North American race *hornemanni*, then one or other of these birds could well have been the first Irish record of the race *exilipies*. It is worth noting that Tony Lancaster was involved in the finding of all three birds. The Irish total has now risen to twelve birds, of which only five were identified to race.[503]

Crossbill (Red Crossbill)

Loxia curvirostra curvirostra | Crosghob

IRRUPTIVE VISITOR AND BREEDER
INCREASING

Crossbills inhabit the conifer forests of Europe, Asia and North America. They are resident, but will undertake partial migrations when population levels are high at the same time as spruce and pine seed crops are low. For many years, it was only after such irruptions that they could be seen in Ireland. Now, with frequent irruptions, and conifer forests here to detain them, they are continuously present in Ireland, and in Donegal.[504]

Hart saw three birds for the first time in 1890 – the first known sighting

Crossbill, Lough Sallagh.

JOHN CROMIE

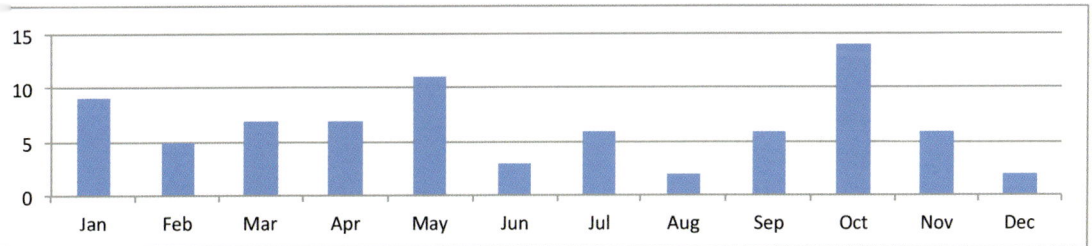

Fig. 18.69: Monthly distribution of Crossbill sightings

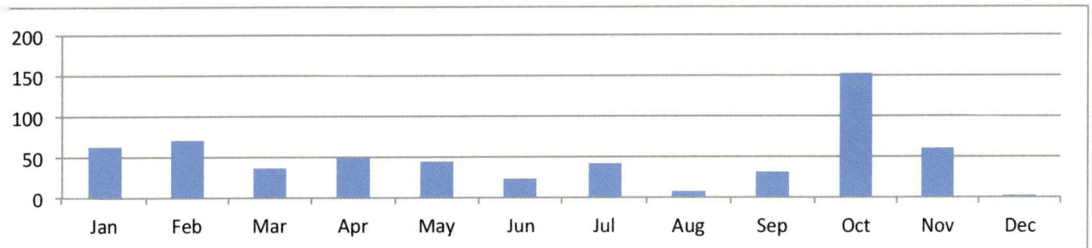

Fig. 18.70: Monthly distribution of individual Crossbills

in the county.[505] They were not recorded then until 1953 when a large flock was seen at Glenties in early July. The first evidence of breeding was in 1981, when adults were seen at Glenveagh from March, and then three juveniles in April and May. Proof of breeding is now obtained somewhere in the county every few years, and given the extent of our largely unexplored conifer forests, it is tempting to assume that there is a resident population. But it could still be the case that Donegal's population would fade away if not renewed every few years by fresh irruptions. Either way, the evidence is still lacking.

Sightings take place in all months (see Figures 18.69 and 18.70). Records at migration hotspots have been relatively few, but all the likely places have had a few, and they have certainly been responsible for boosting the October totals.

Goldfinch (European Goldfinch)

Carduelis carduelis Britannica | Lasair choille

RESIDENT AND PARTIAL MIGRANT
INCREASING

The Goldfinch is a common resident, and after Blackcap, the most successful in recent years. It is essentially a bird of open, weedy habitats, with scrub and trees used for breeding. Abandoned farmland, derelict cottage gardens and building sites are all suitable. Donegal would seem ideal, and to Ussher and Warren (1900), it was the commonest finch between Glenties and Dawros, where trees were almost absent. At that time it bred on Arranmore and at Dungloe and Bunbeg, and large

Goldfinch, Tory.

parties were seen in west Donegal.[506] Yet its presence in the county remained patchy up until the last fifty years, or thereabouts.[507] Being one of our few really showy species, and with a good song, it was trapped as a cage bird in days gone by, to an extent that limited its wild population. With legal protection now, its numbers are on the rise. Its distribution has filled out,[508] and the Countryside Bird Survey has recorded a 32 per cent increase in its national distribution over the forty-four years between the first breeding atlas in 1972 and the third in 2016.[509]

Goldfinches are regular migrants at Tory, but always in small numbers. So the immigration of birds in winter from places like Scotland seems to be on a small scale, and not enough to compensate for their tendency to move south in winter. However, the provision of food in gardens must help to retain birds through the year, and in winter they will also be found in alders with roaming flocks of Siskins and Redpolls.

Siskin (Eurasian Siskin)

Spinus spinus | Siscín

RESIDENT AND WINTER VISITOR

INCREASING

Siskins have been present in the county for as long as records go back, in mixed woods, and gardens with ornamental conifers. They expanded to some degree in the first half of the last century, but then declined in mid-century.[510] The spread of conifer plantations since the 1960s has enabled them to consolidate their distribution.

Our resident birds are joined in winter by visitors from Scandinavia and central Europe. The presence of these immigrant Siskins at garden bird-feeding stations has shown that the move west across Britain and Ireland is progressive, with birds only reaching Donegal at the very end of winter. Donegal-bred Siskins are also found throughout the winter, feeding with their close relatives, Goldfinches and Redpolls, on the seed cones of alder trees. These grow naturally along river banks and in damp marshy ground, but are now increasingly grown as part of grant-aided tree plantations.

Very few Siskins turn up at Tory and the other migration stations around the coast.

Lapland Bunting

Calcarius lapponicus lapponicus | Gealóg Laplannach

UNCOMMON MIGRANT

DECREASING

Lapland Buntings are circumpolar inhabitants of the tundra zone. The north European populations migrate to the steppes of central Eurasia for the winter, although small numbers stay around the North Sea coasts, where they are joined by small numbers of the

Lapland Bunting and Short-toed Lark, Tory.

GRACE MEENAN

birds from Greenland and northern Canada. It is quite likely that these are the birds that pass through Donegal. This was indicated in 1955 when an analysis of the meteorological conditions, and a comparison with events at Fair Isle to the south of Shetland, suggested that Greenland rather than arctic Eurasia was indeed the origin of the birds arriving at Tory.[511]

Lapland Bunting was known to Kennedy et al. as an extreme rarity in Ireland, with only seven records. That 1954 publication just missed the discovery of about 300 birds in north Donegal in the autumn of 1953.[512] The expedition that found these birds recorded up to sixty arriving daily in September at Inishtrahull. At the same time, they were moving west from Malin Head along rough ground up to two miles inland. This was followed on Tory in the autumns from 1954 to 1959, where there were daily numbers, mostly in single figures. Then in 1960, 302 bird/days were recorded between 9 September and 23 October, with seventy on 11 September. In 1962, between 27 August and 14 October there were 747 bird/days. In 1963, they were present daily from 3 September to 5 October, with a maximum of only seventy. And in 1964 the maximum was eight birds on 20 October, in the only month they were recorded.[513]

Since 1964, the last fully operational year of the Tory Bird Observatory, there has been a lull in observations. Good years have included 1993, when about 190 passed through Malin Head between 23 September and 20 October. In 1994, the highest daily total from Malin Head was sixty, with a possible cumulative total of 140. In 1997, small numbers were distributed all along the north coast of the Fanad peninsula. Figure 18.71 shows the bird/days since 2000. As it is not known how many days each bird was logged, this graph is only an indication of the variation between years. It reflects the major influx of birds throughout Britain and Ireland in 2010,[514] but it also suggests that numbers were greater, and more consistent in the 1950s and '60s when there was better coverage of the migration hotspots.

The autumn movements tail off in early November, but there have been two genuine winter records, on very different habitat at Lough Swilly. On 19 December 2020, five birds were seen on the shingly shore of Ballymoney,

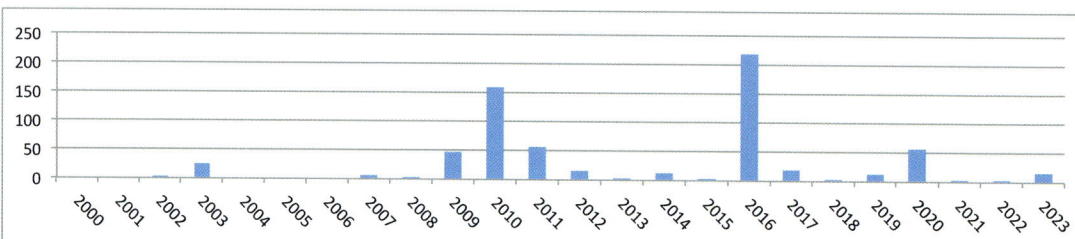

Fig. 18.71: Lapland Bunting bird/days since 2000

which backs onto extensive arable fields, and one bird was on the arable farmland at Big Isle on 20 November 2023.

Spring records are few, with most indicating return visits following the years of major autumn influxes. There were three individuals in 2011, at Tory on 22 April, Bloody Foreland on 24 April and Arranmore on 19 May. Four birds were at Tory on 9 April 2013. Tory had a passage from 19 to 26 April 2017 peaking at four birds, and likely involving at least ten. And there was a bird on Tory on 17 April 2020.

A few Lapland Buntings are now occasionally recorded as autumn migrants or winter visitors on the south and east coasts of Ireland, but Donegal remains the only county where significant numbers can be seen, albeit irregularly.

Snow Bunting

Plectrophenax nivalis nivalis | Gealóg shneachta

WINTER VISITOR AND PASSAGE MIGRANT

DECREASING

Like the Lapland Bunting, Snow Bunting breeds in the circumpolar tundra zone. This nominate race breeds in North America and the western part of Eurasia. They move south for the winter on a broad front.

Ussher and Warren report three summer records on Arranmore – two in summer plumage shot on the first week of May 1883, and an adult male shot on 28 July 1883. An adult male was seen on 18 August 1883.[515] There is a report of a bird seen with food for young on several dates in late May 1977 but with no further evidence of breeding.[516] This record should probably be disregarded. Birds in summer plumage continue to be seen in May, but these would be lingering winter visitors or spring migrants.

The Snow Bunting has two distinct habitats in Donegal – the shore and the mountain summits. Most records come from the shore, where they like shingle, decomposing lines of seaweed and salt marshes. The highest count was of 250 birds at Glashagh Bay on 8 November 1998, in a winter when there were several good flocks along the north coast, at Malin Head (30), Melmore Head (13), Ballyness Bay (43) and Bloody Foreland (15). The peak count on Inishtrahull in the autumn of 1965 was 176 on 4 October. More typical records are of less than ten birds. Although many fewer counts

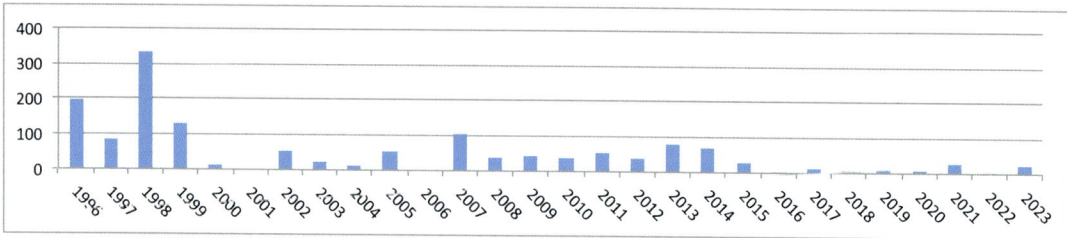

Fig. 18.72: The annual totals of the 1,442 Snow Buntings recorded since 1996

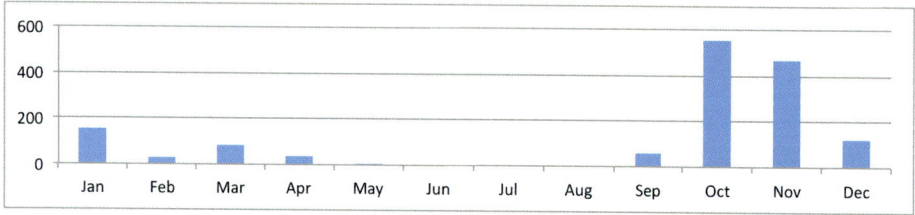

Fig. 18.73: Monthly distribution since 1996 of 1,498 Snow Buntings

come from the mountain summits, they include the third largest gathering, of 100 birds, which was seen on the summit of Croaghgorm (674 m) in the Blue Stack Mountains on 26 January 2007. The impact of two of these exceptional counts on Figure 18.72 can be seen, but they don't alter the general decline indicated by the chart, which is almost certainly attributable to climate change.

On Tory, autumn migration in the 1950s and '60s was recorded daily from late September until the end of October, mostly in single figures. In 1959 'a number of male birds, all exceptionally tame, was a major feature of the spring'; the peak count was sixty-seven on 12 April. Only two females were recorded.[517] Snow Bunting is now regarded on Tory as a scarce migrant and rare winter visitor.[518] The high October and November totals in Figure 18.73 include only five birds on Tory, few in the uplands, and the vast majority in many coastal flocks.

Subspecies

Snow Bunting (Icelandic)

Plectrophenax nivalis insulae | Gealóg shneachta (Íoslannach)

WINTER VISITOR AND PASSAGE
MIGRANT

This island race of the Snow Bunting migrates south to Ireland and Britain. It is not known what proportion of our birds belong to each of the two races, and it is likely that they mingle together in the wintering flocks.

Right: Snow Bunting (male in breeding plumage), Tory.

ANTON MEENAN

Corn Bunting

Emberiza calandra calandra | Gealóg bhuachair

EXTINCT RESIDENT

The story of the Corn Bunting is salutary. It was lost to Donegal, and to Ireland, well before the full impact of changing agricultural practices was appreciated. And we were not alone. The first breeding atlas lists widespread declines throughout Britain, starting between the 1920s and 1940s, when the species had already been eliminated from Wales. By the end of the century it was declining in twenty-two out of thirty-four European countries, and increasing in only two.[519]

Corn Buntings are birds of agricultural grasslands and tilled ground, nesting in low scrub – typically brambles and gorse. Taylor and O'Halloran say that the most important factors in their demise are thought to be the decline in mixed farming and the loss of temporary grasslands, hay meadows and under-sown cereals, especially spring cereals and overwinter cereal stubbles.[520]

At the end of the nineteenth century, Hart wrote that the Corn Bunting was 'not infrequent in the roughly cultivated bare districts in Donegal. Several pairs breed annually in my neighbourhood' (near Portsalon in the Fanad peninsula).[521] Ussher and Warren confirmed its breeding presence on Tory and Arranmore, and mentioned the neighbourhood of Dungloe as an example of the Irish coastal districts where it was common. They added that individuals arrive occasionally on distant islands like Inishtrahull and remain some time.[522]

Fifty years later, Kennedy et al. reported that in Donegal 'the only place it can be found with any certainty is the coastline between Gortahork and Bloody Foreland'.[523] Their correspondent, C.V. Stoney, reported ten pairs on Carrickfinn in 1935, where none remained in 1948, at which time there were still a few breeding on Arranmore between Aphort and Ballintra. Stoney also reported that it had totally disappeared from inland haunts near Raphoe many years previously, and that the decrease in north-west Donegal was of long standing. But there were at least four birds with nests on Gola in 1953, and Philip Redman found three singing birds on Inishbofin in 1954, but none on Tory at the same time.[524]

A resident population of about twelve pairs of Corn Buntings held on through the 1960s at Malin Head.[525] After that, one or two birds were seen in most years up until 1971. In 1975, Ruttledge mentioned one (unnamed) district in Donegal where it was plentiful.[526] This could be an out-of-date reference to Malin Head. But there is the intriguing possibility that he was referring to Inch Island, where four birds were reported on 27 January 1980. Inch would have had good habitat and may well have held a small breeding population through the 1960s and 1970s.

There are two very late sightings, of one bird at Dungloe on 26 June 1993 and two at Malin Head on 27 October of the same year. Evidence of migration is slim, and birds striking the lights at Tory and Inishtrahull were more likely to have been local.[527] But it is just possible that the two Malin birds were migrants, which would leave the Dungloe bird as the last known survivor of the resident population in Donegal, and almost the last in Ireland.

The Corn Bunting fell foul of the mechanisation of agriculture, and the arrival of supermarkets. Together, these factors undermined the self-sufficient culture of organic cottage gardens around the coast, where wildflowers and seed-eating birds still had their place among the vegetables and chickens.

Yellowhammer

Emberiza citrinella caliginosa | Buíóg
RESIDENT
DECREASING
RED-LISTED BOCCI-4

Yellowhammers are resident throughout much of Europe. In the past they were to be found almost anywhere throughout Donegal. Hart said it was one of the most characteristic species in the breeding season at Fanad, and was scattered through the county in the winter.[528] On the islands, Ussher and Warren recorded it as a breeding species on both Tory and Arranmore,[529] but a half-century later it had gone from Tory.[530]

The Yellowhammer suffers from similar pressures to those that led its relative, the Corn Bunting, to extinction. Those pressures are mostly related to the efficiency of modern agricultural practices, which leaves little food for seed-eating birds through the winter. However, unlike the Corn Bunting, the Yellowhammer still survives.

The three atlases are illuminating. Yellowhammers were virtually ubiquitous at the time of the first atlas, from 1968 to 1972.[531] By 1988–91 a retreat from the south of the county can be seen.[532] The final survey was from 2007 to 2011, when virtually the only presence of the species was around the east bank of Lough Swilly.[533] Nationally the species retreated over the same period, from being found throughout the island, to being largely confined to the arable zone south-east of a line from Belfast to County Cork – with a sizeable outlier in the north-west from Lough Swilly to the Lower Bann valley in County Antrim.

The Donegal situation seems to have stabilised since 2013. A garden feeding station on Inch Island has had numbers peaking at forty-three on 23 February 2013. Smaller numbers are found elsewhere around Inch Lough and Inch Levels, where feeding is also provided. And supplementary feeding retains a small number near Burtonport, on the west coast. How long we would retain a population in the county without supplementary winter feeding is questionable. A modification of standard farming practice to include measures like unharvested margins around tillage fields is clearly the only long-term solution.

As a migrant, the Yellowhammer is not very frequent. But there have been exceptional falls, as on 26 September 1963 when 500 were estimated at Malin Head.

The Yellowhammer is Red-listed in BirdWatch Ireland's 'Birds of Conservation Concern in Ireland 4: 2020–2026' due to a >50 per cent long-term decline in its population.[534]

Ortolan Bunting

Emberiza hortulana | Gealóg gharraí

RARE VAGRANT

The Ortolan is a summer migrant from mountain areas on the southern fringe of the Sahara, to most of continental Europe.

There have been nine birds recorded in Donegal. The first was at Inishtrahull, from 13 to 14 September 1953 (P.S. Redman) – the second (or third) occurrence in Ireland. An immature was present on Tory on 1 September 1960. A third was also on Tory, from 25 September to 1 October

Ortolan Bunting, Tory.

ANTON MEENAN

1992. A rare spring bird was seen at Malin Beg on 25 May 1998. Two turned up in 2000 – a first-winter bird on Arranmore from 11 to 12 September, and a female/first-winter bird on Tory from 16 to 17 September. Further birds on Tory were seen on 16 September 2003, 10 June 2020 and 9 September 2023. That June bird was a female, and can probably be treated as our second spring migrant.

Little Bunting

Emberiza pusilla | Gealóg bheag

RARE VAGRANT

The Little Bunting breeds across the Eurasian taiga zone, and winters in south-east Asia.

There have been eleven records so far. Arrivals have all been in late autumn or winter, with the earliest date being 30 September. The first bird to arrive in Donegal was found inside a house on Tory, where it had apparently fallen down the chimney. This was in the first week of November 1952. It was the fourth Irish record.[535] Tory's next bird had a more conventional welcome (in a bird net) on 6 October 1961.[536] Arranmore had one on 7 December 1997. The next five birds were all on Tory – 30 September 1998, 2 October 2005, 3 October 2011, 5 October 2015 and 16 January 2016. One bird was present at Malin Beg on 19 and 20 October 2018. Tory had yet another on 8 October 2020, and the most recent was from 6 to 12 November 2023.

Yellow-breasted Bunting

Emberiza aureola | Gealóg bhroinnbhuí

RARE VAGRANT

The Yellow-breasted Bunting breeds east from European Russia, and winters in south Asia. With two records, Donegal has had more than its fair share of the five birds recorded in Ireland. Sadly, this species is now Critically Endangered globally, so we would be very lucky to see any more turning up in Donegal.

Both our birds were recorded on Tory. The first one, and also the first for Ireland, was a female, on 18 September 1959 (F. Cooke).[537] The second was a female or immature on 21 September 1998 (J. Dowdall, A. Lancaster et al.).

Black-headed Bunting

Emberiza melanocephala | Gealóg cheanndubh

RARE VAGRANT

The Black-headed Bunting breeds in south-east Europe and the Middle East. It winters in India.

The two Donegal records are both from Tory. The first was an adult male from 21 to 27 July 1997 (N. O'Neill et al.). The second bird was on 1 October 2003 (E. Dempsey et al.). These birds were the sixth and seventh Irish records.

Red-headed Bunting

Emberiza bruniceps | Gealóg cheannrua

ESCAPE

This central Asian species is a potential vagrant, but as it is widely kept as a cage bird, all records in Britain and Ireland are regarded as more likely to be escapes. Two birds were recorded in 1953. The first was on Tory on 26 April and is in the National Museum. The second was an adult male recorded on Inishtrahull by Ian Nisbet on 10 September. These two, and another at Belmullet, arriving at migration outposts in 1953 are highly suggestive of natural vagrancy, but the possibility of them being escapes cannot be ruled out.

KIM PERIERA

Reed Bunting, St John's Point.

Reed Bunting (Common Reed Bunting)

Emberiza schoeniclus schoeniclus | Gealóg ghiolcaí

RESIDENT

DECREASING

As its name implies, the Reed Bunting is the counterpart of Yellowhammer in damper terrain. It likes tall vegetation and prefers reedbeds, but can be found in scrubby habitat, and even in conifer re-stock sites.

It is to be found throughout the county, wherever its habitat requirements are met. One or two pairs breed on Tory.[538] There are also breeding season records from Arranmore and Inishfree Lower.

Reed Buntings have suffered from agricultural intensification in a different way from Yellowhammer – through the loss of habitat to land drainage. This may have thinned out the population, but not the geographical range, so the atlas surveys don't show any significant change between the 1960s and 2010s.[539] Neither does the Countryside Bird Survey, which suggests that adaptation to drier habitats may have compensated for loss of wetland sites.[540]

Flocking in winter was more frequent in the recent past, often at cattle-feeding sites. But with indoor feeding now being the norm, these have largely disappeared. Numbers were usually less than a dozen, with thirty being the maximum recorded, on 21 February 1993, near Manorcunningham. A few birds can still be seen together on Inch Levels in winter.

On Tory, there are usually only one or two birds that can be seen on spring or autumn passage, with fifteen on 26 September 1999 being easily the highest total recorded there. The 350 birds seen at Malin Head on 28 September 1963 were part of an exceptional irruption that included 500 Yellowhammers.

Bobolink

Dolichonyx oryzivorus | Bobóilinc

RARE VAGRANT

This bunting-like bird belongs to an exclusively American family, the American Orioles *Icteridae*. The only one so far seen in Donegal was near Fanad Head, between 20 and 24 September 2020, and is the fourth Irish record. It was initially found by Wilton Farrelly, but over the five days it was also enjoyed by a large number of birdwatchers.

VICTOR CASCHERA

Bobolink, Fanad.

Baltimore Oriole, Tory.

Baltimore Oriole

Icterus galbula | Óiréal tuaisceartach

RARE VAGRANT

From the same family as the Bobolink (above), the Baltimore Oriole has only three Irish records, of which Donegal's was the third. It was a second-calendar-year bird, and was seen on Tory on 15 May 2017 by Anton Meenan.

It is remarkable that these last two individual birds, the Bobolink and the Baltimore Oriole, are from the same family, and are the only North American vagrants to reach Donegal from that great order of birds, the passerines, which includes all the songbird families. A number of passerine vagrants from America are expected each year in the south-west of Ireland, but in Donegal, our vagrant songbirds have always come from the north, east and south, but not the west – at least until now.

APPENDIX 1

Locations (with Irish OS grid reference for 1 x 1 km squares)

Aghla Beg	B9624	Bo Island	B7825
Aghla More	B9523	Bogay	C3516
Akibbon Lough	C0618	Breaghy Head	C0538
Annagary	B7919	Brownhall	G9369
Aphort	B6614	Bullaba River	C0013
Ardara	G7391	Bunaninver	B8130
Ardee	C2816	Bunbeg	B8023
Ardnageer	G9690	Buncrana	C3431
Ardnamona Nature Reserve	G9684	Bundoran	G8158
Ards	C0734	Bunlin	C1828
Arranmore (Illanaran)	B6315	Burtonport	B7115
Arranmore (Leabgarrow)	B6815	Carlan Isles	C2033
Arranmore (Rinrawros Point)	B6418	Carndonagh	C4745
Arranmore (Torneady Point)	B6519	Carnowen	H2399
Arranmore (Torries)	B6514	Carraghblagh	C2441
Auchterlinn (Oughterlin)	C2629	Carrickabracky (Doagh Isle)	C3952
Ball Hill	G8976	Carrickfin	B7821
Ballbane Head	C6645	Carrigart	C1237
Ballindrait	H3099	Castle Shanaghan	C2316
Ballintra (Arranmore)	B6714	Castlewray	C2214
Ballintra (south Donegal)	G9270	Clonkillybeg	C1219
Ballyarr Nature Reserve	C1820	Clonleigh	C3300
Ballybofey	H1394	Clonmass Isle	C0836
Ballyhoorisky Point	C1545	Clooney Lough	G7299
Ballyliffin	C3848	Convoy	C2101
Ballymoney	C3121	Coolmore (Rossnowlagh)	G8666
Ballyness Bay	B9032	Crana River	C3933
Ballyshannon	G8761	Creeslough	C0530
Barnesmore Gap	H0285	Croaghgorm	G9489
Barnesmore Lakes	H0483	Croaghturr	C0840
Big Isle	C2313	Crockglass	C4731
Blanket Nook	C3019	Crohy Head	B7008
Bloody Foreland	B8134	Culdaff	C5349
Blue Stack Mountains	G9490	Culmore	C4724

Islands off Gweedore	B7825	Lough Salt Mountain	C1326
Isle Burn	C2310	Lough Swilly	C2921
Keadew	B7417	Lough Ultan	H0769
Kerrykeel	C2032	Loughros More Bay	G7093
Kildoney Point	G8264	Maghera	G6590
Kilmacrenan	C1420	Magheragallon	B8025
Kiltooris Lough	G6896	Magheraroarty	B8834
Kincaslough	B7419	Maghery	B7109
Kincrum	G7999	Malin Beg	G4980
Kindrum Lough	C1943	Malin Head	C3959
Kinnagoe Bay	C6347	Malin More	G4983
Knockalla	C2435	Malin Town	C4649
Lagacurry (Doagh Isle)	C4251	Manorcunningham	C2411
Leannan Estuary	C2422	Marblehill	C0535
Leannan River	C2020	Meenbannad	B7416
Lenan Head	C2944	Meenglass	H1191
Letterkenny Golf Course	C2013	Meenlaragh	B8832
Lifford	H3398	Melmore Head	C1345
Lisfannan	C3328	Mill Bay	C3123
Lough Alaan	H1596	Milford	C1927
Lough Anarget	G5283	Mill River	C3631
Lough Beagh/Veagh	C0221	Mintiaghs Lough	C3840
Lough Derg	H0774	Mongorry Hill	C2404
Lough Eske	G9782	Mountcharles	G8776
Lough Fad	C3943	Mourne Beg River	H0988
Lough Fern	C1824	Moville	C6138
Torglass (Gola)	B7526	Moylenav Mountain	B9513
Lough Foyle (west bank)	C5232	Muckish	B9928
Lough Golagh	G9666	Muckross Head	G6273
Lough Greenan	C1126	Mullaghderg Lough	B7620
Lough Hanane	C2246	Mulroy Bay	C2034
Lough Keel	C1524	Mirraghmullan	B7604
Lough Meela	B7413	Naran	G7199
Lough Nafullanrany	B7617	Newtowncunningham	C3116
Lough Namafin	G7983	Nougherwole	B8746
Lough Naminn	C3941	O'Boyle's Island	G7699
Lough Roshin	G9264	Owenea River	G7892
Lough Sallagh	H0591	Owengarve River	G8992

Owey Island	B7123	Saldanha Head	C2638
Pettigo	H1066	Sessiagh Lough	C0436
Pettigo Plateau	H0373	Shalwy	G6574
Poisoned Glen	B9416	Sheep Haven Bay	C0836
Pollan Reservoir	C4435	Shellfield	C2619
Port	G5489	Slieve League	G5278
Port Lough	C3416	Slievetooey	G6190
Portsalon	C2440	Speenoge	C3521
Porthall	C3403	St John's Point	G7069
Portnablagh	C0437	St Johnstown	C3409
Portnoo	B7099	St Peter's Lough	G8678
Quigley's Point	C5131	Stookaruddan	C4558
Ramelton	C2221	Swilly Burn	C0304
Raphoe	C2503	Swilly Estuary	C2112
Rath Lough	G9668	Tawny	G6074
Rathgory	C2038	Teelin	G5975
Rathlin O'Birne	G4679	Tormore (Melmore)	C1043
Rathmullan	C2927	Tormore (Slievetooey)	G5590
Ray	C2625	Tormore (Tory)	B8746
Rinboy Lough	C1744	Torries	B6514
Rinmore Point	C1946	Tory	B8546
River Deele	H2598	Trabeg	C1644
River Faughan	C4716	Tranarossan Isles	C1043
River Finn	H2594	Trawbreaga Bay	C4549
River Foyle	C3510	Tremone Bay	C5947
River Swilly	C0810	Trumman Lough	G9472
Roaninish	B6502	Tullagh Bay	C3548
Rockhill	C1409	Tullagh Point	C3350
Rockstown Harbour	C3249	Umfin Island	B7628
Rocky Point	G4883	Urris Cliffs	C2941
Rosapenna Lough	C1138	Urris Hills	C3041
Roshin Point	G7699	Vances Point	C5333
Rossan Point	G4884	Whitecastle	C5533
Rossbeg	G6697		
Rossguill	C1040		
Rossnowlagh (Lower)	G8568		
Rossylongan	G9177		
Rutland Island	B7014		

APPENDIX 2

Names of Plants and Other Animals

VERNACULAR NAME	SCIENTIFIC NAME
Plants	
Alder	Alnus glutinosa
Birch	Betula pubescens
Bird's-foot trefoil	Lotus corniculatus
Eelgrass	Zostera marina
Gorse	Ulex europaeus
Hawthorn	Crategus monogyna
Heather	Calluna vulgaris
Holly	Ilex aquifolium
Marram	Ammophila arenaria
Oak (sessile)	Quercus petraea
Pine (Corsican)	Pinus ngra var. corsicana
Pine (lodgepole)	Pinus contorta
Pine (maritime)	Pinus pinaster
Pine (Scots)	Pinus sylvestris
Reed	Phragmites australis
Rhododendron	Rhododendron ponticum
Rowan	Sorbus aucuparia
Spruce (Sitka)	Picea sitchensis
Wild cherry	Prunus avium
Willow	Salix sp.

VERNACULAR NAME	SCIENTIFIC NAME
Invertebrates	
Common oyster	Ostrea eulis
Mussels	Mytilus / Modiolus species
Pacific oyster	Ostrea gigas

VERNACULAR NAME	SCIENTIFIC NAME
Fish	
Sprat	Sprattus edulis
Birds	
Marsh Tit	Poecile palustris
Eurasian Nuthatch	Sitta europaea
Mammals	
Bank vole	Myodes glareolus
Greater white-toothed shrew	Crocidura russula
Grey seal	Halichoerus grypus
Grey wolf	Canis lupus
Mink (American)	Mustela vison
Pygmy shrew	Sorex minutus
Rabbit	Oryctolagus cuniculus
Red deer	Cervus elaphus
Red fox	Vulpes vulpes

APPENDIX 3

Irish Firsts

Donegal records verified as the first in Ireland for the species

In date order

Additional subspecies (in brackets)

No subsequent Irish record **(in bold)**

No other Donegal record (*)

Not confirmed (†)

SPECIES	YEAR	LOCATION
Black-crowned Night Heron*	1834	Letterkenny
Blue-headed Wagtail	1954	Tory
Common Rosefinch	1954	Tory
Yellow-breasted Bunting	1959	Tory
Caspian Tern*	1959	Tory
Eastern Olivaceous **/Sykes's Warbler†**	**1959**	**Tory**
Arctic Warbler	1960	Tory
Rock Pipit (Scandinavian race)	1961	Tory
Black / White-crowned Wheatear*	**1964**	**Portnoo**
Booted Warbler*	2003	Tory
Semipalmated Plover*	2003	Arranmore
Bullfinch (Northern race)*	2004	Tory
Dresser's Eider*	**2010**	**Glashagh Bay**
Collared Flycatcher*	**2012**	**Tory**
Eastern Yellow Wagtail*	2013	Tory
Egyptian Vulture*	**2021**	**Dunfanaghy**

APPENDIX 4

Donegal Singletons

The fifty-two species or races with only a single verified record in Donegal

Additional subspecies (in brackets)

No other Irish record **(in bold)**

Subsequent records yet to be verified exist for these species (*)

Escapes excluded

An additional five species await verification (see Appendix 6)

Longest-standing record is the Black-crowned Night Heron, from 1834

SPECIES	YEAR	LOCATION
Ferruginous Duck*	1990	Inch
Dresser's Eider	**2010**	**North Fanad**
Hooded Merganser	2015	Tory
Yellow-billed Cuckoo	1989	Tory
Pallas's Sandgrouse	1863	Naran and Killybegs
Pied-billed Grebe	1988	L. Anarget
Stone Curlew	1903	Gweedore
Black-winged Stilt	1916	Tory
Ringed Plover (Northern)	1960	Tory
Semipalmated Plover	2003	Arranmore
Little Ringed Plover*	2021	Inch
Kildeer	2013	Killybegs
Stilt Sandpiper	2018	Inch
Temminck's Stint	2007	Blanket Nook
Long-billed Dowicher	1962	Tory
Spotted Sandpiper	2013	Tory
Greater Yellowlegs	1964	Tory
Collared Pratincole	2019	Blanket Nook
Black-winged Pratincole	2023	Blanket Nook
Ivory Gull	1913	Teelin
Ross's Gull	1983	Killybegs
Caspian Gull	1998	Killybegs

Slaty-backed Gull	2015	Killybegs
Caspian Tern	1959	Tory
American Bittern	1974	Malin Beg
Little Bittern	1908	Owey
Black-crowned Night Heron	1834	Letterkenny
Honey Buzzard	2001	Ards
Scops Owl	1911	Ballyliffin
Hobby	2020	Drumkeen
Lesser Grey Shrike	1990	Sheskinmore
Woodchat Shrike	2020	Dunfanaghy
Eastern Skylark	1915	Inishtrahull
Shore Lark	2009	Arranmore
Red-rumped Swallow	2007	Tory
Pallas's Warbler*	2016	Malin Beg
Radde's Warbler	2003	Tory
Greenish Warbler*	1998	Tory
Paddyfield Warbler	1998	Tory
Booted Warbler	2003	Tory
Eastern Olivaceous / **Sykes's Warbler**	1959	Tory
Eastern Subalpine Warbler	2018	Tory
Firecrest*	1997	Malin Beg
Nightingale	2019	Tory
Collared Flycatcher	**2012**	**Tory**
Black / White-crowned Wheatear	**1964**	**Portnoo**
Eastern Yellow Wagtail	2013	Tory
Pechora Pipit	2001	Tory
Red-throated Pipit	2015	Tory
Bullfinch (Northern)	2004	Tory
Baltimore Oriole	2017	Tory
Bobolink	**2020**	**Fanad Head**

APPENDIX 5

Checklist and Status

Species *(and additional subspecies in grey)*	Scientific Names	Status	Trend	Check
Fulvous Whistling Duck	Dendrocygna bicolor	escape		
Pale-bellied Brent Goose	Branta bernicla hrota	winter	up	
Dark-bellied Brent Goose	Branta bernicla bernicla	rarity		
Canada Goose	Branta canadensis canadensis	all-year	up	
Todd's Canada Goose	Branta canadensis interior/parvipes	rarity		
Barnacle Goose	Branta leucopsis	winter	up	
Cackling Goose	Branta hutchinsii	rarity		
Bar-headed Goose	Anser indicus	escape		
Snow Goose	Anser caerulescens caerulescens	rarity		
Greylag Goose	Anser anser anser	all-year	mixed	
Taiga Bean Goose	Anser fabilis fabilis	rarity		
Pink-footed Goose	Anser brachyrhynchus	winter	up	
Tundra Bean Goose	Anser serrirostris rossicus	rarity		
White-fronted Goose (Greenland)	Anser albifrons flavirostris	winter	stable	
White-fronted Goose (Russian)	Anser albifrons albifrons	rarity		
Black Swan	Cygnus atratus	escape		
Mute Swan	Cygnus olor	all-year	stable	
Bewick's Swan	Cygnus columbianus bewickii	formerly winter	down	
Whooper Swan	Cygnus cygnus	winter/passage	mixed	
Egyptian Goose	Alopochen aegyptiaca	rarity		
Shelduck	Tadorna tadorna	all-year	stable	
Ruddy Shelduck	Tadorna ferruginea	rarity		
Mandarin Duck	Aix galericulata	escape		
Garganey	Spatula querquedula	rarity		
Blue-winged Teal	Spatula discors	rarity		
Shoveler	Anas clypeata	all-year	up	
Gadwall	Mareca strepera	winter	stable	
Wigeon	Mareca penelope	winter	up	
American Wigeon	Mareca americana	rarity		
Mallard	Anas platyrhynchos platyrhyncos	all-year	stable	
American Black Duck	Anas rubripes	rarity		
Pintail	Anas acuta acuta	winter	up	
Teal	Anas crecca crecca	all-year	stable	
Green-winged Teal	Anas carolinensis	rarity		

447

Species	Scientific Names	Status	Trend	Check
Red-crested Pochard	Netta rufina	vagrant		
Pochard	Aythya ferina	winter		
Ferruginous Duck	Aythya nyroca	rarity		
Ring-necked Duck	Aythya collaris	rarity	up	
Tufted Duck	Aythya fuligula	all-year	stable	
Scaup	Aythya marila marila	winter	down	
Lesser Scaup	Aythya affinis	rarity		
King Eider	Somateria spectabilis	rarity		
Eider	Somateria mollissima mollissima	all-year		
Northern Eider	Somateria mollissima borealis	rarity		
Dresser's Eider	Somateria mollissima dall-yearseri	rarity		
Surf Scoter	Melanitta perspicillata	rarity	down	
Velvet Scoter	Melanitta fusca	winter	up	
Common Scoter	Melanitta nigra	winter		
Long-tailed Duck	Clangula hyemalis	winter	down	
Goldeneye	Bucephala clangula clangula	winter	down	
Smew	Mergus albellus	rarity	down	
Hooded Merganser	Lophodytes cucullatus	rarity		
Goosander	Mergus merganser merganser	winter	stable	
Red-breasted Merganser	Mergus serrator	all-year	stable	
Ruddy Duck	Oxyura jamaicensis jamaicensis	rarity		
Capercaillie	Tetrao urogallus urogallus	extinct		
Red Grouse	Lagopus lagopus scotica	all-year	stable	
Grey Partridge	Perdix perdix perdix	extinct	down	
Pheasant	Phasianus colchicus ssp.	all-year	down	
Quail	Coturnix coturnix coturnix	formerly summer	down	
Nightjar	Caprimulgus europaeus europaeus	extinct	down	
Alpine Swift	Tachymarptis melba	rarity		
Swift	Apus apus apus	summer	down	
Yellow-billed Cuckoo	Coccyzus americanus	rarity		
Cuckoo	Cuculus canorus canorus	summer	stable	
Pallas's Sandgrouse	Syrrhaptes paradoxus	rarity		
Rock Dove / Feral Pigeon	Columba livia livia	all-year	down	
Stock Dove	Columba oenas oenas	all-year	down	
Wood Pigeon	Columba palumbus palumbus	all-year	stable	
Turtle Dove	Streptopelia turtur turtur	rarity	stable	
Collared Dove	Streptopelia decaocto decaocto	all-year	stable	
Water Rail	Rallus aquaticus aquaticus	all-year		

Species	Scientific Names	Status	Trend	Check
Icelandic Water Rail	Rallus aquaticus hibernicus	winter		
Corncrake	Crex crex	summer	stable	
Spotted Crake	Porzana porzana	rarity		
Moorhen	Gallinula chloropus chloropus	all-year	down	
Coot	Fulica atra atra	all-year	down	
Crane	Grus grus	rarity		
Little Grebe	Tachybaptus ruficollis ruficollis	all-year	stable	
Pied-billed Grebe	Podilymbus podiceps	rarity		
Red-necked Grebe	Podiceps grisigena grisigena	rarity		
Great Crested Grebe	Podiceps cristatus cristatus	all-year	stable	
Slavonian Grebe	Podiceps auritus auritus	winter	down	
Black-necked Grebe	Podiceps nigricollis nigricollis	rarity		
Chilean Flamingo	Phoenicopterus chilensis	escape		
Stone Curlew	Burhinus oedicnemus oedicnemus	rarity		
Oystercatcher	Haematopus ostralegus ostralegus	all-year	stable	
Black-winged Stilt	Himantopus himantopus	rarity		
Avocet	Recurvirostra avosetta	rarity		
Lapwing	Vanellus vanellus	all-year	mixed	
Golden Plover	Pluvialis apricaria	all-year	down	
American Golden Plover	Pluvialis dominica	rarity		
Grey Plover	Pluvialis squatarola squatarola	winter	stable	
Ringed Plover	Charadrius hiaticula hiaticula	all-year	stable	
Ringed Plover (Greenland)	Charadrius hiaticula psammodromus	passage		
Ringed Plover (Northern)	Charadrius hiaticula tundrae	rarity		
Semipalmated Plover	Charadrius semipalmatus	rarity		
Little Ringed Plover	Charadrius dubius curanicus	rarity		
Killdeer	Charadrius vociferus vociferus	rarity		
Dotterel	Charadrius morinellus	rarity		
Whimbrel	Numenius phaeopus islandicus	passage	stable	
Curlew	Numenius arquata arquata	all-year	mixed	
Bar-tailed Godwit	Limosa lapponica lapponica	winter	up	
Black-tailed Godwit	Limosa limosa islandica	winter	up	
Turnstone	Arenaria interpres interpres	winter	stable	
Knot	Calidris canutus islandica	winter	up	
Ruff	Calidris pugnax	passage	up	
Stilt Sandpiper	Calidris himantopus	rarity		
Curlew Sandpiper	Calidris ferruginea	passage	stable	
Temminck's Stint	Calidris temminckii	rarity		

Species	Scientific Names	Status	Trend	Check
Sanderling	Calidris alba alba	winter	stable	
Dunlin	Calidris alpina alpina	winter	down	
Dunlin (Arctic)	Calidris alpina arctica	passage		
Dunlin (British)	Calidris alpina schinzii	summer	down	
Purple Sandpiper	Calidris maritima	winter	stable	
Baird's Sandpiper	Calidris bairdii	rarity		
Little Stint	Calidris minuta	passage	stable	
White-rumped Sandpiper	Calidris fuscicollis	rarity		
Buff-breasted Sandpiper	Calidris subruficollis	rarity		
Pectoral Sandpiper	Calidris melanotos	rarity		
Semipalmated Sandpiper	Calidris pusilla	rarity		
Long-billed Dowitcher	Limnodromus scolopaceus	rarity		
Woodcock	Scolopax rusticola	all-year	down	
Jack Snipe	Lymnocryptes minimus	winter		
Common Snipe	Gallinago gallinago gallinago	all-year	stable	
Faroe Snipe	Gallinago gallinago faeroeensis	winter		
Wilson's Phalarope	Phalaropus tricolor	rarity		
Red-necked Phalarope	Phalaropus lobatus	rarity		
Grey Phalarope	Phalaropus fulicarius	passage	stable	
Common Sandpiper	Actitis hypoleucos	summer	stable	
Spotted Sandpiper	Actitis macularius	rarity		
Green Sandpiper	Tringa ochropus	rarity		
Lesser Yellowlegs	Tringa flavipes	rarity		
Redshank	Tringa totanus totanus	all-year	stable	
Redshank (Icelandic)	Tringa totanus robusta	winter	stable	
Wood Sandpiper	Tringa glareola	rarity		
Spotted Redshank	Tringa erythropus	rarity	up	
Greenshank	Tringa nebularia	winter	up	
Greater Yellowlegs	Tringa melanoleuca	rarity		
Collared Pratincole	Glareola pratincola pratincola	rarity		
Black-winged Pratincole	Glareola nordmanni	rarity		
Kittiwake	Rissa tridactyla tridactyla	all-year	down	
Ivory Gull	Pagophila eburnea	rarity		
Sabine's Gull	Xema sabini	passage	stable	
Bonaparte's Gull	Chroicocephalus philadelphia	rarity		
Black-headed Gull	Chroicocephalus ridibundus	all-year	up	
Little Gull	Hydrocoloeus minutus	rarity	stable	
Ross's Gull	Rhodostethia rosea	rarity		

Species	Scientific Names	Status	Trend	Check
Laughing Gull	Leucophaeus atricilla	rarity		
Mediterranean Gull	Ichthyaetus melanocephalus	winter	up	
Common Gull	Larus canus canus	all-year	stable	
Ring-billed Gull	Larus delawarensis	rarity	stable	
Great Black-backed Gull	Larus marinus	all-year	stable	
Glaucous Gull	Larus hyperboreus hyperboreus	winter	down	
Iceland Gull	Larus glaucoides glaucoides	winter	down	
Kumlien's Gull	Larus glaucoides kumlieni	rarity		
Thayer's Gull	Larus glaucoides thayeri	rarity		
Herring Gull	Larus argentatus argenteus	all-year	mixed	
Herring Gull (Scandinavian)	Larus argentatus argentatus	rarity		
American Herring Gull	Larus smithsonianus	rarity		
Caspian Gull	Larus cachinnans	rarity		
Yellow-legged Gull	Larus michahellis michahellis	rarity		
Slaty-backed Gull	Larus schistisagus	rarity		
Lesser Black-backed Gull	Larus fuscus	all-year	up	
Baltic Gull	Larus fuscus fuscus	rarity		
Gull-billed Tern	Gelochelidon nilotica	rarity		
Caspian Tern	Hydroprogne caspia	rarity		
Sandwich Tern	Thalasseus sandvicensis	summer	up	
Little Tern	Sterna albifrons albifrons	summer	down	
Roseate Tern	Sterna dougalli dougalli	rarity		
Common Tern	Sterna hirundo hirundo	summer	down	
Arctic Tern	Sterna paradisaea	summer	stable	
White-winged Tern	Chlidonias leucopterus	rarity		
Black Tern	Chlidonias niger niger	passage		
Great Skua	Stercorarius skua	summer	mixed	
Pomarine Skua	Stercorarius pomarinus	passage	stable	
Arctic Skua	Stercorarius parasiticus	passage	stable	
Long-tailed Skua	Stercorarius longicaudus longicaudus	rarity		
Little Auk	Alle alle alle	winter	stable	
Guillemot	Uria aalge albionis	summer	down	
Razorbill	Alca torda islandica	summer	up	
Great Auk	Pinguinus impennis	extinct		
Black Guillemot	Cepphus grylle arcticus	all-year		
Puffin	Fratercula arctica	summer	down	
Red-throated Diver	Gavia stellata	all-year	stable	
Black-throated Diver	Gavia arctica arctica	winter	stable	

Species	Scientific Names	Status	Trend	Check
Great Northern Diver	Gavia immer	winter	stable	
White-billed Diver	Gavia adamsii	rarity		
Wilson's Petrel	Oceanites oceanicus ssp.	rarity		
Albatross sp. (Thalassarche sp.)	Thalassarche sp.	rarity		
European Storm-petrel	Hydrobates pelagicus pelagicus	summer	stable	
Leach's Storm-petrel	Oceanodroma leucorhous leucorhous	passage	stable	
Fulmar	Fulmarus glacialis glacialis	summer	mixed	
Gadfly Petrel sp.	Pterodroma madeira / feae / deserta	rarity		
Cory's Shearwater	Calonectris borealis	rarity		
Sooty Shearwater	Ardenna grisea	passage	stable	
Great Shearwater	Ardenna gravis	passage		
Manx Shearwater	Puffinus puffinus	passage	stable	
Balearic Shearwater	Puffinus mauretanicus	rarity		
White Stork	Ciconia ciconia ciconia	rarity		
Gannet	Morus bassanus	passage	stable	
Cormorant	Phalacrocorax carbo carbo	all-year	mixed	
Shag	Gulosus aristotelis aristotelis	all-year	down	
Glossy Ibis	Plegadis falcinellus	rarity	up	
Spoonbill	Platalea leucordia leucordia	rarity		
Bittern	Botaurus stellaris stellaris	rarity		
American Bittern	Botaurus lentiginosus	rarity		
Little Bittern	Ixobrychus minutus minutus	rarity		
Black-crowned Night Heron	Nycticorax nycticorax nycticorax	rarity		
Cattle Egret	Bubulcus ibis	rarity		
Grey Heron	Ardea cinerea cinerea	all-year	stable	
Great White Egret	Ardea alba alba	rarity	up	
Little Egret	Egretta garzetta garzetta	all-year	up	
Osprey	Pandion haliaetus haliaetus	rarity		
Honey Buzzard	Pernis apivoris	rarity		
Egyptian Vulture	Neophron percnopterus	rarity		
Golden Eagle	Aquila chrysaetos chrysaetos	re-introduction		
Sparrowhawk	Accipiter nisus nisus	all-year	stable	
Marsh Harrier	Circus aeruginosus aeruginosus	rarity		
Hen Harrier	Circus cyaneus	all-year	up	
Pallid Harrier	Circus macrourus	rarity		
Red Kite	Milvus milvus milvus	rarity		
White-tailed Eagle	Haliaetus albicilla albicilla	rarity	up	
Rough-legged Buzzard	Buteo lagopus lagopus	rarity		

Species	Scientific Names	Status	Trend	Check
Buzzard	Buteo buteo buteo	all-year	stable	
Barn Owl	Tyto alba alba	all-year	down	
Scops Owl	Otis scops scops	rarity		
Long-eared Owl	Asio otus otus	all-year	down	
Short-eared Owl	Asio flammeus flammeus	rarity	down	
Snowy Owl	Bubo scandiaca	rarity		
Hoopoe	Upupa epops epops	rarity		
Roller	Coracias garrulus garrulus	rarity		
Kingfisher	Alcedo atthis ispida	all-year	stable	
Bee-eater	Merops apiaster	rarity		
Wryneck	Jynx torquilla torquilla	rarity		
Great Spotted Woodpecker	Dendrocopus major anglicus	all-year	up	
Green Woodpecker	Picus viridis viridis	uncertain		
Kestrel	Falco tinnunculus tinnunculus	all-year	down	
Red-footed Falcon	Falco vespertinus	rarity		
Merlin	Falco columbarius aesalon	all-year	stable	
Icelandic Merlin	Falco columbarius subaesalon	passage		
Hobby	Falco subbuteo subbuteo	rarity		
Lanner Falcon	Falco biarmicus	escape		
Saker Falcon	Falco churrug	escape		
Gyrfalcon	Falco rusticolus	rarity	stable	
Peregrine Falcon	Falco peregrinus peregrinus	all-year	stable	
Red-backed Shrike	Lanius collurio	rarity		
Lesser Grey Shrike	Lanius minor	rarity		
Great Grey Shrike	Lanius excubitor excubitor	rarity		
Woodchat Shrike	Lanius senator senator	rarity		
Golden Oriole	Oriolus oriolus	rarity		
Jay	Garrulus glandarius hibernicus	all-year	up	
Magpie	Pica pica pica	all-year	stable	
Chough	Pyrrhocorax pyrrhocorax pyrrhocorax	all-year	stable	
Jackdaw	Coloeus monedula spermologus	all-year	stable	
Eastern Jackdaw	Coloeus monedula/sommerringii	rarity		
Rook	Corvus frugilegus frugilegus	all-year	stable	
Carrion Crow	Corvus corone corone	rarity		
Hooded Crow	Corvus cornix cornix	all-year	down	
Raven	Corvus corax corax	all-year	stable	
Waxwing	Bombycilla garrulus garrulus	winter		
Coal Tit	Periprus ater hibernicus	all-year	up	

453

Species	Scientific Names	Status	Trend	Check
Blue Tit	Cyanistes caeruleus obscurus	all-year	up	
Great Tit	Parus major newtoni	all-year	up	
Skylark	Alauda arvensis arvensis	all-year	down	
Eastern Skylark	Alauda arvensis intermedia	rarity		
Shore Lark	Eremophila alpestris flava	rarity		
Short-toed Lark	Calandrella brachydactyla brachydactyla	rarity		
Sand Martin	Riparia riparia riparia	summer	down	
Swallow	Hirundo rustica rustica	summer	down	
House Martin	Delichon urbicum urbicum	summer	stable	
Red-rumped Swallow	Cecropis daurica rufula	rarity		
Long-tailed Tit	Aegithalos caudatus rosaceus	all-year	stable	
Wood Warbler	Phylloscopus sibilatrix	summer	down	
Yellow-browed Warbler	Phylloscopus inornatus	rarity	up	
Pallas's Warbler	Phylloscopus proregulus	rarity		
Radde's Warbler	Phylloscopus schwarzi	rarity		
Dusky Warbler	Phylloscopus fuscatus	rarity		
Willow Warbler	Phylloscopus trochilus trochilus	summer	up	
Chiffchaff	Phylloscopus collybita collybita	summer	up	
Siberian Chiffchaff	Phylloscopus collybita tristis	rarity		
Greenish Warbler	Phylloscopus trochiloides viridanus	rarity		
Arctic Warbler	Phylloscopus borealis borealis	rarity		
Sedge Warbler	Acrocephalus schoenobaenus	summer	stable	
Paddyfield Warbler	Acrocephalus agricola	rarity		
Blyth's Reed Warbler	Acrocephalus dumetorum	rarity		
Reed Warbler	Acrocephalus scirpaceus scirpaceus	rarity		
Marsh Warbler	Acrocephalus palustris	rarity		
Booted Warbler	Iduna caligata	rarity		
Eastern Olivaceous / Sykes's Warbler	Iduna pallida / rama	rarity		
Melodious Warbler	Hippolais polyglotta	rarity		
Grasshopper Warbler	Locustella naevia naevia	summer	stable	
Blackcap	Sylvia atricapilla atricapilla	summer	up	
Garden Warbler	Sylvia borin borin	rarity		
Barred Warbler	Curruca nisoria	rarity		
Lesser Whitethroat	Curruca curruca curruca	rarity	up	
Eastern Subalpine Warbler	Curruca cantillans albistriata	rarity		
Whitethroat	Curruca communis communis	summer	stable	
Firecrest	Regulus ignicapilla ignicapilla	rarity		
Goldcrest	Regulus regulus regulus	all-year	stable	

Species	Scientific Names	Status	Trend	Check
Wren	Troglodytes troglodytes indigennus	all-year		
Treecreeper	Certhia familiaris britannica	all-year	up	
Rosy Starling	Pastor roseus	rarity		
Starling	Sturnus vulgaris vulgaris	all-year	down	
Song Thrush	Turdus philomelos clarkei	all-year	stable	
Hebridean Song Thrush	Turdus philomelos hebridensis	uncertain		
Mistle Thrush	Turdus viscivorus viscivorus	all-year	down	
Redwing	Turdus iliacus iliacus	winter	stable	
Icelandic Redwing	Turdus iliacus coburni	winter	stable	
Blackbird	Turdus merula merula	all-year	stable	
Fieldfare	Turdus pilaris	winter	stable	
Ring Ouzel	Turdus torquatus torquatus	summer	down	
Spotted Flycatcher	Musicapa striata striata	summer	down	
Robin	Erithacus rubecula melophilus	all-year	stable	
Bluethroat	Luscinia svecica svecica	rarity		
Nightingale	Luscinia megarhynchos megarhynchos	rarity		
Red-breasted Flycatcher	Ficedula parva	rarity		
Pied Flycatcher	Ficedula hypoleuca hypoleuca	rarity	stable	
Collared Flycatcher	Ficedula albicollis	rarity		
Black Redstart	Phoenicurus ochruros gibraltariensis	rarity		
Redstart	Phoenicurus phoenicurus phoenicurus	rarity	down	
Whinchat	Saxola rubetra	summer	down	
Stonechat	Saxicola torquata hibernans	all-year	stable	
Wheatear	Oenanthe oenanthe oenanthe	summer	stable	
Greenland Wheatear	Oenanthe oenanthe leucorrhoa	passage		
Black / White-crowned Wheatear	Oenanthe leucura / leucopyga	rarity		
Dipper	Cinclus cinclus hibernicus	all-year	stable	
House Sparrow	Passer domesticus domesticus	all-year	stable	
Tree Sparrow	Passer montanus montanus	all-year	down	
Dunnock	Prunella modularis hebridium	all-year	up	
Western Yellow Wagtail	Motacilla flava flavissima	rarity	down	
Blue-headed Wagtail	Motacilla flava flava	rarity		
Grey-headed Wagtail	Motacilla flava thunbergi	rarity		
Eastern Yellow Wagtail	Motacilla tschutschensis tschutschensis / plexa	rarity		
Citrine Wagtail	Motacilla citreola	rarity		
Grey Wagtail	Motacilla cinerea cinerea	all-year	down	
Pied Wagtail	Motacilla alba yarrellii	all-year	up	
White Wagtail	Motacilla alba alba	passage		

Species	Scientific Names	Status	Trend	Check
Richard's Pipit	Anthus richardi	rarity		
Meadow Pipit	Anthus pratensis	all-year	down	
Tree Pipit	Anthus trivialis trivialis	rarity		
Pechora Pipit	Anthus gustavi	rarity		
Red-throated Pipit	Anthus cervinus	rarity		
Water Pipit	Anthus spinoletta spinoletta	rarity		
Rock Pipit	Anthus petrosus petrosus	all-year	stable	
Rock Pipit (Scandinavian)	Anthus petrosus littoralis	rarity		
Chaffinch	Fringilla coelebs gengleri	all-year	mixed	
Chaffinch (continental)	Fringilla coelebs coelebs	winter	down	
Brambling	Fringilla montifringilla	winter	down	
Hawfinch	Coccothraustes coccothraustes coccothraustes	rarity		
Bullfinch	Pyrrhula pyrrhula pileata	all-year	up	
Bullfinch (Northern)	Pyrrhula pyrrhula pyrrhula	rarity		
Common Rosefinch	Carpodacus erythrinus	rarity	stable	
Greenfinch	Chloris chloris harrisoni	all-year	down	
Greenfinch (Continental)	Chloris chloris chloris	uncertain		
Twite	Linaria flavirostris pipilans	all-year	down	
Linnet	Linaria cannabina cannabina	all-year	down	
Mealy Redpoll	Acanthis flammea flammea	rarity		
Greenland Redpoll	Acanthis flammea rostrata	passage		
Lesser Redpoll	Acanthis cabaret	all-year	up	
Arctic Redpoll	Acanthis hornemanni	rarity		
Crossbill	Loxia curvirostra curvirostra	all-year	up	
Goldfinch	Carduelis carduelis britannica	all-year	up	
Siskin	Spinus spinus	all-year	up	
Lapland Bunting	Calcarius lapponicus lapponicus	passage	down	
Snow Bunting	Plectrophenax nivalis nivalis	winter	down	
Corn Bunting	Emberiza calandra calandra	extinct	down	
Yellowhammer	Emberiza citrinella caliginosa	all-year	down	
Ortolan Bunting	Emberiza hortulana	rarity		
Little Bunting	Emberiza pusilla	rarity		
Yellow-breasted Bunting	Emberiza aureola	rarity		
Black-headed Bunting	Emberiza melanocephala	rarity		
Red-headed Bunting	Emberiza bruniceps	escape		
Reed Bunting	Emberiza schoeniclus schoeniclus	all-year	down	
Bobolink	Dolichonyx oryzivorus	rarity		
Baltimore Oriole	Icterus galbula	rarity		

APPENDIX 6

Taxa Totals

Categories	No.	Taxa
All taxa	376	
All full species	345	
All additional subspecies	31	

Species validated by IRBC or other authority – grouped by AERC categories

A	Naturally occurring in the county since 1950	310	
A	Either/or	4	Albatross sp. , Gadfly Petrel sp Eastern Olivaceous / Sykes's Warbler Black White-crowned Wheatear
B	Naturally occurring, but not recorded since 1950	11	Pallas's Sandgrouse, Stone Curlew Black-winged Stilt, Little Bittern Black-crowned Night Heron, Ivory Gull Great Auk, Rough-legged Buzzard Capercaillie, Scops Owl, Roller
C1	Self-sustaining introductions in Ireland	4	Mute Swan, Canada Goose Ruddy Duck, Pheasant
C2	From self-sustaining introductions outside Ireland		(Red Kite, below)
B / C*		2	Red Kite B, C1 and C2 White-tailed Eagle B, C1
		331	

Additional species

Full species not yet assessed by IRBC		5	Egyptian Goose, Alpine Swift Black-winged Pratincole, Pallid Harrier Dusky Warbler
D1 Escapes		8	Fulvous Whistling Duck, Bar-headed Goose, Black Swan, Mandarin Duck, Chilean Flamingo Lanner Falcon, Saker Falcon Red-headed Bunting

Inconclusive historical record	1	Green Woodpecker
	14	

Additional Subspecies

Validated by IRBC or other authorities	27	Dark-bellied Brent Goose
		Todd's Canada Goose
		White-fronted Goose (Russian)
		Northern Eider, Dresser's Eider
		Ringed Plover (Greenland)
		Ringed Plover (Northern), Dunlin (Arctic)
		Dunlin (British), Faroe Snipe
		Redshank (Icelandic), Kumlien's Gull
		Thayer's Gul, Herring Gull (Scandinavian)
		Baltic Gull, Icelandic Merlin
		Eastern Jackdaw, Eastern Skylark
		Siberian Chiffchaff, Icelandic Redwing
		Greenland Wheatear
		Blue-headed Wagtail, White Wagtail
		Rock Pipit (Scandinavian)
		Chaffinch (Continental)
		Bullfinch (Northern), Greenland Redpoll
Subspecies not yet validated	4	Icelandic Water Rail
		Hebridean Song Thrush
		Grey-headed Wagtail
		Continental Greenfinch
	31	

* Note that recent records of White-tailed Eagle are from the Irish re-introduction project which is now probably self-sustaining. If that is the case, the recent Donegal records qualify the eagle for the C1 category. Recent Red Kite records are thought to be from both of the successful re-introduction projects, in Ireland and Scotland, qualifying the species for both C1 and C2.

The Donegal Ruddy Ducks are assumed to have been from a naturalised population in Northern Ireland. The species has since been virtually exterminated throughout western Europe, so any new records would more likely be vagrants from North America, and therefore Category A.

For your personal list, you can make up your own rules. But if what you want is a friendly competition, or just to be in line with most other birdwatchers, then this breakdown of the Donegal list should be useful. The totally validated list of full species for the county is 331.

NOTES

Chapter 2: Early Days

1. G. D'Arcy, *Ireland's Lost Birds* (Dublin: Four Courts Press, 1999).

2. A. Day and P. McWilliams (eds), *Ordnance Survey Memoirs of Ireland. Volume 38: Parishes of County Donegal I, 1833–5, North-east Donegal* (Belfast and Dublin: The Institute of Irish Studies in association with The Royal Irish Academy, 1997).

3. R.J. Ussher and R. Warren, *The Birds of Ireland* (London: Gurney & Jackson, 1900).

4. H.C. Hart, *The Flora of the County Donegal, or, List of the Flowering Plants and Ferns, with their Localities and Distribution* (Dublin: Sealy, Bryers & Walker, 1898).

5. H.C. Hart, 'Notes on the Birds of Donegal', *The Zoologist*, 1891–2.

6. J.S. Elliot, 'Additional Notes on the Birds of Donegal', *The Zoologist*, 3rd series, vol. XVI, 1892.

7. Ussher and Warren, *The Birds of Ireland*.

8. G.R. Humphries, *A List of Irish Birds* (Dublin: Stationery Office, 1937).

9. P.G. Kennedy, R.F. Ruttledge and C.F. Scroope, *Birds of Ireland* (Edinburgh and London: Oliver & Boyd, 1954).

10. R.S.R. Fitter, 'Birds on Roaninish', *The Irish Naturalists' Journal*, vol. 9, 1948, p. 128.

11. N.P. Cummins and I.M. Goodbody, 'Storm Petrel on Roaninish', *Irish Naturalists' Journal*, vol. 9, 1948, p. 129.

Chapter 3: Recent Times

1. K.W. Perry, *The Birds of the Inishowen Peninsula* (privately published, 1975).

Chapter 5: Farmland

1. L.J. Lewis, D. Coombes, B. Burke, J. O'Halloran, A. Walsh, T.D. Tierney and S. Cummins, 'Countryside Bird Survey: Status and trends of common and widespread breeding birds 1998–2016', *Irish Wildlife Manuals*, no. 115 (Dublin: National Parks and Wildlife Service, 2019).

2. G. Gilbert, A. Stanbury and L. Lewis, 'Birds of Conservation Concern in Ireland 4: 2020–2026', *Irish Birds*, vol. 43, 2021, pp. 1–22.

3. J.T.R. Sharrock, *The Atlas of Breeding Birds in Britain and Ireland* (Tring: British Trust for Ornithology, 1976).

4. D.W. Gibbons, J.B. Reid and R.A. Chapman (eds), *The New Atlas of Breeding Birds in Britain and Ireland* (London: Poyser, 1993).

5. A.J. Taylor and J. O'Halloran, 'The Decline of the Corn Bunting, *Miliaria calandra*, in the Republic of Ireland', *Biology and Environment: Proceedings of the Royal Irish Academy*, vol. 102B, 2002, pp. 165–75.

Chapter 6: Woods and Forests

1. S. Ó Gaoithin, *The Ancient Woodlands of Glenveagh* (Donegal: Glenveagh National Park, 2021).

Chapter 7: Uplands

1. C. Mac Lochlainn, 'Breeding and Wintering Bird Communities of Glenveagh National Park, Co. Donegal', *Irish Birds*, vol. 2, 1984, pp. 482–500.

Chapter 8: Lakes, Rivers and Streams

1. L. Campbell, *Room for the River: The Foyle River catchment landscape. Connecting people, place and nature* (Buncrana: Merdog Books, 2021).

Chapter 10: The Coast

1. S. Cramp, W.R.P. Bourne and D. Saunders, *The Seabirds of Britain and Ireland* (London: Collins, 1974).

2. C.S. Lloyd, M.L. Tasker and K. Partridge, *The Status of Seabirds in Britain and Ireland* (London: Poyser, 1991).

3. P.I. Mitchell, N. Ratcliffe, S. Newton and T.E. Dunn, *Seabird Populations of Britain and Ireland: Results of the Seabird 2000 census (1998–2002)* (London: T. & A.D. Poyser, 2004).

4. S. Cummins, C. Lauder, A. Lauder and T.D. Tierney, 'The Status of Ireland's Breeding Seabirds: Birds Directive Article 12 Reporting 2013–2018', *Irish Wildlife Manuals*, no. 114 (Dublin: National Parks and Wildlife Service, 2019); J. Roller and D. Tierney, pers. comm.

5. Ussher and Warren, *The Birds of Ireland*.

6. P. Phillips, C. Ingram and R. Salter, *Tory Island Bird Report 2016* (privately published, 2017).

7. P. Phillips, C. Ingram and R. Salter, *Tory Island Bird Report 2014* (privately published, 2015).

8. M.K. Bell, 'Breeding Wader Survey at Selected Sites in Cos Donegal and Sligo', BirdWatch Ireland, unpublished report, 2022.

9. Inishtrahull Bird Observatory, https://inishtrahullbirdobs436659775.wordpress.com.

10. R. Nairn and J.R. Sheppard, 'Breeding Waders of Sand Dune Machair in North-west Ireland', *Irish Birds*, vol. 6, 1998, pp. 177–90; D. Suddaby, T. Nelson and J. Veldman, 'Resurvey and Comparative Changes of Breeding Wader Populations', *Birds*, vol. 3, 1985, pp. 53–70; B. Madden, T. Cooney, A. O'Donoghue, D.W. Norris and O.J. Merne, 'Breeding Waders of Machair Systems in Ireland in 1996', *Irish Birds*, vol. 6, 1998, pp. 177–90; D. Suddaby, T. Nelson and J. Veldman, 'Resurvey and Comparative Changes of Breeding Wader Populations of Irish Machair and Associated Wet Grasslands in 2009', *Irish Birds*, vol. 8, 2009, pp. 533–42; C. McMonagle, M. Bell and A. Donaghy, 'Survey of Breeding Wader Populations at Machair and Offshore Islands in North-West Ireland, 2017, CAAB Cooperating Across Borders for Biodiversity Project, an INTERREG V Project under the European Regional Development Fund', BirdWatch Ireland, unpublished report, 2017; M.K. Bell, 'Breeding Wader Survey at Selected Sites in Cos Donegal and Sligo', BirdWatch Ireland, unpublished report, 2018; M.K. Bell, 'Breeding Wader Survey at Selected Sites in Cos Donegal and Sligo', BirdWatch Ireland, unpublished report, 2019; M.K. Bell and K. Flynn, 'Breeding Wader Survey at Selected Sites in Cos Donegal and Sligo', BirdWatch Ireland, unpublished report, 2020; Bell, 'Breeding Wader Survey at Selected Sites in Cos Donegal and Sligo', 2022.

Chapter 11: Top Twenty Sites

1. E. Dempsey and M. O'Clery, *Finding Birds in Ireland: The complete guide*, 2nd edn (Dublin: Gill &Macmillan, 2014); P. Milne and C. Hutchinson, *Where to Watch Birds: Ireland* (London: Christopher Helm, 2009).

Chapter 12: Lough Swilly

1. J.R. Leebody, 'Notes on the Birds of Lough Swilly', *The Irish Naturalist*, vol. 1, 1892, pp. 173–7; Perry, *The Birds of the Inishowen Peninsula*; R. Sheppard, 'The Wintering Waterbirds of Lough Swilly, County Donegal', *Irish Birds*, vol. 7, 2002, pp. 65–78.

2. A. Lauder and C. Lauder, 'Identification of Breeding Waterbird Hotspots in Ireland', *Irish Wildlife Manuals*, no. 129 (Dublin: National Parks and Wildlife Service, 2020).

3. Leebody, 'Notes on the Birds of Lough Swilly'.

4. C.D. Hutchinson, *Ireland's Wetlands and Their Birds* (Dublin: Irish Wildbird Conservancy, 1979).

5. R. Sheppard, *Ireland's Wetland Wealth* (Dublin: Irish Wildbird Conservancy, 1993).

6. O. Crowe, *Ireland's Wetlands and Their Waterbirds: Status and distribution* (Newcastle: BirdWatch Ireland, 2005); L.J Lewis, B. Burke, N. Fitzgerald, T.D. Tierney and S. Kelly, 'Irish Wetland Bird Survey: Waterbird status and distribution 2009/10–2015/16', *Irish Wildlife Manuals*, no. 106 (Dublin: National Parks and Wildlife Service, 2019).

7. B. Burke, L.J. Lewis, N. Fitzgerald, T. Frost, G. Austin and T.D. Tierney, 'Estimates of Waterbird Numbers Wintering in Ireland 2011/12 – 2015/16', *Irish Birds*, no. 41, 2018, pp. 1–12; Lewis, Burke, Fitzgerald, Tierney and Kelly, 'Irish Wetland Bird Survey: Waterbird status and distribution 2009/10–2015/16'.

Chapter 13: Migrant Hotspots

1. R.M. Barrington, *The Migration of Birds* (London: Porter, and Dublin: Ponsonby, 1900).

2. E.A.S. Baynes, 'The *Lepidoptera* of Tory Island, Co. Donegal', *The Entomologist*, vol. 90, 1957, pp. 310–13.

3. A. Gibbs, I.C.T. Nisbet and P. Redman, 'Birds of North Donegal in Autumn, 1953', *British Birds*, vol. 47, 1954 pp. 217–28.

4. P.S. Redman, 'Birds of Tory Island, County Donegal in 1954', *Tory Island Bird Report 2015* (privately published, 2016).

5. Tory Bird Observatory, unpublished annual reports, 1959 to 1964.

6. R.G. Pettitt, 'Tory Island, 1965', pp. 1–5, unpublished report.

7. J. Hobbs, *First Irish Records, 1800 to 2010*, version 1.3 (privately published, 2021).

8. P. Phillips, C. Ingram and R. Salter, *Tory Island Bird Reports*, 2011 to 2016 (privately published).

9. Gibbs, Nisbet and Redman, 'Birds of North Donegal in Autumn, 1953'.

10. Kennedy, Ruttledge and Scroope, *Birds of Ireland*.

11. O.J. Merne, 'Malin Head Observatory', *IWC News*, no. 2, 1974.

12. T.R.E. Devlin and O.J. Merne, 'Malin Head Observatory Report 1961–1964', pp. 1–38, unpublished report, 1964; T.R.E. Devlin and O.J. Merne, 'Malin Head Observatory Report 1965', pp. 1–32, unpublished report, 1965.

13. Perry, *The Birds of the Inishowen Peninsula*.

14. D.I.M. Wallace, A. McGeehan and D. Allen, 'Autumn Migration in Westernmost Donegal', *British Birds*, vol. 94, 2001, pp. 103–20.

15. M. Mac Gloinn, *Éin Árainn Mhór: Birds of Arranmore* (n.p.: Shearwater Publishing, 2022).

Chapter 15: Seawatching

1. Gibbs, Nisbet and Redman, 'Birds of North Donegal in Autumn, 1953'; P.S. Redman, 'Birds of Tory Island, County Donegal in 1954', Tory *Island Bird Report 2015* (privately published, 2016); Devlin and Merne, 'Malin Head Observatory Report 1965'.

Chapter 16: Looking Ahead

1. K.V. Rosenberg, A.M. Dokter, P.J. Blancher, J.R. Sauer, A.C. Smith, A. Paul, J.C. Stanton, A. Panjabi, L. Helft and P.P. Marra, 'Decline of North American Avifauna', *Science*, vol. 366, no. 6,461, 19 September 2019, pp. 120–4.

2. BirdLife International, *State of the World's Birds 2022: Insights and solutions for the biodiversity crisis* (Cambridge: BirdLife International, 2022).

3. Gilbert, Stanbury and Lewis, 'Birds of Conservation Concern in Ireland 4: 2020–2026'.

4. M. Hine and R. Malin, 'News and Comment: Concern for Scotland's seabirds as avian influenza continues to take hold', *British Birds*, vol. 115, 2022, p. 423.

5. Gilbert, Stanbury and Lewis, 'Birds of Conservation Concern in Ireland 4: 2020–2026'.

6. Burke, Lewis, Fitzgerald, Frost, Austin and Tierney, 'Estimates of Waterbird Numbers Wintering in Ireland 2011/12 – 2015/16'.

7. Inishowen Upland Farmers Project, https://inishoweneip.com.

8. Inishowen Rivers Trust, https://inishowenriverstrust.com.

9. Inch Wildfowl Reserve, https://inchwildfowlreserve.ie.

Chapter 17: Introduction to Species Accounts

1. F. Gill, D. Donsker and P. Rasmussen (eds), *IOC World Bird List*, v. 11.1, 2021, http://www.worldbirdnames.org.

2. C. Ó Caomhánaigh, *Dictionary of Bird Names in Irish*, http://gofree.indigo.ie/~cocaomh/HomePage.htm, 2002.

3. W. Thompson, *The Natural History of Ireland. Vol. 3: Birds* (London: Reeve, Benham & Reeve, 1851).

4. Ussher and Warren, *The Birds of Ireland*.

5. Kennedy, Ruttledge and Scroope, *Birds of Ireland*.

6. R.F. Ruttledge, *Ireland's Birds* (London: Witherby, 1966).

7. R.F. Ruttledge, *A List of the Birds of Ireland* (Dublin: Stationery Office, 1975).

8. C.D. Hutchinson, *Birds in Ireland* (Calton: Poyser, 1989).

9. S. Delany, D. Scott, T. Dodman and D. Stroud, *An Atlas of Wader Populations in Africa and Western Eurasia* (Wageningen, The Netherlands: Wetlands International, 2009).

10. D.A. Scott and P.M Rose, *Atlas of Anatidae Populations in Africa and Western Eurasia* (Wageningen: Wetlands International, 1996).

11. J. del Hoyo, A. Elliott and J. Sargatal (eds), *Handbook of the Birds of the World*, vols 1–7; J. del Hoyo, A. Elliott and D.A. Christie (eds), *Handbook of the Birds of the World*, vols 8–16 (Barcelona: Lynx Edicions, 1992–2011).

12. Sharrock, *The Atlas of Breeding Birds in Britain and Ireland*.

13. Kennedy, Ruttledge and Scroope, *Birds of Ireland*.

14. J.P. Hillis, 'Rare Irish Breeding Birds, 1992–2001', *Irish Birds*, vol. 7, 2003, pp. 157–72.

15. J.P. Hillis, 'Annual Reports of the Irish Rare Breeding Birds Panel', *Irish Birds*, vols 7–9, 2002–12.

16. K.W. Perry, 'Rare Breeding Birds in Ireland in 2012 – including a review of significant population changes over the past decade', *Irish Birds*, vol. 9, 1993, pp. 563–76; K.W. Perry and S.F. Newton, 'Rare Breeding Birds in Ireland in 2013', *Irish Birds*, vol. 10, 2014, pp. 63–79; S.F. Newton, 'Rare Breeding Birds in Ireland in 2016', *Irish Birds*, vol. 10, 2016, pp. 383–90; S.F. Newton, 'An Overview of Rare Breeding Birds in Ireland in 2017', *Irish Birds*, vol. 10, 2018, pp. 541–4; B. Burke, O. Crowe and S.F. Newton, 'Rare and Scarce Breeding Birds in Ireland in 2017 and 2018', *Irish Birds*, no. 42, 2020, pp. 63–70; O. Crowe, T.D. Tierney and B. Burke, 'Status of Rare Breeding Birds Across the Island of Ireland 2013–2018', *Irish Birds*, no. 43, 2021, pp. 29–38.

17. Cramp, Bourne and Saunders, *The Seabirds of Britain and Ireland*.

18. C.S. Lloyd, 'Inventory of Seabird Breeding Colonies in the Republic of Ireland', Forest and Wildlife Service, Bray, unpublished report, 1982.

19. Mitchell, Ratcliffe, Newton and Dunn, *Seabird Populations of Britain and Ireland: Results of the Seabird 2000 census (1998–2002)*.

20. Cummins, Lauder, Lauder and Tierney, 'The Status of Ireland's Breeding Seabirds: Birds Directive Article 12 Reporting 2013–2018'.

21. National Parks and Wildlife Service, https://www.npws.ie/publications.

22. Gilbert, Stanbury and Lewis, 'Birds of Conservation Concern in Ireland 4: 2020–2026'.

Chapter 18: Species Accounts

1. Irish Rare Birds Committee, 'Irish Rare Birds Committee Review of Presumed Vagrant Canada Goose *Branta canadensis* Records to Ascertain the Occurrence of Cackling Goose *Branta hutchinsii*', *Irish Birds*, vol. 9, 2013, pp. 613–22.

2. J. Hobbs, *A List of Irish Birds* (version 12.0) (www.southdublinbirds.com) (Dublin: South Dublin Branch of BirdWatch Ireland, 2022).

3. S. Doyle, A. Walsh, B.J. McMahon and D.T. Tierney, 'Barnacle Geese *Branta leucopsis* in Ireland: Results of the 2018 census', *Irish Birds*, vol. 11, 2018, pp. 23–38.

4. Irish Rare Birds Committee, 'Irish Rare Birds Committee Review of Presumed Vagrant Canada Goose *Branta canadensis* Records to Ascertain the Occurrence of Cackling Goose *Branta hutchinsii*'.

5. C. Mitchell, R. Heard and D. Stroud, 'The Merging of Populations of Greylag Goose Breeding in Britain', *British Birds*, vol. 105, 2012, pp. 498–505.

6. Lewis, Burke, Fitzgerald, Tierney and Kelly, 'Irish Wetland Bird Survey: Waterbird status and distribution 2009/10–2015/16'.

7. H. Boland and A. Speer, pers. comm.

8. Hart, 'Notes on the Birds of Donegal'.

9. Ussher and Warren, *The Birds of Ireland*.

10. J.F. Dowdall and E. Larrissey, 'Tundra Bean Goose in County Louth: A race new to Ireland', *Irish Birds*, vol. 6, 1999, pp. 432–3.

11. R.F. Ruttledge, 'Re-assessment of the Record of Long-tailed Skuas in the Shannon Valley in 1860', *Irish Birds*, vol. 2, 1982 pp. 197–8.

12. D.W. Norriss and H.J. Wilson, 'Disturbance and Flock Size Changes in Greenland White-fronted Geese Wintering in Ireland', *Wildfowl*, vol. 39, 1988, pp. 63–70.

13. Hart, 'Notes on the Birds of Donegal'.

14. E. Rees, 'Northwest European Bewick's Swans: A national and flyway perspective', *Waterbirds in the UK 2019/20, the Annual Report of the Wetland Bird Survey*, 2021.

15. R. Sheppard, 'Whooper and Bewick's Swans in North-west Ireland', *Irish Birds*, vol. 2, 1981, pp. 48–59.

16. Gilbert, Stanbury and Lewis, 'Birds of Conservation Concern in Ireland 4: 2020–2026'.

17. Sheppard, 'Whooper and Bewick's Swans in North-west Ireland'.

18. B. Burke, J.G. McElwaine, N. Fitzgerald, S.B.A. Kelly, N. McCulloch, A.J. Walsh and L.J. Lewis, 'Population Size, Breeding Success and Habitat Use of Whooper Swan *Cygnus cygnus* and Bewick's Swan *Cygnus columbianus bewickii* in Ireland: Results of the 2020 International Swan Census', *Irish Birds*, no. 43, 2021, pp. 57–70.

19. Sheppard, 'Whooper and Bewick's Swans in North-west Ireland'.

20. Sheppard, *Ireland's Wetland Wealth*.

21. Crowe, *Ireland's Wetlands and Their Waterbirds: Status and distribution*.

22. Lewis, Burke, Fitzgerald, Tierney and Kelly, 'Irish Wetland Bird Survey: Waterbird status and distribution 2009/10–2015/16'.

23. S. Cropper, 'Whooper Swans in Co. Donegal in August', *British Birds*, vol. 31, 1937, p. 151.

24. Hobbs, *A List of Irish Birds* (version 12.0).

25. Leebody, 'Notes on the Birds of Lough Swilly'.

26. Gilbert, Stanbury and Lewis, 'Birds of Conservation Concern in Ireland 4: 2020–2026'.

27. Lewis, Burke, Fitzgerald, Tierney and Kelly, 'Irish Wetland Bird Survey: Waterbird status and distribution 2009/10–2015/16'.

28. Crowe, *Ireland's Wetlands and Their Waterbirds: Status and distribution*.

29. Leebody, 'Notes on the Birds of Lough Swilly'.

30. Sharrock, *The Atlas of Breeding Birds in Britain and Ireland*; Gibbons, Reid and Chapman, *The New Atlas of Breeding Birds in Britain and Ireland*; Balmer, Gillings, Caffrey, Swann, Downie and Fuller, *Bird Atlas 2007–11*.

31. Ussher and Warren, *The Birds of Ireland*.

32. Scott and Rose, *Atlas of Anatidae Populations in Africa and Western Eurasia*.

33. Gilbert, Stanbury and Lewis, 'Birds of Conservation Concern in Ireland 4: 2020–2026'.

34. K.W. Perry, pers. comm.

35. Scott and Rose, *Atlas of Anatidae Populations in Africa and Western Eurasia*.

36. Anon., 'Scaup on the Slide', *Waterbirds in the UK 2019/20, the Annual Report of the Wetland Bird Survey*, 2021.

37. Leebody, 'Notes on the Birds of Lough Swilly'.

38. Sheppard, *Ireland's Wetland Wealth*.

39. Gilbert, Stanbury and Lewis, 'Birds of Conservation Concern in Ireland 4: 2020–2026'.

40. Ussher and Warren, *The Birds of Ireland*.

41. Kennedy, Ruttledge and Scroope, *Birds of Ireland*.

42. Gilbert, Stanbury and Lewis, 'Birds of Conservation Concern in Ireland 4: 2020–2026'.

43. W. Farrelly and D. Charles, 'The Dresser's Eider in County Donegal: A new western Palearctic bird', *Birding World*, vol. 23, pp. 62–4; J. Hobbs, *First Records for the Western Palearctic 1800 to 2015* (version 1.6) (privately published, 2023).

44. Gilbert, Stanbury and Lewis, 'Birds of Conservation Concern in Ireland 4: 2020–2026'.

45. Balmer, Gillings, Caffrey, Swann, Downie and Fuller, *Bird Atlas 2007–11*.

46. Scott and Rose, *Atlas of Anatidae Populations in Africa and Western Eurasia*.

47. Lewis, Coombes, Burke, O'Halloran, Walsh, Tierney and Cummins, 'Countryside Bird Survey: Status and trends of common and widespread breeding birds 1998–2016'.

48. Gilbert, Stanbury and Lewis, 'Birds of Conservation Concern in Ireland 4: 2020–2026'.

49. Ibid.

50. Scott and Rose, *Atlas of Anatidae Populations in Africa and Western Eurasia*.

51. Leebody, 'Notes on the Birds of Lough Swilly'.

52. R. Sheppard, 'The Breeding of the Goosander in Ireland', *Irish Birds*, vol. 1, 1978, pp. 224–8.

53. Hart, 'Notes on the Birds of Donegal'.

54. Sharrock, *The Atlas of Breeding Birds in Britain and Ireland*.

55. Day and McWilliams (eds), *Ordnance Survey Memoirs of Ireland. Volume 38: Parishes of County Donegal I, 1833–5, North-east Donegal*.

56. D'Arcy, *Ireland's Lost Birds*.

57. S. Cummins, A. Bleasdale, C. Douglas, S.F. Newton, J. O'Halloran and H.J. Wilson, 'Densities and Population Estimates of Red Grouse *Lagopus lagopus scotica* in Ireland Based on the 2006–2008 National Survey', *Irish Birds*, vol. 10, 2015, pp. 197–210.

58. Sharrock, *The Atlas of Breeding Birds in Britain and Ireland*.

59. Gilbert, Stanbury and Lewis, 'Birds of Conservation Concern in Ireland 4: 2020–2026'.

60. W. Thompson, *The Natural History of Ireland. Vol. 2: Birds* (London: Reeve, Benham & Reeve, 1850).

61. B. Bryce, pers. comm.

62. Gilbert, Stanbury and Lewis, 'Birds of Conservation Concern in Ireland 4: 2020–2026'.

63. Thompson, *The Natural History of Ireland. Vol. 2: Birds*.

64. Ussher and Warren, *The Birds of Ireland*.

65. Gilbert, Stanbury and Lewis, 'Birds of Conservation Concern in Ireland 4: 2020–2026'.

66. Ruttledge, *Ireland's Birds*.

67. Sharrock, *The Atlas of Breeding Birds in Britain and Ireland*.

68. Gilbert, Stanbury and Lewis, 'Birds of Conservation Concern in Ireland 4: 2020–2026'.

69. Ibid.

70. Balmer, Gillings, Caffrey, Swann, Downie and Fuller, *Bird Atlas 2007–11*; Lewis, Coombes, Burke, O'Halloran, Walsh, Tierney and Cummins, 'Countryside Bird Survey: Status and trends of common and widespread breeding birds 1998–2016'.

71. Ussher and Warren, *The Birds of Ireland*.

72. Ibid.

73. Lewis, Coombes, Burke, O'Halloran, Walsh, Tierney and Cummins, 'Countryside Bird Survey: Status and trends of common and widespread breeding birds 1998–2016'.

74. Gilbert, Stanbury and Lewis, 'Birds of Conservation Concern in Ireland 4: 2020–2026'.

75. R.K. Murton, *The Woodpigeon* (London: Collins, 1965).

76. Lewis, Coombes, Burke, O'Halloran, Walsh, Tierney and Cummins, 'Countryside Bird Survey: Status and trends of common and widespread breeding birds 1998–2016'.

77. Gilbert, Stanbury and Lewis, 'Birds of Conservation Concern in Ireland 4: 2020–2026'.

78. Sharrock, *The Atlas of Breeding Birds in Britain and Ireland*.

79. Ussher and Warren, *The Birds of Ireland*.

80. del Hoyo, Elliott and Sargatal (eds), *Handbook of the Birds of the World*.

81. Sharrock, *The Atlas of Breeding Birds in Britain and Ireland*.

82. R. Sheppard and R.E. Green, 'Status of the Corncrake in Ireland in 1993', *Irish Birds*, vol. 5, 1994, pp. 125–38.

83. M. Duffy, *The Corncrake Conservation Project: Annual report* (Dublin: National Parks and Wildlife Service, 2018).

84. Gilbert, Stanbury and Lewis, 'Birds of Conservation Concern in Ireland 4: 2020–2026'.

85. Thompson, *The Natural History of Ireland. Vol. 2: Birds*.

86. Balmer, Gillings, Caffrey, Swann, Downie and Fuller, *Bird Atlas 2007–11*.

87. Sheppard, *Ireland's Wetland Wealth*; Crowe, *Ireland's Wetlands and Their Waterbirds: Status and distribution*.

88. Irish Rare Birds Committee, 'Irish Rare Bird Report 2016', *Irish Birds*, vol. 10, p. 573.

89. W.K. Bigger, pers. comm.

90. Lewis, Burke, Fitzgerald, Tierney and Kelly, 'Irish Wetland Bird Survey: Waterbird status and distribution 2009/10–2015/16'.

91. Hart, 'Notes on the Birds of Donegal'.

92. Leebody, 'Notes on the Birds of Lough Swilly'.

93. Lewis, Burke, Fitzgerald, Tierney and Kelly, 'Irish Wetland Bird Survey: Waterbird status and distribution 2009/10–2015/16'.

94. Leebody, 'Notes on the Birds of Lough Swilly'.

95. Gilbert, Stanbury and Lewis, 'Birds of Conservation Concern in Ireland 4: 2020–2026'.

96. Ussher and Warren, *The Birds of Ireland*.

97. Kennedy, Ruttledge and Scroope, *Birds of Ireland*.

98. R. Nairn and J. O'Halloran, *Bird Habitats in Ireland* (Cork: The Collins Press, 2012).

99. R. Sheppard, 'The Winter and Spring Feeding Ecology of Oystercatchers *Haematopus ostralegus* on the Edible Mussel *Mytilus edulus*', MSc thesis, University of Wales, Bangor, 1971.

100. Crowe, *Ireland's Wetlands and Their Waterbirds: Status and distribution*.

101. Gilbert, Stanbury and Lewis, 'Birds of Conservation Concern in Ireland 4: 2020–2026'.

102. Nairn and Sheppard, 'Breeding Waders of Sand Dune Machair in North-west Ireland'.

103. Madden, Cooney, O'Donoghue, Norris and Merne, 'Breeding Waders of Machair Systems in Ireland in 1996'.

104. Suddaby, Nelson and Veldman, 'Resurvey and Comparative Changes of Breeding Wader Populations of Irish Machair and Associated Wet Grasslands in 2009'.

105. Bell, 'Breeding Wader Survey at Selected Sites in Cos Donegal and Sligo', 2018.

106. Bell, 'Breeding Wader Survey at Selected Sites in Cos Donegal and Sligo', 2019.

107. Bell and Flynn, 'Breeding Wader Survey at Selected Sites in Cos Donegal and Sligo'.

108. Crowe, *Ireland's Wetlands and Their Waterbirds: Status and distribution*.

109. Gilbert, Stanbury and Lewis, 'Birds of Conservation Concern in Ireland 4: 2020–2026'.

110. Hart, 'Notes on the Birds of Donegal'.

111. Sharrock, *The Atlas of Breeding Birds in Britain and Ireland*.

112. Ibid.

113. Cox, Eddleston and Newton, 'Upland Bird Survey Report 2002: Donegal', BirdWatch Ireland Conservation Report no. 02/4.

114. Bell, 'Upland Breeding Wader Survey Report Cos Donegal and Sligo'.

115. Crowe, *Ireland's Wetlands and Their Waterbirds: Status and distribution*.

116. Gilbert, Stanbury and Lewis, 'Birds of Conservation Concern in Ireland 4: 2020–2026'.

117. Ibid.

118. Madden, Cooney, O'Donoghue, Norris and Merne, 'Breeding Waders of Machair Systems in Ireland in 1996'.

119. Lewis, Burke, Fitzgerald, Tierney and Kelly, 'Irish Wetland Bird Survey: Waterbird status and distribution 2009/10–2015/16'.

120. Crowe, *Ireland's Wetlands and Their Waterbirds: Status and distribution*.

121. Ussher and Warren, *The Birds of Ireland*.

122. B.G. O'Donoghue, A. Donaghy and S.B.A. Kelly, 'National Survey of Breeding Eurasian Curlew *Numenius arquata* in the Republic of Ireland, 2015–2017', *Wader Study*, vol. 126, 2019, pp. 43–8.

123. Sharrock, *The Atlas of Breeding Birds in Britain and Ireland*.

124. D. Moloney, 'Curlew Report, 2019, CAAB Cooperating Across Borders for Biodiversity Project, an INTERREG V Project under the European Regional Development Fund', BirdWatch Ireland, unpublished report, 2019.

125. Crowe, *Ireland's Wetlands and Their Waterbirds: Status and distribution*.

126. Gilbert, Stanbury and Lewis, 'Birds of Conservation Concern in Ireland 4: 2020–2026'.

127. Crowe, *Ireland's Wetlands and Their Waterbirds: Status and distribution*.

128. Gilbert, Stanbury and Lewis, 'Birds of Conservation Concern in Ireland 4: 2020–2026'.

129. Sheppard, *Ireland's Wetland Wealth.*

130. Hutchinson, *Ireland's Wetlands and Their Birds.*

131. Gilbert, Stanbury and Lewis, 'Birds of Conservation Concern in Ireland 4: 2020–2026'.

132. A. McGeehan, *To the Ends of the Earth: Ireland's place in bird migration* (Cork: The Collins Press, 2018).

133. Crowe, *Ireland's Wetlands and Their Waterbirds: Status and distribution.*

134. L.J. Lewis, G. Austin, H. Boland, T. Frost, O. Crowe and T.D. Tierney, 'Waterbird Populations on Non-Estuarine Coasts in Ireland: Results of the 2015/16 Non-Estuarine Coastal Waterbird Survey (NEWS-III)', *Irish Birds*, vol. 10, 2017, pp. 511–22.

135. Lack, *The Atlas of Wintering Birds in Britain and Ireland.*

136. Gilbert, Stanbury and Lewis, 'Birds of Conservation Concern in Ireland 4: 2020–2026'.

137. Ibid.

138. Lewis, Burke, Fitzgerald, Tierney and Kelly, 'Irish Wetland Bird Survey: Waterbird status and distribution 2009/10–2015/16'.

139. Crowe, *Ireland's Wetlands and Their Waterbirds: Status and distribution.*

140. Gilbert, Stanbury and Lewis, 'Birds of Conservation Concern in Ireland 4: 2020–2026'.

141. Bell, 'Upland Breeding Wader Survey Report Cos Donegal and Sligo', 2020; Cox, Eddleston and Newton, 'Upland Bird Survey Report 2002: Donegal', BirdWatch Ireland Conservation Report no. 02/4.

142. M. Bell, pers. comm.

143. Gilbert, Stanbury and Lewis, 'Birds of Conservation Concern in Ireland 4: 2020–2026'.

144. S. Foster, H. Boland, K. Colhoun, B. Etheridge and R. Summers, 'Flock Composition of Purple Sandpipers *Calidris maritima* in the west of Ireland', *Irish Birds*, vol. 9, 2010, pp. 31–4.

145. Lewis, Austin, Boland, Frost, Crowe and Tierney, 'Waterbird Populations on Non-Estuarine Coasts in Ireland: Results of the 2015/16 Non-Estuarine Coastal Waterbird Survey (NEWS-III)'.

146. Gilbert, Stanbury and Lewis, 'Birds of Conservation Concern in Ireland 4: 2020–2026'.

147. Delany, Scott, Dodman and Stroud, *An Atlas of Wader Populations in Africa and Western Eurasia.*

148. Wallace, McGeehan and Allen, 'Autumn Migration in Westernmost Donegal'.

149. Sharrock, *The Atlas of Breeding Birds in Britain and Ireland.*

150. Gibbons, Reid and Chapman, *The New Atlas of Breeding Birds in Britain and Ireland.*

151. Balmer, Gillings, Caffrey, Swann, Downie and Fuller, *Bird Atlas 2007–11.*

152. Gilbert, Stanbury and Lewis, 'Birds of Conservation Concern in Ireland 4: 2020–2026'.

153. C. Wernham, M. Toms, J. Marchant, J. Clark, G. Sitiwardena and S. Baillie, *The Migration Atlas* (London: Poyser 2000).

154. Nairn and Sheppard, 'Breeding Waders of Sand Dune Machair in North-west Ireland'.

155. Madden, Cooney, O'Donoghue, Norris and Merne, 'Breeding Waders of Machair Systems in Ireland in 1996'.

156. Suddaby, Nelson and Veldman, 'Resurvey and Comparative Changes of Breeding Wader Populations of Irish Machair and Associated Wet Grasslands in 2009'.

157. McMonagle, Bell and Donaghy, 'Survey of Breeding Wader Populations at Machair and Offshore Islands in North-West Ireland'.

158. Bell, 'Breeding Wader Survey at Selected Sites in Cos Donegal and Sligo', 2018.

159. Bell, 'Breeding Wader Survey at Selected Sites in Cos Donegal and Sligo', 2019.

160. Bell, 'Upland Breeding Wader Survey Report Cos Donegal and Sligo'.

161. D. Moloney, pers. comm.

162. Gilbert, Stanbury and Lewis, 'Birds of Conservation Concern in Ireland 4: 2020–2026'.

163. Delany, Scott, Dodman and Stroud, *An Atlas of Wader Populations in Africa and Western Eurasia*.

164. Humphries, *A List of Irish Birds*.

165. Wallace, McGeehan and Allen, 'Autumn Migration in Westernmost Donegal'.

166. Hobbs, *A List of Irish Birds* (version 12.0).

167. Hutchinson, *Birds in Ireland*.

168. Ussher and Warren, *The Birds of Ireland*.

169. C.V. Stoney, 'Red-necked Phalarope Breeding in Co. Donegal', *The Irish Naturalist*, vol. 33, 1924, p. 109.

170. Kennedy, Ruttledge and Scroope, *Birds of Ireland*.

171. F. Egginton, pers. comm.

172. R.S.A. van Bemmelen et al., 'A Migratory Divide Among Red-necked Phalaropes in the Western Palearctic Reveals Contrasting Migration and Wintering Movement Strategies', *Frontiers in Ecology and Evolution*, 2019.

173. Gilbert, Stanbury and Lewis, 'Birds of Conservation Concern in Ireland 4: 2020–2026'.

174. Kennedy, Ruttledge and Scroope, *Birds of Ireland*.

175. Cox, Eddleston and Newton, 'Upland Bird Survey Report 2002: Donegal', BirdWatch Ireland Conservation Report no. 02/4.

176. Bell, 'Upland Breeding Wader Survey Report Cos Donegal and Sligo', 2020.

177. M. Bell, pers. comm.

178. Nairn and Sheppard, 'Breeding Waders of Sand Dune Machair in North-west Ireland'.

179. Madden, Cooney, O'Donoghue, Norris and Merne, 'Breeding Waders of Machair Systems in Ireland in 1996'.

180. Suddaby, Nelson and Veldman, 'Resurvey and Comparative Changes of Breeding Wader Populations of Irish Machair and Associated Wet Grasslands in 2009'.

181. McMonagle, Bell and Donaghy, 'Survey of Breeding Wader Populations at Machair and Offshore Islands in North-West Ireland'.

182. Bell, 'Breeding Wader Survey at Selected Sites in Cos Donegal and Sligo', 2018.

183. Bell, 'Breeding Wader Survey at Selected Sites in Cos Donegal and Sligo', 2019.

184. Bell and Flynn, 'Breeding Wader Survey at Selected Sites in Cos Donegal and Sligo'.

185. Gilbert, Stanbury and Lewis, 'Birds of Conservation Concern in Ireland 4: 2020–2026'.

186. Delany, Scott, Dodman and Stroud, *An Atlas of Wader Populations in Africa and Western Eurasia*.

187. Crowe, *Ireland's Wetlands and Their Waterbirds: Status and distribution*.

188. Delany, Scott, Dodman and Stroud, *An Atlas of Wader Populations in Africa and Western Eurasia.*

189. Crowe, *Ireland's Wetlands and Their Waterbirds: Status and distribution.*

190. Ussher and Warren, *The Birds of Ireland.*

191. Cramp, Bourne and Saunders, *The Seabirds of Britain and Ireland.*

192. P.S. Watson and D.J. Radford, 'Census of Breeding Seabirds at Horn Head, County Donegal, in June 1980', *Seabird Report 1977–81*, vol. 6, 1982.

193. Gilbert, Stanbury and Lewis, 'Birds of Conservation Concern in Ireland 4: 2020–2026'.

194. Kennedy, Ruttledge and Scroope, *Birds of Ireland.*

195. Ussher and Warren, *The Birds of Ireland.*

196. A. O'Donnell, 'Scarce Migrants in Ireland, 2007 and 2008', *Irish Birds*, vol. 9, 2012, pp. 421–46.

197. A. Whilde, 'A Survey of Gulls Breeding Inland in the West of Ireland in 1977 and 1978 and a Review of the Inland Breeding Habit in Ireland and Britain', *Irish Birds*, vol. 1, 1978, pp. 134–60; A. Whilde, D.C.F. Cotton and R. Sheppard, 'A Repeat Survey of Gulls Breeding Inland in Counties Donegal, Sligo, Mayo and Galway, with Recent Counts from Leitrim and Fermanagh', *Irish Birds*, vol. 5, 1993, pp. 67–72.

198. Lewis, Burke, Fitzgerald, Tierney and Kelly, 'Irish Wetland Bird Survey: Waterbird status and distribution 2009/10–2015/16'.

199. Whilde, 'A Survey of Gulls Breeding Inland in the West of Ireland in 1977 and 1978 and a Review of the Inland Breeding Habit in Ireland and Britain'.

200. Whilde, Cotton and Sheppard, 'A Repeat Survey of Gulls Breeding Inland in Counties Donegal, Sligo, Mayo and Galway, with Recent Counts from Leitrim and Fermanagh'.

201. Cummins, Lauder, Lauder and Tierney, 'The Status of Ireland's Breeding Seabirds: Birds Directive Article 12 Reporting 2013–2018'.

202. D. Cabot, 'Birds on Inishduff, Co. Donegal', *Irish Naturalists' Journal*, vol. 14, 1962, pp. 36–7.

203. Cramp, Bourne and Saunders, *The Seabirds of Britain and Ireland.*

204. Ussher and Warren, *The Birds of Ireland.*

205. Ibid.

206. Kennedy, Ruttledge and Scroope, *Birds of Ireland.*

207. A. McGeehan and R. Millington, 'The Adult Thayer's Gull in Donegal', *Birding World*, vol. 11, 1998, pp. 102–8.

208. Cramp, Bourne and Saunders, *The Seabirds of Britain and Ireland.*

209. Lloyd, 'Inventory of Seabird Breeding Colonies in the Republic of Ireland'.

210. Mitchell, Ratcliffe, Newton and Dunn, *Seabird Populations of Britain and Ireland: Results of the Seabird 2000 census (1998–2002).*

211. Ussher and Warren, *The Birds of Ireland.*

212. Kennedy, Ruttledge and Scroope, *Birds of Ireland.*

213. Whilde, 'A Survey of Gulls Breeding Inland in the West of Ireland in 1977 and 1978 and a Review of the Inland Breeding Habit in Ireland and Britain'; Whilde, Cotton and Sheppard, 'A Repeat Survey of Gulls Breeding Inland in Counties Donegal, Sligo, Mayo and Galway, with Recent Counts from Leitrim and Fermanagh'.

214. G.R. Hosey and F. Goodridge, 'Establishment of Territories in Two Species of Gull on Walney Island, Cumbria', *Bird Study*, vol. 27, 1980, pp. 73–80.

215. Hobbs, *A List of Irish Birds* (version 12.0).

216. Hart, 'Notes on the Birds of Donegal'.

217. Ussher and Warren, *The Birds of Ireland*.

218. Kennedy, Ruttledge and Scroope, *Birds of Ireland*.

219. C. Hannon, S.D. Berrow and S.F. Newton, 'The Status and Distribution of Breeding Sandwich *Sterna sandvicensis*, Roseate *S. dougallii*, Common *S. hirundo*, Arctic *S. paradisaea* and Little Terns *S. albifrons* in Ireland in 1995', *Irish Birds*, vol. 6, 1997, pp. 1–22.

220. Ussher and Warren, *The Birds of Ireland*.

221. Hart, 'Notes on the Birds of Donegal'.

222. R.K. Norman and D.R. Saunders, 'Status of Little Terns in Great Britain and Ireland in 1967', *British Birds*, vol. 62, 1969, pp. 4–13.

223. Cramp, Bourne and Saunders, *The Seabirds of Britain and Ireland*.

224. A. Whilde, 'The 1984 All Ireland Tern Survey', *Irish Birds*, vol. 3, 1985, pp. 1–32.

225. Hannon, Berrow and Newton, 'The Status and Distribution of Breeding Sandwich *Sterna sandvicensis*, Roseate *S. dougallii*, Common *S. hirundo*, Arctic *S. paradisaea* and Little Terns *S. albifrons* in Ireland in 1995'.

226. B.W. Tucker, 'Roseate Terns in Donegal', *British Birds*, vol. 34, 1941, p. 245.

227. Hannon, Berrow and Newton, 'The Status and Distribution of Breeding Sandwich *Sterna sandvicensis*, Roseate *S. dougallii*, Common *S. hirundo*, Arctic *S. paradisaea* and Little Terns *S. albifrons* in Ireland in 1995'.

228. Kennedy, Ruttledge and Scroope, *Birds of Ireland*.

229. Cramp, Bourne and Saunders, *The Seabirds of Britain and Ireland*; Whilde, 'The 1984 All Ireland Tern Survey'; Hannon, Berrow and Newton, 'The Status and Distribution of Breeding Sandwich *Sterna sandvicensis*, Roseate *S. dougallii*, Common *S. hirundo*, Arctic *S. paradisaea* and Little Terns *S. albifrons* in Ireland in 1995'.

230. Ussher and Warren, *The Birds of Ireland*.

231. Kennedy, Ruttledge and Scroope, *Birds of Ireland*.

232. Whilde, 'The 1984 All Ireland Tern Survey'; Hannon, Berrow and Newton, 'The Status and Distribution of Breeding Sandwich *Sterna sandvicensis*, Roseate *S. dougallii*, Common *S. hirundo*, Arctic *S. paradisaea* and Little Terns *S. albifrons* in Ireland in 1995'.

233. Ussher and Warren, *The Birds of Ireland*.

234. Kennedy, Ruttledge and Scroope, *Birds of Ireland*.

235. Ibid.

236. Thompson, *The Natural History of Ireland. Vol. 3: Birds*.

237. Ussher and Warren, *The Birds of Ireland*.

238. Ruttledge, *Ireland's Birds*; R.F. Ruttledge, 'Re-assessment of the Record of Long-tailed Skuas in the Shannon Valley in 1860', *Irish Birds*, vol. 2, 1982 pp. 197–8.

239. Hutchinson, *Birds in Ireland*.

240. Ussher and Warren, *The Birds of Ireland*.

241. Ibid.

242. Kennedy, Ruttledge and Scroope, *Birds of Ireland*.

243. Cramp, Bourne and Saunders, *The Seabirds of Britain and Ireland*.

244. Gilbert, Stanbury and Lewis, 'Birds of Conservation Concern in Ireland 4: 2020–2026'.

245. D'Arcy, *Ireland's Lost Birds*.

246. Lack, *The Atlas of Wintering Birds in Britain and Ireland*.

247. Ussher and Warren, *The Birds of Ireland*.

248. Redman, 'Birds of Tory Island, County Donegal in 1954', *Tory Island Bird Report 2015*.

249. Gilbert, Stanbury and Lewis, 'Birds of Conservation Concern in Ireland 4: 2020–2026'.

250. Ussher and Warren, *The Birds of Ireland*.

251. J. Cromie, 'Breeding Status of Red-throated Diver *Gavia stellata* in Ireland', *Irish Birds*, vol. 7, 2002, pp. 13–20.

252. D.R. Wilson, 'The Storm Petrel Colony on Roaninish', *Bird Study*, vol. 6, 1959, pp. 73–6; D. Cabot, 'Birds on Roaninish, Co. Donegal', *Irish Naturalists' Journal*, vol. 13, 1961, pp. 238–9.

253. D. Moloney, pers. comm.

254. Ussher and Warren, *The Birds of Ireland*.

255. Redman, 'Birds of Tory Island, County Donegal in 1954', *Tory Island Bird Report 2015*.

256. Cramp, Bourne and Saunders, *The Seabirds of Britain and Ireland*.

257. Lloyd, Tasker and Partridge, *The Status of Seabirds in Britain and Ireland*.

258. B. Bryce, pers. comm.

259. Hutchinson, *Birds in Ireland*.

260. R. Patterson, 'Fork-tailed Petrels in North of Ireland', *The Zoologist*, vol. 15, 1891, pp. 468–9.

261. H. Boyd, 'The "Wreck" of Leach's Petrels in the Autumn of 1952', *British Birds*, vol. 47, 1954, pp. 137–63.

262. B. Bryce, pers. comm.

263. Gilbert, Stanbury and Lewis, 'Birds of Conservation Concern in Ireland 4: 2020–2026'.

264. Kennedy, Ruttledge and Scroope, *Birds of Ireland*.

265. Cummins, Lauder, Lauder and Tierney, 'The Status of Ireland's Breeding Seabirds: Birds Directive Article 12 Reporting 2013–2018'.

266. Hobbs, *A List of Irish Birds* (version 12.0).

267. Gibbs, Nisbet and Redman, 'Birds of North Donegal in Autumn, 1953'; P.S. Redman, 'Status of the Sooty Shearwater in British Waters', unpublished report, 1955.

268. Thompson, *The Natural History of Ireland. Vol. 3: Birds*.

269. Redman, 'Birds of Tory Island, County Donegal in 1954', *Tory Island Bird Report 2015*.

270. Devlin and Merne, 'Malin Head Observatory Report 1961–1964'.

271. K. Colhoun, pers. comm.

272. Redman, 'Birds of Tory Island, County Donegal in 1954', *Tory Island Bird Report 2015*.

273. Gilbert, Stanbury and Lewis, 'Birds of Conservation Concern in Ireland 4: 2020–2026'.

274. Hart, 'Notes on the Birds of Donegal'.

275. Ussher and Warren, *The Birds of Ireland*.

276. R.A. Macdonald, 'The Breeding Population and Distribution of the Cormorant in Ireland', *Irish Birds*, vol. 3, 1987, pp. 405–16.

277. Lewis, Burke, Fitzgerald, Tierney and Kelly, 'Irish Wetland Bird Survey: Waterbird status and distribution 2009/10–2015/16'.

278. Crowe, *Ireland's Wetlands and Their Waterbirds: Status and distribution*.

279. Cramp, Bourne and Saunders, *The Seabirds of Britain and Ireland*.

280. Kennedy, Ruttledge and Scroope, *Birds of Ireland*.

281. Thompson, *The Natural History of Ireland. Vol. 2: Birds*.

282. Ussher and Warren, *The Birds of Ireland*.

283. Kennedy, Ruttledge and Scroope, *Birds of Ireland*.

284. Ussher and Warren, *The Birds of Ireland*.

285. R.M. Barrington, 'Little Bittern in Co. Donegal', *Irish Naturalist*, vol. 17, no. 3, 1908, p. 59.

286. Ussher and Warren, *The Birds of Ireland*.

287. P. Smiddy and B. Duffy, 'Little Egret *Egretta garzetta*: A new breeding bird for Ireland', *Irish Birds*, vol. 6, 1997, pp. 55–6.

288. Ussher and Warren, *The Birds of Ireland*.

289. Kennedy, Ruttledge and Scroope, *Birds of Ireland*.

290. Anon., 'Rare Buzzard Tracked', *Wings*, no. 23, 2001, p. 13.

291. W. Thompson, *The Natural History of Ireland. Vol. 1: Birds* (London, Reeve, Benham & Reeve, 1849).

292. Golden Eagle Trust, pers. comm.

293. Hart, 'Notes on the Birds of Donegal'.

294. Ussher and Warren, *The Birds of Ireland*.

295. C.J. Carroll, 'Extermination of the Golden Eagle in Ireland', *British Birds*, vol. 9, 1915, pp. 251–2; W.J. Williams, 'Golden Eagle and Marsh-Harrier in Ireland', *British Birds*, vol. 20, 1927, pp. 107–8.

296. L. O'Toole, pers. comm.

297. Gilbert, Stanbury and Lewis, 'Birds of Conservation Concern in Ireland 4: 2020–2026'.

298. Sharrock, *The Atlas of Breeding Birds in Britain and Ireland*; Gibbons, Reid and Chapman, *The New Atlas of Breeding Birds in Britain and Ireland*; Balmer, Gillings, Caffrey, Swann, Downie and Fuller, *Bird Atlas 2007–11*.

299. Ussher and Warren, *The Birds of Ireland*.

300. Ibid.

301. D.W. Norriss, J. Marsh, D. McMahon and G.A. Oliver, 'A National Survey of Breeding Hen Harriers *Cirus cyaneus* in Ireland 1998–2000', *Irish Birds*, vol. 7, 2002, pp. 1–12.

302. C. Barton, C. Pollock, D.W. Norriss, T. Nagle, G.A. Oliver and S. Newton, 'The Second National Survey of Breeding Hen Harriers *Circus cyaneus* in Ireland', *Irish Birds*, vol. 8, 2005, pp. 1–20.

303. M. Ruddock, A. Mee, J. Lusby, A. Nagle, S. O'Neill and L. O'Toole, 'The 2015 National Survey of Breeding Hen Harrier in Ireland', *Irish Wildlife Manuals*, no. 93 (Dublin: National Parks and Wildlife Service, 2016).

304. A. Day and D. McWilliams (eds), *Ordnance Survey Memoirs of Ireland. Volume 39: Parishes of County Donegal II, 1835–6, Mid, West and South Donegal* (Belfast and Dublin: The Institute of Irish Studies in association with The Royal Irish Academy 1997).

305. Day and McWilliams (eds), *Ordnance Survey Memoirs of Ireland. Volume 38: Parishes of County Donegal I, 1833–5, North-east Donegal.*

306. Gilbert, Stanbury and Lewis, 'Birds of Conservation Concern in Ireland 4: 2020–2026'.

307. D'Arcy, *Ireland's Lost Birds.*

308. Ussher and Warren, *The Birds of Ireland.*

309. Gilbert, Stanbury and Lewis, 'Birds of Conservation Concern in Ireland 4: 2020–2026'.

310. Ussher and Warren, *The Birds of Ireland.*

311. Kennedy, Ruttledge and Scroope, *Birds of Ireland.*

312. Thompson, *The Natural History of Ireland. Vol. 1: Birds.*

313. Hart, 'Notes on the Birds of Donegal'.

314. Sharrock, *The Atlas of Breeding Birds in Britain and Ireland.*

315. D.W. Norriss, 'The Status of the Buzzard as a Breeding Species in the Republic of Ireland, 1977–1991', *Irish Birds*, vol. 4, 1991, pp. 291–8.

316. Balmer, Gillings, Caffrey, Swann, Downie and Fuller, *Bird Atlas 2007–11.*

317. Irish Raptor Study Group, '2001 Annual Roundup', unpublished report.

318. Sharrock, *The Atlas of Breeding Birds in Britain and Ireland.*

319. Gibbons, Reid and Chapman, *The New Atlas of Breeding Birds in Britain and Ireland*; Balmer, Gillings, Caffrey, Swann, Downie and Fuller, *Bird Atlas 2007–11.*

320. Gilbert, Stanbury and Lewis, 'Birds of Conservation Concern in Ireland 4: 2020–2026'.

321. Hart, 'Notes on the Birds of Donegal'; Ussher and Warren, *The Birds of Ireland.*

322. Thompson, *The Natural History of Ireland. Vol. 1: Birds.*

323. Gilbert, Stanbury and Lewis, 'Birds of Conservation Concern in Ireland 4: 2020–2026'.

324. Ussher and Warren, *The Birds of Ireland.*

325. Kennedy, Ruttledge and Scroope, *Birds of Ireland.*

326. O. Crowe, S. Cummins, N. Gilligan, P. Smiddy and T.D. Tierney, 'An Assessment of the Current Distribution and Status of the Kingfisher *Alcedo atthis* in Ireland', *Irish Birds*, vol. 9, 2010, pp. 41–54.

327. Ussher and Warren, *The Birds of Ireland.*

328. Hutchinson, *Birds in Ireland.*

329. Hobbs, *A List of Irish Birds* (version 12.0).

330. Anon., *Natural History Review*, vol. 1, 1854, p. 91.

331. Ibid., p. 95.

332. Ussher and Warren, *The Birds of Ireland.*

333. Kennedy, Ruttledge and Scroope, *Birds of Ireland.*

334. Sharrock, *The Atlas of Breeding Birds in Britain and Ireland*; Gibbons, Reid and Chapman, *The New Atlas of Breeding Birds in Britain and Ireland.*

335. Balmer, Gillings, Caffrey, Swann, Downie and Fuller, *Bird Atlas 2007–11*.

336. Gilbert, Stanbury and Lewis, 'Birds of Conservation Concern in Ireland 4: 2020–2026'.

337. Hobbs, *A List of Irish Birds* (version 12.0).

338. Cox, Eddleston and Newton, 'Upland Bird Survey Report 2002: Donegal', BirdWatch Ireland Conservation Report no. 02/4.

339. D.W. Norriss, B. Haran, J. Hennigan, A. McElheron, D.J. McLaughlin, V. Swan and A. Walsh, 'Breeding Biology of Merlins *Falco columbarius* in Ireland, 1986–1992', *Irish Birds*, vol. 9, 2010, pp. 23–30.

340. Ussher and Warren, *The Birds of Ireland*.

341. J. Temple Lang, 'Peregrine Survey – Republic of Ireland, 1967–1968', *Irish Bird Report*, no. 16, 1968, pp. 8–12; J. Temple Lang, 'Peregrine', *Irish Bird Report*, no. 17, 1969, p. 33; J. Temple Lang, 'Peregrine', *Irish Bird Report*, no. 18, 1970, p. 28.

342. D.W. Norriss and H.J. Wilson, 'Survey of the Peregrine *Falco peregrinus* breeding population in the Republic of Ireland 1981', *Bird Study*, vol. 30, 1983, pp. 91–101.

343. D.W. Norriss, 'The 1991 Survey and Weather Impacts on the Peregrine *Falco peregrinus* Breeding Population in the Republic of Ireland', *Bird Study*, vol. 42, 1995, pp. 20–30; B. Madden, J. Hunt and D. Norriss, 'The 2002 Survey of the Peregrine *Falco peregrinus* Breeding Population in the Republic of Ireland', *Irish Birds*, vol. 8, 2009, pp. 543–8.

344. J. Baird, P. Harrison, reported by E. Masterson, 'Irish Rare Bird Report 2005', *Irish Birds*, vol. 8, 2008, p. 390.

345. Ussher and Warren, *The Birds of Ireland*.

346. Ibid.

347. Kennedy, Ruttledge and Scroope, *Birds of Ireland*.

348. Gibbons, Reid and Chapman, *The New Atlas of Breeding Birds in Britain and Ireland*.

349. Ussher and Warren, *The Birds of Ireland*.

350. D. Cabot, 'The Status and Distribution of the Chough, *Pyrrhocorax pyrrhocorax* (L.) in Ireland, 1960–65', *Irish Naturalists' Journal*, vol. 15, 1965, pp. 95–100; I.D. Bullock, D.R. Drewett and S.P. Mickleburgh, 'The Chough in Ireland', *Irish Birds*, vol. 2, 1983, pp. 257–71; S.D. Berrow, K.L. Mackie, O. O'Sullivan, K.B. Shepherd, C. Mellon and J.A. Coveney, 'The Second International Chough Survey in Ireland, 1992', *Irish Birds*, vol. 5, 1993, pp. 1–10; N. Gray, G. Thomas, M. Trewby and S.F. Newton, 'The Status and Distribution of Choughs *Pyrrhocorax pyrrhocorax* in the Republic of Ireland 2002/03', *Irish Birds*, vol. 7, 2003, pp. 147–56.

351. N. Gray, M. Trewby, S. Cummins and S. Newton, 'The Distribution and Feeding Ecology of the Chough *Pyrrhocorax pyrrhocorax* in Co. Donegal: September 2004 – August 2005', *Conservation Report,* BirdWatch Ireland, 2005.

352. Ussher and Warren, *The Birds of Ireland*.

353. Wernham, Toms, Marchant, Clark, Sitiwardena and Baillie, *The Migration Atlas*.

354. Lewis, Coombes, Burke, O'Halloran, Walsh, Tierney and Cummins, 'Countryside Bird Survey: Status and trends of common and widespread breeding birds 1998–2016'.

355. Kennedy, Ruttledge and Scroope, *Birds of Ireland*.

356. Lewis, Coombes, Burke, O'Halloran, Walsh, Tierney and Cummins, 'Countryside Bird Survey: Status and trends of common and widespread breeding birds 1998–2016'.

357. Balmer, Gillings, Caffrey, Swann, Downie and Fuller, *Bird Atlas 2007–11*.

358. Lack, *The Atlas of Wintering Birds in Britain and Ireland*.

359. Lewis, Coombes, Burke, O'Halloran, Walsh, Tierney and Cummins, 'Countryside Bird Survey: Status and trends of common and widespread breeding birds 1998–2016'.

360. Ibid.

361. A.S. Copland, O. Crowe, M.W. Wilson and J. O'Halloran, 'Habitat Associations of Eurasian Skylarks *Alauda arvensis* breeding on Irish Farmland and Implications for Agri-environment Planning', *Bird Study*, vol. 59, 2012, pp. 155–65.

362. Cox, Eddleston and Newton, 'Upland Bird Survey Report 2002: Donegal', BirdWatch Ireland Conservation Report no. 02/4.

363. Kennedy, Ruttledge and Scroope, *Birds of Ireland*.

364. Hobbs, *A List of Irish Birds* (version 12.0).

365. Balmer, Gillings, Caffrey, Swann, Downie and Fuller, *Bird Atlas 2007–11*.

366. Ibid.

367. Ussher and Warren, *The Birds of Ireland*.

368. Lewis, Coombes, Burke, O'Halloran, Walsh, Tierney and Cummins, 'Countryside Bird Survey: Status and trends of common and widespread breeding birds 1998–2016'.

369. Ibid.

370. Gibbons, Reid and Chapman, *The New Atlas of Breeding Birds in Britain and Ireland*.

371. Lewis, Coombes, Burke, O'Halloran, Walsh, Tierney and Cummins, 'Countryside Bird Survey: Status and trends of common and widespread breeding birds 1998–2016'.

372. Gilbert, Stanbury and Lewis, 'Birds of Conservation Concern in Ireland 4: 2020–2026'.

373. Ibid.

374. K. Mullarney (on behalf of the Irish Records Panel), 'Review of Irish Records of Greenish Warblers', *Irish Birds*, vol. 2, 1984, pp. 536–45.

375. Hobbs, *A List of Irish Birds* (version 12.0).

376. Ibid.

377. Wallace, McGeehan and Allen, 'Autumn Migration in Westernmost Donegal'.

378. Lewis, Coombes, Burke, O'Halloran, Walsh, Tierney and Cummins, 'Countryside Bird Survey: Status and trends of common and widespread breeding birds 1998–2016'.

379. Ussher and Warren, *The Birds of Ireland*.

380. Sharrock, *The Atlas of Breeding Birds in Britain and Ireland*.

381. Perry, *The Birds of the Inishowen Peninsula*.

382. Gibbons, Reid and Chapman, *The New Atlas of Breeding Birds in Britain and Ireland*; Balmer, Gillings, Caffrey, Swann, Downie and Fuller, *Bird Atlas 2007–11*.

383. Lewis, Coombes, Burke, O'Halloran, Walsh, Tierney and Cummins, 'Countryside Bird Survey: Status and trends of common and widespread breeding birds 1998–2016'.

384. Lack, *The Atlas of Wintering Birds in Britain and Ireland*.

385. N. Doherty, pers. comm.

386. I.J. Herbert, 'The Status of Garden Warbler at Crom Estate, Co. Fermanagh, and a Review of Its Status in Ireland', *Irish Birds*, vol. 4, 1991, pp. 369–76.

387. Ussher and Warren, *The Birds of Ireland*.

388. D. Winstanley, R. Spencer and K. Williamson, 'Where Have All the Whitethroats Gone?', *Bird Study*, vol. 21, 1974, pp. 1–14.

389. Sharrock, *The Atlas of Breeding Birds in Britain and Ireland*.

390. Balmer, Gillings, Caffrey, Swann, Downie and Fuller, *Bird Atlas 2007–11*.

391. Lewis, Coombes, Burke, O'Halloran, Walsh, Tierney and Cummins, 'Countryside Bird Survey: Status and trends of common and widespread breeding birds 1998–2016'.

392. Hutchinson, *Birds in Ireland*.

393. Lewis, Coombes, Burke, O'Halloran, Walsh, Tierney and Cummins, 'Countryside Bird Survey: Status and trends of common and widespread breeding birds 1998–2016'.

394. Cox, Eddleston and Newton, 'Upland Bird Survey Report 2002: Donegal', BirdWatch Ireland Conservation Report no. 02/4.

395. Lewis, Coombes, Burke, O'Halloran, Walsh, Tierney and Cummins, 'Countryside Bird Survey: Status and trends of common and widespread breeding birds 1998–2016'.

396. Balmer, Gillings, Caffrey, Swann, Downie and Fuller, *Bird Atlas 2007–11*.

397. Lewis, Coombes, Burke, O'Halloran, Walsh, Tierney and Cummins, 'Countryside Bird Survey: Status and trends of common and widespread breeding birds 1998–2016'.

398. Ussher and Warren, *The Birds of Ireland*.

399. Lewis, Coombes, Burke, O'Halloran, Walsh, Tierney and Cummins, 'Countryside Bird Survey: Status and trends of common and widespread breeding birds 1998–2016'.

400. Lack, *The Atlas of Wintering Birds in Britain and Ireland*.

401. Ibid.

402. B. Madden, 'A Remarkable Nocturnal Thrush Migration', *Irish East Coast Bird Report*, 1992, pp. 71–3.

403. Hobbs, *A List of Irish Birds* (version 12.0).

404. Devlin and Merne, 'Malin Head Observatory Report 1961–1964'.

405. Ussher and Warren, *The Birds of Ireland*.

406. Lewis, Coombes, Burke, O'Halloran, Walsh, Tierney and Cummins, 'Countryside Bird Survey: Status and trends of common and widespread breeding birds 1998–2016'.

407. Gilbert, Stanbury and Lewis, 'Birds of Conservation Concern in Ireland 4: 2020–2026'.

408. Lack, *The Atlas of Wintering Birds in Britain and Ireland*.

409. Hutchinson, *Birds in Ireland*.

410. J. Wann and R. Sheppard, *Pilot Ecological Study of Two Donegal Islands: Inishfree Upper and Inishmeane*, unpublished report for Donegal County Council: Aulino Wann and Associates, 2010.

411. P. Phillips, 'The Breeding Birds of Tory', *Tory Island Bird Report 2013* (privately published, 2014).

412. Lewis, Coombes, Burke, O'Halloran, Walsh, Tierney and Cummins, 'Countryside Bird Survey: Status and trends of common and widespread breeding birds 1998–2016'.

413. Lack, *The Atlas of Wintering Birds in Britain and Ireland*.

414. Balmer, Gillings, Caffrey, Swann, Downie and Fuller, *Bird Atlas 2007–11*.

415. Kennedy, Ruttledge and Scroope, *Birds of Ireland*.

416. Hart, 'Notes on the Birds of Donegal'.

417. Ussher and Warren, *The Birds of Ireland*.

418. Sharrock, *The Atlas of Breeding Birds in Britain and Ireland*.

419. Balmer, Gillings, Caffrey, Swann, Downie and Fuller, *Bird Atlas 2007–11*.

420. Cox, Eddleston and Newton, 'Upland Bird Survey Report 2002: Donegal', BirdWatch Ireland Conservation Report no. 02/4.

421. A. Mee, 'The Status and Ecology of a Remnant Population of Ring Ouzel *Turdus torquatus* in the MacGillyguddy's Reeks, Kerry', *Irish Birds*, vol. 11, 2018, pp. 13–22.

422. A. McGeehan and J. Wyllie, *Birds through Irish Eyes* (Cork: The Collins Press, 2012).

423. A. Whilde, *Threatened Mammals, Birds, Amphibians and Fish in Ireland. Irish Red Data Book 2: Vertebrates* (Belfast: Her Majesty's Stationery Office, 1993)

424. S.F. Newton, A. Donaghy, D. Allen and D. Gibbons, 'Birds of Conservation Concern in Ireland', *Irish Birds*, vol. 6, 1999, pp. 333–42; P. Lynas, S.F. Newton and J.A. Robinson, 'The Status of Birds in Ireland: An analysis of conservation concern 2008–2013', *Irish Birds*, vol. 8, 2007, pp. 149–66; K. Colhoun and S.T. Cummins, 'Birds of Conservation Concern in Ireland 2014–2019', *Irish Birds*, vol. 9, pp. 523–44.

425. Gilbert, Stanbury and Lewis, 'Birds of Conservation Concern in Ireland 4: 2020–2026'.

426. Lewis, Coombes, Burke, O'Halloran, Walsh, Tierney and Cummins, 'Countryside Bird Survey: Status and trends of common and widespread breeding birds 1998–2016'.

427. Gilbert, Stanbury and Lewis, 'Birds of Conservation Concern in Ireland 4: 2020–2026'.

428. Sharrock, *The Atlas of Breeding Birds in Britain and Ireland*; Balmer, Gillings, Caffrey, Swann, Downie and Fuller, *Bird Atlas 2007–11*.

429. S.E. Newson, D.I. Leech, C.M Hewson, H.Q.P. Crick and P.V. Grice, 'Potential Impact of Grey Squirrels *Sciurus carolinensis* on Woodland Bird Populations in England', *Journal of Ornithology*, vol. 151, 2010, pp. 211–18; D.K. Stevens, G.Q.A. Anderson, P.V. Grice, K. Norris and N. Butcher, 'Predators of Spotted Flycatcher *Muscicapa striata* Nests in Southern England as Determined by Digital Nest-cameras', *Bird Study*, vol. 55, 2008, pp. 179–87.

430. Lewis, Coombes, Burke, O'Halloran, Walsh, Tierney and Cummins, 'Countryside Bird Survey: Status and trends of common and widespread breeding birds 1998–2016'.

431. Kennedy, Ruttledge and Scroope, *Birds of Ireland*.

432. Gilbert, Stanbury and Lewis, 'Birds of Conservation Concern in Ireland 4: 2020–2026'.

433. Hart, 'Notes on the Birds of Donegal'.

434. Sharrock, *The Atlas of Breeding Birds in Britain and Ireland*; Gibbons, Reid and Chapman, *The New Atlas of Breeding Birds in Britain and Ireland*; Balmer, Gillings, Caffrey, Swann, Downie and Fuller, *Bird Atlas 2007–11*.

435. Gilbert, Stanbury and Lewis, 'Birds of Conservation Concern in Ireland 4: 2020–2026'.

436. Cox, Eddleston and Newton, 'Upland Bird Survey Report 2002: Donegal', BirdWatch Ireland Conservation Report no. 02/4.

437. Lewis, Coombes, Burke, O'Halloran, Walsh, Tierney and Cummins, 'Countryside Bird Survey: Status and trends of common and widespread breeding birds 1998–2016'.

438. Cox, Eddleston and Newton, 'Upland Bird Survey Report 2002: Donegal', BirdWatch Ireland Conservation Report no. 02/4.

439. Lewis, Coombes, Burke, O'Halloran, Walsh, Tierney and Cummins, 'Countryside Bird Survey: Status and trends of common and widespread breeding birds 1998–2016'.

440. K.W. Perry and P. Agnew, 'Breeding Dipper Populations in North-west Ireland', *Irish Birds*, vol. 5, 1993, pp. 45–8.

441. K.W. Perry, 'Population Changes of Dippers in North-West Ireland', *Irish Birds*, vol. 2, 1983, pp. 272–7.

442. Sharrock, *The Atlas of Breeding Birds in Britain and Ireland*; Gibbons, Reid and Chapman, *The New Atlas of Breeding Birds in Britain and Ireland*; Balmer, Gillings, Caffrey, Swann, Downie and Fuller, *Bird Atlas 2007–11*.

443. Lack, *The Atlas of Wintering Birds in Britain and Ireland*.

444. Balmer, Gillings, Caffrey, Swann, Downie and Fuller, *Bird Atlas 2007–11*.

445. Lewis, Coombes, Burke, O'Halloran, Walsh, Tierney and Cummins, 'Countryside Bird Survey: Status and trends of common and widespread breeding birds 1998–2016'.

446. Hutchinson, *Birds in Ireland*.

447. W.B. Alexander, 'The Tree-Sparrow in Arranmore Co. Donegal', *British Birds*, vol. 34, 1940, pp. 107–8; Kennedy, Ruttledge and Scroope, *Birds of Ireland*.

448. T. Mackie, *Frank Egginton* (privately published, 2022), p. 52.

449. R. Sheppard, 'Hybrid Tree x House Sparrows in County Donegal', *Irish Birds*, vol. 5, 1995, pp. 319–20.

450. Lewis, Coombes, Burke, O'Halloran, Walsh, Tierney and Cummins, 'Countryside Bird Survey: Status and trends of common and widespread breeding birds 1998–2016'.

451. Redman, 'Birds of Tory Island, County Donegal in 1954', *Tory Island Bird Report 2015*; Irish Rare Birds Committee, 'Irish Rare Bird Report 2016'.

452. Cox, Eddleston and Newton, 'Upland Bird Survey Report 2002: Donegal', BirdWatch Ireland Conservation Report no. 02/4.

453. Lewis, Coombes, Burke, O'Halloran, Walsh, Tierney and Cummins, 'Countryside Bird Survey: Status and trends of common and widespread breeding birds 1998–2016'.

454. Lack, *The Atlas of Wintering Birds in Britain and Ireland*; Balmer, Gillings, Caffrey, Swann, Downie and Fuller, *Bird Atlas 2007–11*.

455. P. Smiddy and J. O'Halloran, 'Breeding Biology of the Grey Wagtail *Motacilla cinerea* in Southwest Ireland', *Bird Study*, vol. 45, 1998, pp. 331–6.

456. Gilbert, Stanbury and Lewis, 'Birds of Conservation Concern in Ireland 4: 2020–2026'.

457. Balmer, Gillings, Caffrey, Swann, Downie and Fuller, *Bird Atlas 2007–11*.

458. Lewis, Coombes, Burke, O'Halloran, Walsh, Tierney and Cummins, 'Countryside Bird Survey: Status and trends of common and widespread breeding birds 1998–2016'.

459. E.M. Nicholson, 'Richard's Pipit in Donegal', *British Birds*, vol. 49, 1956, pp. 44–5.

460. Phillips, 'The Breeding Birds of Tory', *Tory Island Bird Report 2013*.

461. Cox, Eddleston and Newton, 'Upland Bird Survey Report 2002: Donegal', BirdWatch Ireland Conservation Report no. 02/4.

462. Lewis, Coombes, Burke, O'Halloran, Walsh, Tierney and Cummins, 'Countryside Bird Survey: Status and trends of common and widespread breeding birds 1998–2016'.

463. Hutchinson, *Birds in Ireland*.

464. Lewis, Coombes, Burke, O'Halloran, Walsh, Tierney and Cummins, 'Countryside Bird Survey: Status and trends of common and widespread breeding birds 1998–2016'.

465. R. Howard and A. Moore, *A Complete Checklist of the Birds of the World* (London: Macmillan, 1984).

466. S. Cramp and K.E.L. Simmons (eds), *Handbook of the Birds of Europe, the Middle East and North Africa: The birds of the western Palearctic* (vols 1–3); S. Cramp (ed.) (vols 4–6); S. Cramp and C.M. Perrins (eds) (vols 7–9) (Oxford: Oxford University Press, 1977–94); del Hoyo, Elliott and Sargatal (cds), *Handbook of the Birds of the World*.

467. Tory Bird Observatory, 'Tory Island 1960', unpublished report, 1960, pp. 1–17.

468. Gilbert, Stanbury and Lewis, 'Birds of Conservation Concern in Ireland 4: 2020–2026'.

469. Redman, 'Birds of Tory Island, County Donegal in 1954', *Tory Island Bird Report 2015*.

470. Tory Bird Observatory, 'Tory Island 1961', unpublished report, 1961, pp. 1–19.

471. Lewis, Coombes, Burke, O'Halloran, Walsh, Tierney and Cummins, 'Countryside Bird Survey: Status and trends of common and widespread breeding birds 1998–2016'.

472. Hart, 'Notes on the Birds of Donegal'.

473. Thompson, *The Natural History of Ireland. Vol. 1: Birds*.

474. Lewis, Coombes, Burke, O'Halloran, Walsh, Tierney and Cummins, 'Countryside Bird Survey: Status and trends of common and widespread breeding birds 1998–2016'.

475. Sharrock, *The Atlas of Breeding Birds in Britain and Ireland*; Gibbons, Reid and Chapman, *The New Atlas of Breeding Birds in Britain and Ireland*; Balmer, Gillings, Caffrey, Swann, Downie and Fuller, *Bird Atlas 2007–11*.

476. K. Fahy, 'Irish Rare Bird Report 2010', *Irish Birds*, vol. 9, 2011, p. 311.

477. Ussher and Warren, *The Birds of Ireland*.

478. Kennedy, Ruttledge and Scroope, *Birds of Ireland*.

479. Lewis, Coombes, Burke, O'Halloran, Walsh, Tierney and Cummins, 'Countryside Bird Survey: Status and trends of common and widespread breeding birds 1998–2016'.

480. Devlin and Merne, 'Malin Head Observatory Report 1961–1964'; Devlin and Merne, 'Malin Head Observatory Report 1965'.

481. Sharrock, *The Atlas of Breeding Birds in Britain and Ireland*.

482. Ussher and Warren, *The Birds of Ireland*.

483. Redman, 'Birds of Tory Island, County Donegal in 1954', *Tory Island Bird Report 2015*.

484. Tory Bird Observatory, 'Tory Island 1962', unpublished report, 1962, pp. 1–18.

485. Devlin, and Merne, 'Malin Head Observatory Report 1965'.

486. Sharrock, *The Atlas of Breeding Birds in Britain and Ireland*.

487. Balmer, Gillings, Caffrey, Swann, Downie and Fuller, *Bird Atlas 2007–11*.

488. D.T. McLoughlin and D. Cotton, 'The Status of Twite *Carduelis flavirostris* in Ireland 2008', *Irish Birds*, vol. 8, 2008, pp. 323–30; D.T. McLoughlin, 'Twite on the Edge', *Wings*, no. 57, 2010, pp. 15–17; D.T. McLoughlin, C. Benson, B. Williams and D.C. Cotton, 'The Movement Patterns of Two Populations of Twites *Carduelis flavirostris* in Ireland', *Ringing and Migration*, vol. 25, 2010, pp. 15–21.

489. D.T. McLoughlin, 'Management Prescriptions for Twite in Ireland', *Irish Wildlife Manuals*, no. 52 (Dublin: National Parks and Wildlife Service, 2011).

490. Tory Bird Observatory, 'Tory Island 1960'.

491. Devlin and Merne, 'Malin Head Observatory Report 1965'.

492. Gilbert, Stanbury and Lewis, 'Birds of Conservation Concern in Ireland 4: 2020–2026'.

493. Ussher and Warren, *The Birds of Ireland*.

494. Lewis, Coombes, Burke, O'Halloran, Walsh, Tierney and Cummins, 'Countryside Bird Survey: Status and trends of common and widespread breeding birds 1998–2016'.

495. Tory Bird Observatory, 'Tory Island 1963 and 1964', unpublished report, 1964, pp. 1–49.

496. Ussher and Warren, *The Birds of Ireland*.

497. Kennedy, Ruttledge and Scroope, *Birds of Ireland*.

498. Ibid.

499. Hobbs, *A List of Irish Birds* (version 12.0).

500. Cox, Eddleston and Newton, 'Upland Bird Survey Report 2002: Donegal', BirdWatch Ireland Conservation Report no. 02/4.

501. M.W. Wilson, J. Pithon, T. Gittins, T.C. Kelly, S.G. Giller and J. O'Halleron, 'Effects of Growth Stage and Tree Species Composition on Breeding Bird Assemblages of Plantation Forests', *Bird Study*, vol. 53, 2006, pp. 225–36.

502. Lewis, Coombes, Burke, O'Halloran, Walsh, Tierney and Cummins, 'Countryside Bird Survey: Status and trends of common and widespread breeding birds 1998–2016'.

503. Hobbs, *A List of Irish Birds* (version 12.0).

504. Hutchinson, *Birds in Ireland*.

505. Hart, 'Notes on the Birds of Donegal'.

506. Ussher and Warren, *The Birds of Ireland*.

507. Sharrock, *The Atlas of Breeding Birds in Britain and Ireland*.

508. Balmer, Gillings, Caffrey, Swann, Downie and Fuller, *Bird Atlas 2007–11*.

509. Lewis, Coombes, Burke, O'Halloran, Walsh, Tierney and Cummins, 'Countryside Bird Survey: Status and trends of common and widespread breeding birds 1998–2016'.

510. Ruttledge, *Ireland's Birds*.

511. Redman, 'Birds of Tory Island, County Donegal in 1954', *Tory Island Bird Report 2015*.

512. Gibbs, Nisbet and Redman, 'Birds of North Donegal in Autumn, 1953'.

513. Tory Bird Observatory, 'Tory Island 1963 and 1964'.

514. M.G. Pennington, R. Riddington and W.T.S. Miles, 'The Lapland Bunting Influx in Britain and Ireland in 2010/11', *British Birds*, vol. 105, 2012, pp. 654–73.

515. Ussher and Warren, *The Birds of Ireland*.

516. Hutchinson, *Birds in Ireland*.

517. Tory Bird Observatory, 'Tory Island 1958 and 1959', unpublished report, 1959, pp. 1–13.

518. Phillips, Ingram, and Salter, *Tory Island Bird Report 2016*.

519. P.F. Donald, J.D. Wilson and M. Shepherd, 'The Decline of the Corn Bunting', *British Birds*, vol. 87, 1994, pp. 106–32.

520. Taylor and O'Halloran, 'The Decline of the Corn Bunting, *Miliaria calandra*, in the Republic of Ireland'.

521. Hart, 'Notes on the Birds of Donegal'.

522. Ussher and Warren, *The Birds of Ireland*.

523. Kennedy, Ruttledge and Scroope, *Birds of Ireland*.

524. Redman, 'Birds of Tory Island, County Donegal in 1954', *Tory Island Bird Report 2015*.

525. Devlin and Merne, 'Malin Head Observatory Report 1965'.

526. Ruttledge, *A List of the Birds of Ireland*.

527. Kennedy, Ruttledge and Scroope, *Birds of Ireland*.

528. Hart, 'Notes on the Birds of Donegal'.

529. Ussher and Warren, *The Birds of Ireland*.

530. Kennedy, Ruttledge and Scroope, *Birds of Ireland*.

531. Sharrock, *The Atlas of Breeding Birds in Britain and Ireland*.

532. Gibbons, Reid and Chapman, *The New Atlas of Breeding Birds in Britain and Ireland*.

533. Balmer, Gillings, Caffrey, Swann, Downie and Fuller, *Bird Atlas 2007–11*.

534. Gilbert, Stanbury and Lewis, 'Birds of Conservation Concern in Ireland 4: 2020–2026'.

535. Kennedy, Ruttledge and Scroope, *Birds of Ireland*.

536. Tory Bird Observatory, 'Tory Island 1961'.

537. H. Cooke, 'Yellow-breasted Bunting in Co. Donegal', *British Birds*, vol. 53, 1960, p. 229.

538. Phillips, 'The Breeding Birds of Tory', *Tory Island Bird Report 2013*.

539. Balmer, Gillings, Caffrey, Swann, Downie and Fuller, *Bird Atlas 2007–11*.

540. Lewis, Coombes, Burke, O'Halloran, Walsh, Tierney and Cummins, 'Countryside Bird Survey: Status and trends of common and widespread breeding birds 1998–2016'.

BIBLIOGRAPHY

Alexander, W.B., 'The Tree-Sparrow in Arranmore Co. Donegal', *British Birds*, vol. 34, 1940, pp. 107–8

Anon., 'Scaup on the Slide', *Waterbirds in the UK 2019/20, the Annual Report of the Wetland Bird Survey*, 2021

——, 'Proceedings of the Dublin University Zoological Association, 11 February 1854', *Natural History Review*, vol. 1, 1854, p. 91

——, 'Proceedings of the Dublin University Zoological Association, 4 March 1854', *Natural History Review*, vol. 1, 1854, p. 95

——, 'Rare Buzzard Tracked', *Wings*, no. 23, 2001, p. 13

Baird, J., Harrison, P., *per.* Masterson, E., 'Irish Rare Bird Report 2005', *Irish Birds*, vol. 8, 2008, p. 390

Balmer, D.E., Gillings, S., Caffrey, B.J., Swann, R.L., Downie, I.S. and Fuller, R.J., *Bird Atlas 2007–11: The breeding and wintering birds of Britain and Ireland* (Thetford: BTO Books, 2013)

Barrington, R.M., *The Migration of Birds* (London: Porter, and Dublin: Ponsonby, 1900)

Barrington, R.M., 'Little Bittern in Co. Donegal', *Irish Naturalist*, vol. 17, no. 3, 1908, p. 59

Barton, C., Pollock, C., Norriss, D.W., Nagle, T., Oliver, G.A. and Newton, S., 'The Second National Survey of Breeding Hen Harriers *Circus cyaneus* in Ireland', *Irish Birds*, vol. 8, 2005, pp. 1–20

Baynes, E.A.S., 'The *Lepidoptera* of Tory Island, Co. Donegal', *The Entomologist*, vol. 90, 1957, pp. 310–13

Bell, M.K., 'Breeding Wader Survey at Selected Sites in Cos Donegal and Sligo', BirdWatch Ireland, unpublished report, 2019

——, 'Upland Breeding Wader Survey Report Cos Donegal and Sligo', BirdWatch Ireland, unpublished report, 2020

——, 'Breeding Wader Survey at Selected Sites in Donegal and Sligo', BirdWatch Ireland, unpublished report, 2021

——, 'Breeding Wader Survey at Selected Sites in Donegal and Sligo', BirdWatch Ireland, unpublished report, 2022

——, and Flynn, K., 'Breeding Wader Survey at Selected Sites in Cos Donegal and Sligo', BirdWatch Ireland, unpublished report, 2020

Berrow, S.D., Mackie, K.L., O'Sullivan, O., Shepherd, K.B., Mellon, C. and Coveney, J.A., 'The Second International Chough Survey in Ireland, 1992', *Irish Birds*, vol. 5, 1993, pp. 1–10

BirdLife International, *State of the World's Birds 2022: Insights and solutions for the biodiversity crisis* (Cambridge: BirdLife International, 2022)

Boyd, H., 'The "Wreck" of Leach's Petrels in the Autumn of 1952', *British Birds*, vol. 47, 1954, pp. 137–63

Bullock, I.D., Drewett, D.R. and Mickleburgh, S.P., 'The Chough in Ireland', *Irish Birds*, vol. 2, 1983, pp. 257–71

Burke, B., Crowe, O. and Newton S.F., 'Rare and Scarce Breeding Birds in Ireland in 2017 and 2018', *Irish Birds*, no. 42, 2020, pp. 63–70

———, Lewis, L.J., Fitzgerald, N., Frost, T., Austin, G. and Tierney, T.D., 'Estimates of Waterbird Numbers Wintering in Ireland 2011/12 – 2015/16', *Irish Birds*, no. 41, 2018, pp. 1–12

———, McElwaine, J.G., Fitzgerald, N., Kelly, S.B.A., McCulloch, N., Walsh, A.J. and Lewis, L.J., 'Population Size, Breeding Success and Habitat Use of Whooper Swan *Cygnus cygnus* and Bewick's Swan *Cygnus columbianus bewickii* in Ireland: Results of the 2020 International Swan Census', *Irish Birds*, no. 43, 2021, pp. 57–70

Cabot, D., 'Birds on Roaninish, Co. Donegal', *Irish Naturalists' Journal*, vol. 13, 1961, pp. 238–9

———, 'Birds on Inishduff, Co. Donegal', *Irish Naturalists' Journal*, vol. 14, 1962, pp. 36–7

———, 'The Status and Distribution of the Chough, *Pyrrhocorax pyrrhocorax* (L.), in Ireland, 1960–65', *Irish Naturalists' Journal*, vol. 15, 1965, pp. 95–100

Campbell, L., *Room for the River – the Foyle River Catchment Landscape: Connecting people, place and nature* (Buncrana: Merdog Books, 2021)

Carroll, C.J., 'Extermination of the Golden Eagle in Ireland', *British Birds*, vol. 9, 1915, pp. 251–2

Colhoun, K. and Cummins, S., 'Birds of Conservation Concern in Ireland 2014–2019', *Irish Birds*, vol. 9, pp. 523–44

Cooke, H., 'Yellow-breasted Bunting in Co. Donegal', *British Birds*, vol. 53, 1960, p. 229

Copland, A.S., Crowe, O., Wilson, M.W. and O'Halloran, J., 'Habitat Associations of Eurasian Skylarks *Alauda arvensis* breeding on Irish Farmland and Implications for Agri-environment Planning', *Bird Study*, vol. 59, 2012, pp. 155–65

Cox, R.B., Eddleston, C.R. and Newton S.F., 'Upland Bird Survey Report 2002: Donegal', BirdWatch Ireland Conservation Report no. 02/4

Cramp, S. and Simmons, K.E.L. (eds), *Handbook of the Birds of Europe, the Middle East and North Africa: The birds of the western Palearctic* (vols 1–3); Cramp, S. (ed.) (vols 4–6); Cramp, S. and Perrins, C.M. (eds) (vols 7–9) (Oxford: Oxford University Press, 1977–94)

———, Bourne, W.R.P. and Saunders, D., *The Seabirds of Britain and Ireland* (London: Collins, 1974)

Cromie, J., 'Breeding Status of Red-throated Diver *Gavia stellata* in Ireland', *Irish Birds*, vol. 7, 2002, pp. 13–20

Cropper, S., 'Whooper Swans in Co. Donegal in August', *British Birds*, vol. 31, 1937, p. 151

Crowe, O., *Ireland's Wetlands and Their Waterbirds: Status and distribution* (Newcastle: BirdWatch Ireland, 2005)

———, Cummins, S., Gilligan, N., Smiddy, P. and Tierney, T.D., 'An Assessment of the Current Distribution and Status of the Kingfisher *Alcedo atthis* in Ireland', *Irish Birds*, vol. 9, 2010, pp. 41–54

———, Tierney, T.D. and Burke, B., 'Status of Rare Breeding Birds Across the Island of Ireland 2013–2018', *Irish Birds*, no. 43, 2021, pp. 29–38

Cummins, S., Bleasdale, A., Douglas, C., Newton, S.F., O'Halloran, J. and Wilson, H.J., 'Densities and Population Estimates of Red Grouse *Lagopus lagopus scotica* in Ireland Based on the 2006–2008 National Survey', *Irish Birds*, vol. 10, 2015, pp. 197–210

———, Corbishley, H. and Newton, S.F., 'Upland Bird Survey Report', BirdWatch Ireland, unpublished report for National Parks and Wildlife Service, 2003

——— and Goodbody, I.M., 'Storm Petrel on Roaninish', *Irish Naturalists' Journal*, vol. 9, 1948, p. 129

———, Lauder, C., Lauder, A. and Tierney, T.D., 'The Status of Ireland's Breeding Seabirds: Birds Directive Article 12 Reporting 2013–2018', *Irish Wildlife Manuals*, no. 114 (Dublin: National Parks and Wildlife Service, 2019)

D'Arcy, G., *Ireland's Lost Birds* (Dublin: Four Courts Press, 1999)

Day, A. and McWilliams, D. (eds), *Ordnance Survey Memoirs of Ireland. Volume 38: Parishes of County Donegal I, 1833–5, North-east Donegal* (Belfast and Dublin: The Institute of Irish Studies in association with The Royal Irish Academy, 1997)

——— and McWilliams, D. (eds), *Ordnance Survey Memoirs of Ireland. Volume 39: Parishes of County Donegal II, 1835–6, Mid, West and South Donegal* (Belfast and Dublin: The Institute of Irish Studies in association with The Royal Irish Academy, 1997)

Delany, S., Scott, D., Dodman, T. and Stroud, D., *An Atlas of Wader Populations in Africa and Western Eurasia* (Wageningen, The Netherlands: Wetlands International, 2009)

del Hoyo, J., Elliott, A. and Sargatal, J. (eds), *Handbook of the Birds of the World* (vols 1–7); del Hoyo, J., Elliott, A. and Christie, D.A. (eds) (vols 8–16) (Barcelona: Lynx Edicions, 1992–2011)

Dempsey, E. and O'Clery, M., *Finding Birds in Ireland: The complete guide*, 2nd edn (Dublin: Gill & Macmillan, 2014)

Devlin, T.R.E. and Merne, O.J., 'Malin Head Observatory Report 1961–1964', unpublished report, 1964, pp. 1–38

——— 'Malin Head Observatory Report 1965', unpublished report, 1965, pp 1–32

Donald, P.F., Wilson J.D. and Shepherd, M., 'The Decline of the Corn Bunting', *British Birds*, vol. 87, 1994, pp. 106–32

Dowdall, J.F. and Larrissey, E., 'Tundra Bean Goose in County Louth: A race new to Ireland', *Irish Birds*, vol. 6, 1999, pp. 432–3

Doyle, S., Walsh, A., McMahon, B.J. and Tierney, D.T., 'Barnacle Geese *Branta leucopsis* in Ireland: Results of the 2018 census', *Irish Birds*, no. 41, 2018, pp. 23–38

Duffy, M., *The Corncrake Conservation Project: Annual report* (Dublin: National Parks and Wildlife Service, 2018)

Elliot, J.S., 'Additional Notes on the Birds of Donegal', *The Zoologist*, 3rd series, vol. XVI, 1892

Fahy, K., 'Irish Rare Bird Report 2010', *Irish Birds*, vol. 9, 2011, p. 311

Farrelly, W. and Charles, D., 'The Dresser's Eider in County Donegal: A new western Palearctic bird', *Birding World*, vol. 23, 2010, pp. 62–4

Fitter, R.S.R., 'Birds on Roaninish', *The Irish Naturalists' Journal*, vol. 9, 1948, p. 128

Foster, S., Boland, H., Colhoun, K., Etheridge, B. and Summers, R., 'Flock Composition of Purple Sandpipers *Calidris maritima* in the West of Ireland', *Irish Birds*, vol. 9, 2010, pp. 31–4

Gibbons, D.W., Reid, J.B. and Chapman, R.A. (eds), *The New Atlas of Breeding Birds in Britain and Ireland* (London: Poyser, 1993)

Gibbs, A., Nisbet, I.C.T. and Redman, P., 'Birds of North Donegal in Autumn, 1953', *British Birds*, vol. 47, 1954 pp. 217–28

Gilbert, G., Stanbury, A. and Lewis, L., 'Birds of Conservation Concern in Ireland 4: 2020–2026', *Irish Birds*, no. 43, 2021, pp. 1–22

Gill, F., Donsker, D. and Rasmussen, P. (eds), *IOC World Bird List*, v. 11.1, 2021, http://www.worldbirdnames.org

Gill F., Donsker, D. and Rasmussen P. (eds), *IOC World Bird List* v. 14.2, 2024, https://www.worldbirdnames.org

Gray, N., Thomas, G., Trewby, M. and Newton, S.F., 'The Status and Distribution of Choughs *Pyrrhocorax pyrrhocorax* in the Republic of Ireland 2002/03', *Irish Birds*, vol. 7, 2003, pp. 147–56

———, Trewby, M., Cummins, S. and Newton, S., 'The Distribution and Feeding Ecology of the Chough *Pyrrhocorax pyrrhocorax* in Co. Donegal: September 2004 – August 2005', *Conservation Report*, BirdWatch Ireland, 2005

Hannon, C., Berrow, S.D. and Newton, S.F., 'The Status and Distribution of Breeding Sandwich *Sterna sandvicensis*, Roseate *S. dougallii*, Common *S. hirundo*, Arctic *S. paradisaea* and Little Terns *S. albifrons* in Ireland in 1995', *Irish Birds*, vol. 6, 1997, pp. 1–22

Hart, H.C. 'Notes on the Birds of Donegal', *The Zoologist*, 1891–2

———, *The Flora of the County Donegal, or, List of the Flowering Plants and Ferns, with their Localities and Distribution* (Dublin: Sealy, Bryers & Walker, 1898)

Herbert, I.J., 'The Status of Garden Warbler at Crom Estate, Co. Fermanagh, and a Review of Its Status in Ireland', *Irish Birds*, vol. 4, 1991, pp. 369–76

Hillis, J.P., 'Rare Irish Breeding Birds, 1992–2001', *Irish Birds*, vol. 7, 2003, pp. 157–72

———, 'Annual Reports of the Irish Rare Breeding Birds Panel', *Irish Birds*, vols 7–9, 2002–12

Hine, M. and Malin, R., 'News and Comment: Concern for Scotland's seabirds as avian influenza continues to take hold', *British Birds*, vol. 115, 2022, p. 423

Hobbs, J., *First Irish Records 1800 to 2010*, version 1.3 (privately published, 2021)

———, *A List of Irish Birds* (version 12.0) (www.southdublinbirds.com) (Dublin: South Dublin Branch of BirdWatch Ireland, 2022)

———, *First Records for the Western Palearctic 1800 to 2015*, version 1.6 (privately published, 2023)

Hosey, G.R. and Goodridge, F., 'Establishment of Territories in Two Species of Gull on Walney Island, Cumbria', *Bird Study*, vol. 27, 1980, pp. 73–80

Howard, R. and Moore, A., *A Complete Checklist of the Birds of the World* (London: Macmillan, 1984)

Humphries, G.R., *A List of Irish Birds* (Dublin: Stationery Office, 1937)

Hutchinson, C.D., *Ireland's Wetlands and Their Birds* (Dublin: Irish Wildbird Conservancy, 1979)

———, *Birds in Ireland* (Calton: Poyser, 1989)

Inch Wildfowl Reserve, https://inchwildfowlreserve.ie

Inishowen Rivers Trust, https://inishowenriverstrust.com

Inishowen Upland Project, https://inishoweneip.com

Inishtrahull Bird Observatory, https://inishtrahullbirdobs436659775.wordpress.com

Irish Rare Birds Committee, 'Irish Rare Birds Committee Review of Presumed Vagrant Canada Goose *Branta canadensis* Records to Ascertain the Occurrence of Cackling Goose *Branta hutchinsii*', *Irish Birds*, vol. 9, 2013, pp. 613–22

Irish Rare Birds Committee, 'Irish Rare Bird Report 2016', *Irish Birds*, vol. 10, no. 4, 2017, p. 573

Irish Raptor Study Group, '2001 Annual Roundup', unpublished report

Kennedy, P.G., Ruttledge, R.F. and Scroope, C.F., *The Birds of Ireland* (Edinburgh and London: Oliver & Boyd, 1954)

Lack, P. (ed.), *The Atlas of Wintering Birds in Britain and Ireland* (Calton: Poyser, 1986)

Lauder, A. and Lauder, C., 'Identification of Breeding Waterbird Hotspots in Ireland', *Irish Wildlife Manuals*, no. 129 (Dublin: National Parks and Wildlife Service, 2020)

Leebody, J.R., 'Notes on the Birds of Lough Swilly', *The Irish Naturalist*, vol. 1, 1892, pp. 173–7

Lewis, L.J., Austin, G., Boland, H., Frost, T., Crowe, O. and Tierney, T.D., 'Waterbird Populations on Non-Estuarine Coasts in Ireland: Results of the 2015/16 Non-Estuarine Coastal Waterbird Survey (NEWS-III)', *Irish Birds*, vol. 10, 2017, pp. 511–22

——, Burke, B., Fitzgerald, N., Tierney, T.D. and Kelly, S., 'Irish Wetland Bird Survey: Waterbird status and distribution 2009/10–2015/16', *Irish Wildlife Manuals*, no. 106 (Dublin: National Parks and Wildlife Service, 2019)

——, Coombes, D., Burke, B., O'Halloran, J., Walsh, A., Tierney, T.D. and Cummins, S., 'Countryside Bird Survey: Status and trends of common and widespread breeding birds 1998–2016', *Irish Wildlife Manuals*, no. 115 (Dublin: National Parks and Wildlife Service, 2019)

Lloyd, C.S., 'Inventory of Seabird Breeding Colonies in the Republic of Ireland', Forest and Wildlife Service, Bray, unpublished report, 1982

——, Tasker, M.L. and Partridge, K., *The Status of Seabirds in Britain and Ireland* (London: Poyser, 1991)

Lynas, P., Newton, S.F. and Robinson J.A., 'The Status of Birds in Ireland: An analysis of conservation concern 2008–2013', *Irish Birds*, vol. 8, 2007, pp. 149–66

Mac Lochlainn, C., 'Breeding and Wintering Bird Communities of Glenveagh National Park, Co. Donegal', *Irish Birds*, vol. 2, 1984, pp. 482–500

Macdonald, R.A., 'The Breeding Population and Distribution of the Cormorant in Ireland', *Irish Birds*, vol. 3, 1987, pp. 405–16

Mac Gloinn, M., *Éin Árainn Mhór: Birds of Arranmore* (n.p.: Shearwater Publishing, 2022)

Mackie, T., *Frank Egginton* (privately published, 2022), p. 52

Madden, B., 'A Remarkable Nocturnal Thrush Migration', *Irish East Coast Bird Report*, 1992, pp. 71–3

——, Cooney, T., O'Donoghue, A., Norris, D.W. and Merne, O.J., 'Breeding Waders of Machair Systems in Ireland in 1996', *Irish Birds*, vol. 6, 1998, pp. 177–90

——, Hunt, J. and Norriss, D., 'The 2002 Survey of the Peregrine *Falco peregrinus* Breeding Population in the Republic of Ireland', *Irish Birds*, vol. 8, 2009, pp. 543–8

McGeehan, A., *To the Ends of the Earth: Ireland's place in bird migration* (Cork: The Collins Press, 2018)

———and Millington, R. 'The Adult Thayer's Gull in Donegal', *Birding World*, vol. 11, 1998, pp. 102–8

———and Wyllie, J., *Birds through Irish Eyes* (Cork: The Collins Press, 2012)

McLoughlin, D.T., 'Management Prescriptions for Twite in Ireland', *Irish Wildlife Manuals*, no. 52 (Dublin: National Parks and Wildlife Service, 2011)

———, 'Twite on the Edge', *Wings*, no. 57, 2010, pp. 15–17

——— and Cotton, D., 'The Status of Twite *Carduelis flavirostris* in Ireland, 2008', *Irish Birds*, vol. 8, 2008, pp. 323–30

———, Benson, C., Williams, B. and Cotton, D.C., 'The Movement Patterns of Two Populations of Twites *Carduelis flavirostris* in Ireland', *Ringing and Migration*, vol. 25, 2010, pp. 15–21

McMonagle, C., Bell, M. and Donaghy, A., 'Survey of Breeding Wader Populations at Machair and Offshore Islands in North-West Ireland, 2017, CAAB Cooperating Across Borders for Biodiversity Project, an INTERREG V Project under the European Regional Development Fund', BirdWatch Ireland, unpublished report, 2017

Mee, A., 'The Status and Ecology of a Remnant Population of Ring Ouzel *Turdus torquatus* in the MacGillyguddy's Reeks, Kerry', *Irish Birds*, vol. 11, 2018, pp. 13–22

Merne, O.J., 'Malin Head Observatory', *IWC News*, no. 2, 1974

Milne, P. and Hutchinson, C., *Where To Watch Birds: Ireland* (London: Christopher Helm, 2009)

Mitchell, C., Heard, R. and Stroud, D., 'The Merging of Populations of Greylag Goose Breeding in Britain', *British Birds*, vol. 105, 2012, pp. 498–505

Mitchell, P.I., Ratcliffe, N., Newton, S. and Dunn, T.E., *Seabird Populations of Britain and Ireland: Results of the Seabird 2000 census (1998–2002)*(London: T. & A.D. Poyser, 2004)

Moloney, D., 'Curlew Report, 2019, CAAB Cooperating Across Borders for Biodiversity Project, an INTERREG V Project under the European Regional Development Fund', Birdwatch Ireland, unpublished report, 2019

Mullarney, K. (on behalf of the Irish Records Panel), 'Review of Irish Records of Greenish Warblers', *Irish Birds*, vol. 2, 1984, pp. 536–45

Murton, R.K., *The Woodpigeon* (London: Collins, 1965)

Nairn, R. and O'Halloran, J., *Bird Habitats in Ireland* (Cork: The Collins Press, 2012)

——— and Sheppard, J.R., 'Breeding Waders of Sand Dune Machair in North-west Ireland', *Irish Birds*, vol. 3, 1985, pp. 53–70

National Parks and Wildlife Service, https://www.npws.ie/publications

Newson, S.E., Leech, D.I., Hewson, C.M., Crick, H.Q.P. and Grice, P.V., 'Potential Impact of Grey Squirrels *Sciurus carolinensis* on Woodland Bird Populations in England', *Journal of Ornithology*, vol. 151, 2010, pp. 211–18

Newton S.F., 'An Overview of Rare Breeding Birds in Ireland in 2017', *Irish Birds*, vol. 10, 2018, pp. 541–4

———, 'Rare Breeding Birds in Ireland in 2016', *Irish Birds*, vol. 10, 2016, pp. 383–90

———, Donaghy, A., Allen, D. and Gibbons, D., 'Birds of Conservation Concern in Ireland', *Irish Birds*, vol. 6, 1999, pp. 333–42

Nicholson, E.M., 'Richard's Pipit in Donegal', *British Birds*, vol. 49, 1956, pp. 44–5

Norman, R.K. and Saunders, D.R., 'Status of Little Terns in Great Britain and Ireland in 1967', *British Birds*, vol. 62, 1969, pp. 4–13

Norriss, D.W., 'The Status of the Buzzard as a Breeding Species in the Republic of Ireland, 1977–1991', *Irish Birds*, vol. 4, 1991, pp. 291–8

———, 'The 1991 Survey and Weather Impacts on the Peregrine *Falco peregrinus* Breeding Population in the Republic of Ireland', *Bird Study*, vol. 42, 1995, pp. 20–30

——— and Wilson, H.J., 'Survey of the Peregrine *Falco peregrinus* Breeding Population in the Republic of Ireland 1981', *Bird Study*, vol. 30, 1983, pp. 91–101

——— and Wilson, H.J., 'Disturbance and Flock Size Changes in Greenland White-fronted Geese Wintering in Ireland', *Wildfowl*, vol. 39, 1988, pp. 63–70

———, Haran, B., Hennigan, J., McElheron, A., McLaughlin, D.J., Swan, V. and Walsh, A., 'Breeding Biology of Merlins *Falco columbarius* in Ireland, 1986–1992', *Irish Birds*, vol. 9, 2010, pp. 23–30

———, Marsh, J., McMahon, D. and Oliver, G.A., 'A National Survey of Breeding Hen Harriers *Cirus cyaneus* in Ireland 1998–2000', *Irish Birds*, vol. 7, 2002, pp. 1–12

Ó Caomhánaigh, C., *Dictionary of Bird Names in Irish*, http://gofree.indigo.ie/~cocaomh/HomePage.htm, 2002

O'Donnell, A., 'Scarce Migrants in Ireland, 2007 and 2008', *Irish Birds*, vol. 9, 2012, pp. 421–46

O'Donoghue, B.G, Donaghy, A. and Kelly, S.B.A., 'National Survey of Breeding Eurasian Curlew *Numenius arquata* in the Republic of Ireland, 2015–2017', *Wader Study*, vol. 126, 2019, pp. 43–8

Ó Gaoithin, S., *The Ancient Woodlands of Glenveagh* (Glenveagh National Park, 2021)

Patterson, R., 'Fork-tailed Petrels in North of Ireland', *The Zoologist*, vol. 15, 1891, pp. 468–9

Pennington, M.G., Riddington, R. and Miles, W.T.S., 'The Lapland Bunting Influx in Britain and Ireland in 2010/11', *British Birds*, vol. 105, 2012, pp. 654–73

Perry, K.W., *The Birds of the Inishowen Peninsula* (privately published, 1975)

———, 'Population Changes of Dippers in North-West Ireland', *Irish Birds*, vol. 2, 1983, pp. 272–7

———, 'Rare Breeding Birds in Ireland in 2012 – including a review of significant population changes over the past decade', *Irish Birds*, vol. 9, 1993, pp. 563–76

Pettitt, R.G., 'Tory Island, 1965', unpublished report, pp. 1–5.

Phillips, P., 'The Breeding Birds of Tory', *Tory Island Bird Report 2013* (privately published, 2014)

——— and Agnew, P., 'Breeding Dipper Populations in North-west Ireland', *Irish Birds*, vol. 5, 1993, pp. 45–8

——— and Newton, S.F., 'Rare Breeding Birds in Ireland in 2013', *Irish Birds*, vol. 10, 2014, pp. 63–79

——, Ingram, C. and Salter, R., *Tory Island Bird Reports*, 2011 to 2016 (privately published)

——, Ingram, C. and Salter, R., *Tory Island Bird Report 2014* (privately published, 2015)

——, Ingram, C. and Salter, R., *Tory Island Bird Report 2016* (privately published, 2017)

Redman, P.S., 'Status of the Sooty Shearwater in British Waters', unpublished report, 1955

——, 'Birds of Tory Island, County Donegal in 1954', *Tory Island Bird Report 2015* (privately published, 2016)

Rees, E., 'Northwest European Bewick's Swans: A national and flyway perspective', *Waterbirds in the UK 2019/20, the Annual Report of the Wetland Bird Survey*, 2021

Rosenberg, K.V., Dokter, A.M., Blancher, P.J., Sauer, J.R., Smith, A.C., Paul A., Stanton, J.C., Panjabi, A., Helft, L. and Marra, P.P., 'Decline of North American Avifauna', *Science*, vol. 366, no. 6,461, 19 September 2019, pp. 120–4

Ruddock, M., Mee, A., Lusby, J., Nagle, A., O'Neill, S. and O'Toole, L., 'The 2015 National Survey of Breeding Hen Harrier in Ireland', *Irish Wildlife Manuals*, no. 93 (Dublin: National Parks and Wildlife Service, 2016)

Ruttledge, R.F., *Ireland's Birds* (London: Witherby, 1966)

——, *A List of the Birds of Ireland* (Dublin: Stationery Office, 1975)

——, Re-assessment of the Record of Long-tailed Skuas in the Shannon Valley in 1860', *Irish Birds*, vol. 2, 1982 pp. 197–8

—— and Ogilvie, M.A., 'The Past and Current Status of the Greenland White-fronted Goose in Ireland and Britain', *Irish Birds*, vol. 1, 1979, pp. 293–363

Scott, D.A. and Rose, P.M., *Atlas of Anatidae Populations in Africa and Western Eurasia* (Wageningen, The Netherlands: Wetlands International, 1996)

Sharrock, J.T.R., *The Atlas of Breeding Birds in Britain and Ireland* (Tring: British Trust for Ornithology, 1976)

Sheppard R., 'The Winter and Spring Feeding Ecology of Oystercatchers *Haematopus ostralegus* on the Edible Mussel *Mytilus edulis*', MSc thesis, University of Wales, Bangor, 1971

——, 'Whooper and Bewick's Swans in North-west Ireland', *Irish Birds*, vol. 2, 1981, pp. 48–59

——, 'The Breeding of the Goosander in Ireland', *Irish Birds*, vol. 1, 1978, pp. 224–8

——, *Ireland's Wetland Wealth* (Dublin: Irish Wildbird Conservancy, 1993)

——, 'The Wintering Waterbirds of Lough Swilly, County Donegal', *Irish Birds*, vol. 7, 2002, pp. 65–78

——, 'Hybrid Tree x House Sparrows in County Donegal', *Irish Birds*, vol. 5, 1995, pp. 319–20

—— and Green, R.E., 'Status of the Corncrake in Ireland in 1993', *Irish Birds*, vol. 5, 1994, pp. 125–38

Smiddy, P. and O'Halloran, J., 'Breeding Biology of the Grey Wagtail *Motacilla cinerea* in Southwest Ireland', *Bird Study*, vol. 45, 1998, pp. 331–6

—— and Duffy, B., 'Little Egret *Egretta garzetta*: A new breeding bird for Ireland', *Irish Birds*, vol. 6, 1997, pp. 55–6

Speer, A., 'Nature Conservation', in A. Cooper (ed.), *Lough Swilly: A living landscape* (Dublin: Four Courts Press, 2011), pp. 138–48

Stevens, D.K., Anderson, G.Q.A., Grice, P.V., Norris, K. and Butcher, N., 'Predators of Spotted Flycatcher *Muscicapa striata* Nests in Southern England as Determined by Digital Nest-cameras', *Bird Study*, vol. 55, 2008, pp. 179–87

Stoney, C.V., 'Red-necked Phalarope Breeding in Co. Donegal', *The Irish Naturalist*, vol. 33, 1924, p. 109

Suddaby, D., Nelson, T. and Veldman, J., 'Resurvey and Comparative Changes of Breeding Wader Populations of Irish Machair and Associated Wet Grasslands in 2009', *Irish Birds*, vol. 8, 2009, pp. 533–42

——, Nelson, T. and Veldman, J., 'Resurvey of Breeding Wader Populations of Machair and Associated Wet Grasslands in North-west Ireland', *Irish Wildlife Manuals*, no. 44 (Dublin: National Parks and Wildlife Service, 2010)

Taylor, A.J. and O'Halloran, J., 'The Decline of the Corn Bunting, *Miliaria calandra*, in the Republic of Ireland', *Biology and Environment: Proceedings of the Royal Irish Academy*, vol. 102B, 2002, pp. 165–75

Temple Lang, J., 'Peregrine Survey – Republic of Ireland 1967–1968', *Irish Bird Report*, no. 16, 1968, pp. 8–12

——, 'Peregrine', *Irish Bird Report*, no. 17, 1969, p. 33

——, 'Peregrine', *Irish Bird Report*, no. 18, 1970, p. 28

Thompson, W., *The Natural History of Ireland. Vol. 1: Birds* (London: Reeve, Benham & Reeve, 1849)

——, *The Natural History of Ireland. Vol. 2 Birds* (London: Reeve, Benham & Reeve, 1850)

——, *The Natural History of Ireland. Vol. 3: Birds* (London: Reeve, Benham & Reeve, 1851)

Tory Bird Observatory , 'Tory Island 1958 and 1959', unpublished report, 1959, pp. 1–13

——, 'Tory Island 1960', unpublished report, 1960, pp. 1–17

——, 'Tory Island 1961', unpublished report, 1961, pp. 1–19

——, 'Tory Island 1962', unpublished report, 1962, pp. 1–18

——, 'Tory Island 1963 and 1964', unpublished report, 1964, pp. 1–49

Tucker, B.W., 'Roseate Terns in Donegal', *British Birds*, vol. 34, 1941, p. 245

Ussher, R.J. and Warren, R., *The Birds of Ireland* (London: Gurney & Jackson, 1900)

van Bemmelen, R.S.A. et al., 'A Migratory Divide Among Red-Necked Phalaropes in the Western Palearctic Reveals Contrasting Migration and Wintering Movement Strategies', *Frontiers in Ecology and Evolution*, 2019

Wallace, D.I.M., McGeehan, A. and Allen, D., 'Autumn Migration in Westernmost Donegal', *British Birds*, vol. 94, 2001, pp. 103–20

Wann, J. and Sheppard, R., *Pilot Ecological Study of Two Donegal Islands: Inishfree Upper and Inishmeane*, Aulino Wann and Associates, unpublished report for Donegal County Council, 2010

Watson, P.S. and Radford D.J., 'Census of Breeding Seabirds at Horn Head, County Donegal in June 1980', *Seabird Report, 1977–81*, vol. 6, 1982

Wernham C., Toms, M., Marchant, J., Clark, J., Sitiwardena, G. and Baillie S., *The Migration Atlas* (London: Poyser, 2000)

Whilde, A., *Threatened Mammals, Birds, Amphibians and Fish in Ireland. Irish Red Data Book 2: Vertebrates* (Belfast: Her Majesty's Stationery Office, 1993)

———, 'A Survey of Gulls Breeding Inland in the West of Ireland in 1977 and 1978 and a Review of the Inland Breeding Habit in Ireland and Britain', *Irish Birds*, vol. 1, 1978, pp. 134–60

———, 'The 1984 All Ireland Tern Survey', *Irish Birds*, vol. 3, 1985, pp. 1–32

———, Cotton, D.C.F. and Sheppard, R., 'A Repeat Survey of Gulls Breeding Inland in Counties Donegal, Sligo, Mayo and Galway, with Recent Counts from Leitrim and Fermanagh', *Irish Birds*, vol. 5, 1993, pp. 67–72

Williams, W.J., 'Golden Eagle and Marsh-Harrier in Ireland', *British Birds*, vol. 20, 1927, pp. 107–8

Wilson, D.R., 'The Storm Petrel Colony on Roaninish', *Bird Study*, vol. 6, 1959, pp. 73–6

Wilson, M.W., Pithon, J., Gittins, T., Kelly, T.C., Giller, S.G. and O'Halleron, J., 'Effects of Growth Stage and Tree Species Composition on Breeding Bird Assemblages of Plantation Forests', *Bird Study*, vol. 53, 2006, pp. 225–36

Winstanley, D., Spencer, R. and Williamson, K., 'Where Have All the Whitethroats Gone?', *Bird Study*, vol. 21, 1974, pp. 1–14

INDEX

Note: Page locators in italics refer to illustrations, tables and figures. Locators in bold refer to the main entry of that particular species. Other locators refer to mentions elsewhere.